DISTRIBUTED SENSOR SYSTEMS

DISTRIBUTED SENSOR SYSTEMS

PRACTICE AND APPLICATIONS

Habib F. Rashvand
Adcoms, University of Warwick, UK

Jose M. Alcaraz Calero
Hewlett-Packard Laboratories, UK

A John Wiley & Sons, Ltd., Publication

Library of Congress Cataloging-in-Publication Data

Rashvand, Habib F.
    Distributed sensor systems : practice and applications / Habib F. Rashvand, Jose M. Alcaraz Calero.
        p. cm.
    Includes bibliographical references and index.
    ISBN 978-0-470-66124-6 (cloth)
    1. Sensor networks–Design and construction. 2. Sensor networks–Industrial applications.
I. Alcaraz Calero, Jose M. II. Title.
    TK7872.D48R37 2012
    681′.2–dc23

                                                                                    2011051663

A catalogue record for this book is available from the British Library.

Print ISBN: 9780470661246

Set in 11/13 Times by Laserwords Private Limited, Chennai, India
Printed and bound in Singapore by Markono Print Media Pte Ltd

To Liz, Leila, Cyrus and Joan

Contents

List of Figures

List of Tables

List of Tables

Preface

No matter how small a superior technology is, it can lead to new application paradigms and through a series of successful large-scale diffusions, transform societies and provide a better global lifestyle.

This book is complementary to many recent theoretical sensor-networking books and aims to give readers a taste of what the new smart sensing systems can do in their promising real world applications.

Under 'Distributed Sensor Systems', we focus on the applications and uses of smart sensors as the nucleus of an application-based intelligent networking concept. This volume answers many critical questions relating sensors to the very exciting technological developments of this century, which are essential to fulfil our next viable application paradigms. This integrated development should inspire a new technological revolution, similar to those following the Renaissance, such as the great transformation caused by the discovery of the steam engine some 200 years ago and the decentralisation caused by further discoveries in chemistry-electric-information around 100 years ago.

With the recent rapid development of distributed sensors we see it is time to consolidate all the hard work carried out over the decades and open new windows of opportunity for slowly, but steadily, developing new light-weight, ultra-low-power and intelligent sensor devices. This is being achieved by using new technologies such as micro- and nano-electromechanical devices for a wide range of viable products, systems and long awaited solutions in monitoring health, medical, environmental, traffic, tracking, space, underwater and many industrial uses.

Although exploring the basic development of sensors is exciting and essential, due to their long historical development full coverage of the technical details is beyond the scope of this book, we limit our coverage of sensors to sample devices that lend themselves to integrated wireless-enabled solutions and enforce new application momentums. As the communication capability of new sensors is an extremely important requirement of distributed sensing systems, we focus on the class of sensors that can perform sense-and-node or actuate-and-node functions. For this we propose ideal multi-core device models and analyse recent developments, for their suitability and their potential features.

In order to appreciate the distributed capability of these systems we include a brief buffering interface to help the following viewers:

(a) For readers who want to acquire both theories and practice, this book becomes a useful piece of information, helping to consolidate their in-depth understanding of aspects of the theory in an optimised, engineering management approach of applying the technology into projects and deployment phases. That is, the extra half-a-chapter brief becomes a practical interface connecting this book to the many widely available theoretical books.
(b) For practitioners, system developers and the new breed of innovative engineers this book can be used without a need for detailed theories as an informative compact debrief of which the first three chapters should help with a better understanding of the follow-up application chapters. In order to maintain the quality for the volume's target audience all material in this volume is presented in a practical style and most of the latter chapters are provided in the form of case study and application scenarios.
(c) We also envisage a third group of potential users for this book stemming from:

   (i) The many advanced academics, who with an increasing awareness of wireless sensors, could integrate this title into their new courses, particularly one to be studied in the later stage of a bachelor degree course. In either a postgraduate academic course or industrial employment in the field, this book should play a significant role when adopting the technology or applying it for potential products in the near future.
   (ii) The need for trained, skilful, hands-on, able and practical engineers in today's highly competitive industries has never been as high. This book not only helps them with their training courses purely as a textbook, it can also help individuals to acquire a better appreciation of the technology and enable them to understand the potential practicalities of their hard-learned technical abilities.
   (iii) With continuously evolving distributed sensor systems as a response to the growing demand for more intelligent solutions through miniature electronic devices, the need for training and retraining system designers, engineers and technocrats arises. This book can be extensively used throughout the industry and college laboratories for implementing new solutions and new application systems. Whether in the health industry, smart homes, underwater or just monitoring climate change and environmental observations using portable or moving intelligent sensors, all can use this book for various phases of product development.

In general, what clearly appears to be missing in the list provided above is an insightful bridge that will carry motivated young readers from their core undergraduate preparatory learning to the point where they can produce results. Common experience shows that, even amongst this large class, a limited number of students actually overcome the steep transition from advanced mathematics into industrial application-oriented system engineering without the need for training. Where we need to motivate those potentially bright minds at a challenging yet successful pace is into making innovations to derive reality from theory and insightful theoretical analyses. Whilst many books contain useful information that may be readily absorbed by a diligent reader, they do not aim at nurturing the practice by deriving insight into the analysis of simple models. It is this precise gap in the sub-area literature that this title aspires to fill. It is essential that such an attempt is made to motivate the reader to pursue analytical approaches based on models guided by engineering intuition and the resulting path of study accessible to dedicated engineers.

H. F. Rashvand and J. M. Alcaraz Calero

Acknowledgements

The work supported by the Advanced Communication Systems Ltd and its associates.

Jose M. would like to sincerely recognise and thank Carmen for her unconditional support, sailing with us aboard the river of life.

Jose M. would also like to personally thank Professor Pedro M. Ruiz, at the University of Murcia, recognised expert in sensing platforms and wireless sensors because he showed me the incredible world of distributed sensing devices, I couldn't have been able to accomplish my contribution of the book without his help.

Finally, Jose M. would like to thank Fundación Seneca for grant #15714/PD/10.

List of Abbreviations

AAA	Authentication, Authorisation and Accounting
AAL	Ambient Assisted Living
ABF	Array-Based Beamforming
AC	Alternative Current
ACORD	Appliance Coordination
ACK	Acknowledgement
ADC	Analogue to Digital Converter
ADV	Advertisement Message
AES	Advanced Encryption Standard
AFN	Aggregation-and-Forwarding Node
AI	Artificial Intelligence
AM	Ante Meridiem
ANK	Multiple Negative ACK
AMR	Anisotropic Magneto-Resistive
AODV	Ad-hoc On-demand Distance Vector Routing
AOT	Agent Objective Tasks
AOWSN	Application-Oriented Wireless Sensor Network
AP	Access Point
API	Application Program Interface
APS	Application Support
ARM	Advanced RISC Machine
ASIC	Application-Specific Integrated Circuit
ATT	Agent Task Table
AUV	Autonomous underwater Vehicle
BCG	Ballisto-CardioGrams
BIR	Biomedical Imaging Resource
BMS	Building Management Server
BP	Blood Pressure
BS	Base Station
BST	Body Surface Temperature

CAD	Computer Aided Design
CBF	Collaborative Beamforming
CBR	Case-Based Reasoning
CC-MAC	Correlation-based Collaborative MAC
CCA	Cooperative Collision Avoidance
CCD	Charge-Coupled Device
CD	Congestion Degree
CDUA	Convergence, Divergence and Ubiquitous Access
CH	Cluster Head
CMOS	Complementary Metal-Oxide Semiconductor
CNP	Contract Net Protocol
COP	Common Operational Picture
CODA	Congestion Detection and Avoidance
CPU	Central Processing Unit
CRC	Cyclic Redundancy Checking
CSMA	Carrier Sense Multiple Access
CSMA-CA	Carrier Sense Multiple Access – Collision Avoidance
CT	Computer-assisted Tomography
CTP	Collection Tree Protocol
DAC	Digital-to-Analogue Converter
DBF	Distributed Beamforming
DBT	Dry Bulb Temperature
DC	Direct Current
DCO	Digitally Controlled Oscillator
DD	Distributed Devices
DIA	Distributed Intelligent Agent
DIS	Distributed Intelligent System
DLNA	Digital Living Network Alliance
DLUT	Distributed Look-Up-Table
DMA	Direct Memory Access
DMC	Digital Magnetic Compass
DNA	DeoxyriboNucleic Acid
DOA	Direction of Arrival
DoI	Domain of Interest
DOS	Distributed Operating System
DSA	Digital Subtraction Angiogram
DSDV	Destination-Sequenced Distance Vector Routing
DSL	Digital Subscription Line
DSM	Design Sign-off Model
DSS	Distributed Sensor System
DST	Dempster–Shafer Theory
DTB	Distance To Base

ECG	ElectroCardioGram
EEG	Electro-EncephaloGraphy
EEPROM	Electrically-Erasable Programmable Read-Only Memory
EGM	Ventricular Electrogram
EKG	Same as ECG
EM	ElectroMagnetic
EMG	Electromiagram
EMR	Electronic Medical Record
EMS	Emergency Medical Services
EMT	Emergency Medical Technician
EMU	Energy Management Unit
EN	Exposed-Node
EPT	Estimated Propagation Time
ER	Emergency Room
ESSN	Energy-efficient Self-clustering Sensor Network
ETO	Electro−THz−optical
EWM	Embedded Wireless Module
FCC	Federal Communications Commission
FEM	Finite Element method
FFD	Full Function Device
FIFO	First In/First Out
FIR	Finite Impulse Response
FOE	Frequently occurring Events
FOS	Fiber Optic Sensor
FSO	Free Space Optics
FSK	Frequency-Shift Keying
FTSP	Flooding Time Synchronisation Protocol
GAF	Geographic Adaptive Fidelity
GEAR	Geographic and Energy Aware Routing
GFSK	Gaussian Frequency-Shift Keying
GHz	Giga Hertz
GMT	Greenwich Meridian Time
GN	Gateway Node
GP	General Practice
GPIO	General Purpose Input/Output
GPRS	General Packet Radio Service
GPS	Global Positioning System
GPSR	Greedy Perimeter Stateless Routing
GSM	Global System for Mobile Communications
GTS	Guaranteed Time Slots
GW	Gateway
HART	Highway Addressable Remote Transducer Protocol

HF	High Frequency
HIT	Health Information Technology
HMI	Human Machine Interface
HN	Hidden-Node
HPBW	Half-Power Beam Width
HPC	High Performance Computing
HR	Heart Rate
Hz	Hertz
IAP	Inhibitor of the Apoptosis
ICA	Intersection Control Agent
ICD	Intelligent Congestion Detection/Implantable Cardioverter Defibrillators
ICN	Implicit Congestion Notification
ICP	IntraCranial Pressure
ICT	Information and Communication Technology
ICU	Intensive Care Unit
ID	Identifier
IDSQ	Information-Driven Sensor Query
IEC	International Standard for Communications
IFIP	International Federation for Information Processing
IO	Input/Output
IoT	Internet of Things
IP	Internet Protocol
IR	Infrared
IrDA	Infrared Data Association
IROM	Internal ROM
ISA	International Society of Automation
ISFET	Ion Sensitive Field Effect Transistors
ISM	Industrial, Scientific and Medical
ITS	Intelligent Transport System
JSON	JavaScript Object Notation
JTAG	Joint Test Action Group
KB	Kilobytes
Kbits	Kilobits
Kbps	Kilo Bytes per second
KHz	Kilo Hertz
Km	Kilometer
LA	Local Aggregators
LAN	Local Area Network
LCD	Liquid Crystal Display
LEACH	Low-Energy Adaptive Clustering Hierarchy
LED	Light-Emitting Diode
LOC	Lab-On-a-Chip

LOL	Long Objectives List
LPG	Liquefied Petroleum Gas
LQI	Link Quality Indicator
LST	Land Surface Temperature
MA	Master Aggregators
MAC	Medium Access Control
MAS	Multi Agent System
MB	Megabyte
Mbps	Megabyte per second
MCD	Mobile Computing Devices
MCU	Master Control Unit
MDS	Non-Metric multidimensional Scaling
MECN	Minimum Energy Communication Network
MEMS	Micro-Electro-Mechanical Systems
MHz	Mega Hertz
MIMO	Multiple Input/Multiple Output
Min	Minimum
MIT	Massachusetts Institute of Technology
MMX	MultiMedia eXtension
MOSFET	Metal Oxide Semiconductor Field Effect Transistor
MRT	Mean Radiant Temperature
MSN	Micro-Sensor Nodes
MSR	Micro-Sensor Routing
MS	Mobile Sensor
MZI	Mech-Zehnder Interferometer
NACK	Negative Acknowledgment
NASA	National Aeronautics and Space Administration
NEMS	Nano-Electromechanical Systems
NFC	Near Field Communications
NILM	Nonintrusive Load Monitoring
NN	Neuronal Network
NP-hard	non-deterministic polynomial-time hard
NSCLC	Normal/Small Cell Lung Sancer
ODL	On-board Deliberative Layer
OEM	Original Equipment Manufacturer
OODA	Observe, Orient, Decide and Act
OS	Operating System
OSI	Open System Interconnection
P2P	Peer-to-Peer
PBG	Photonic Band Gap
PC	Personal Computer
PCCD	Priority-based Congestion Control Protocol

PCB	Printed Circuit Board
PD	Photo Diode
PDA	Personal Digital Assistant
PEGASIS	Power-Efficient Gathering in Sensor Information Systems
PHR	Personal Health Records
PHY	Physical Layer
PM	Post Merídiem
PMV	Predicted Mean Vote
PMMA	PolyMethyl MethacrylAte
PoI	Point of Interest
PPG	photo-plethysmogram
PRA	Priority-based Rate Adjustment
PSA	Pressure Sensitive Adhesive
PUIP	Programmable Ultrasound Image Processor
PVDF	PolyVinyliDene Fluoride
PWM	Pulse-width Modulation
PZT	Piezoelectric Transducer
QoS	Quality of Service
RAM	Random Access Memory
RBS	Reference Broadcast Synchronisation
RCU	Identify Risks, Critical Bottlenecks and Urgent Tasks
RDT	Reliable Data Transfer
REQ	Request
RF	Radio Frecuency
RFD	Reduce Function Device
RH	Relative Humidity
RFID	Radio Frequency Identification
RISC	Reduced Instruction Set Computer
RoI	Return of Interest
RON	Resilient Overlay Networks
ROM	Read Only Memory
RSSI	Received Signal Strength Indication
RSU	Road Side Unit
S-MAC	Secure MAC
SAR	Sequential Assignment Routing
SARS	Severe Acute Respiratory Syndrome
SASA	Structure-Aware Self-Adaptive
SBL	Sequence-Based Localisation
SCLC	Small Cell Lung Cancer
SCN	Switched Capacitor Network
SDIO	Secure Digital Input/Output
SDRAM	Synchronous Dynamic Random Access Memory

SEAD	Scalable Energy-Efficient Asynchronous Dissemination
SHM	Structural Health Monitoring
SIGOPS	Special Interest Group on Operating Systems
SIS	Self-adapting Intelligent Sensor
SLP	Service Location Protocol
SMA	Sub Miniature version A
SME	Small medium enterprise
SMECN	Small Minimum Energy Communication network
SN	Smart Sensors Node
SNMP	Simple Network Management Protocol
SNR	Signal-to-noise Ratio
SOA	Service Oriented Architecture
SOI	Silicon-On-Insulator
SOS	Save Our Souls
SPI	Serial Peripheral Interface Bus
SaO2	Arterial Blood Oxygen Saturation
SpO2	Oxygen Saturation
SRAM	Synchronous Random Access Memory
SSD	Smart Sensor Device
SSE	Streaming Single instruction, multiple data Extensions
SSL	Secure Socket Layer
TDMA	Time Division Multiple Access
THz	Tera Hertz
TI	Telecom Italia
TOS	Tiny OS
TOSS	Trust, Objectivity, Security and Stability
TPD	Time-Diffusion Synchronisation Protocol
TPSN	Timing-sync Protocol for Sensor Network
TST	Traditional Sensor Transducers
UART	Universal Asynchronous Receiver-Transmitter
UAV	Unmanned Aerial Vehicles
UMB	Ultra Mobile Broadband
UMTS	Universal Mobile Telecommunications System
UOF	U-Object Finder
UPNP	Universal Plug and Play
USA	United States of America
USB	Universal Serial Bus
USN	Ubiquitous Sensor Network
UWB	Ultrawideband
VANET	Vehicular Area Network
VCSEL	Vertical-Cavity Surface-Emitting Laser
Vdd	Supply Voltage

VGA	Virtual Grid Architecture
VI-CMOS	Vertically Integrated CMOS
VLSI	Very Large Scale Integration
VOC	Volatile Organic Compound
VoIP	Voice over IP
VRML	Virtual Reality Modelling Language
VS	Vice Sink
WCS	Wastewater Collection System
WENn	Wireless Electronic Nose Network
WHO	World Health Organisation
WHMS	Wearable Health-Monitoring System
Wi-Fi	IEEE 802.11
WFQ	Weighted Fair Queuing
WiMAX	Worldwide Interoperability for Microwave Access
WLAN	Wireless Local Area Network
WPAN	Wireless Personal Area Networks
WRR	Weighted Round Robin
WSN	Wireless Sensor Network
WSS	Wireless Sensor System
XLM	Cross-layer Module
XML	eXtensible Markable Language
XYZ	Axis x, Axis y and Axis z
ZDO	ZigBee Device Object
ZVE	Zero-Voltage-Threshold
A	Ampere
μTAS	Micro Total Analysis Systems
W	Watt

1

Distributed Sensors

Recent technological discoveries in distributed systems, new clustering and networking protocols in conjunction with progressing wireless, Internet and new smart sensors, have led us to move towards more intelligent sensor solutions for which we expect a growing global demand and new range of superior applications to promote our next stages of technology for a new social and economical development paradigm. In order to achieve this unique opportunity for a maximum global impact using minimum resources we need to harmonise our global, regional and local industrial and academic efforts at all levels under the common goals of sensor enabled smart media distributed systems.

1.1 Primary Objectives

Due to the service nature of sensor technology and its firm market relationships with other technologies in many aspects of our modern life in an ever-growing industrial world, one can visualise the sensitivity, if not dependability, of our social and economical developments to the sensors. Our understanding of this technology can be greatly influenced by the choice of one out of two extreme views of sensors.

First view: they are regarded as transducers, simple converters of a signal or data from one form into another, a more suitable format. The second view: they are regarded as intelligent proactive devices as part of a larger system where sensing plays an important role in bringing in a new layer of control and intelligence over our capabilities in managing various aspects of our life, including health, stability and security for better social and economical prosperity. That is, under the first view, a traditional understanding of the sensors, our poor vision of the sensor technology imposes excessive limitations to smart sensors functionalities. Therefore for their uses in the real world we may fail to justify our existing overwhelming investments on new smart sensor associated research and development projects. The second

Distributed Sensor Systems: Practice and Applications, First Edition.
Habib F. Rashvand and Jose M. Alcaraz Calero.
© 2012 John Wiley & Sons, Ltd. Published 2012 by John Wiley & Sons, Ltd.

view is, however, more acceptable to such an investment as its integration with other systems can add new exciting dimensions to new systems, applications, services and to their applicability to make sensors to help to deploy new breeds of intelligent distributed systems. This would enable further agility, efficacy and product viability to deliver new applications of distributed intelligent sensors providing a new range of versatile solutions enhancing every aspect of our modern life today.

For the second view, we therefore set the book's primary objectives following on from our preliminary discussions. Part of this discussion is associated with a demonstration of the above mentioned practical capabilities of intelligent sensors for which we need to understand some fundamental concepts of these new sensors, appreciate their critical factors for practical deployment whilst we visualise the viability of a new generation of sensors being developed under the protection of innovation where their needs, applications and potential market forces dominate over any lengthy theoretical details of the applied technologies.

In order to make our objectives clearer we review the three main components of a successful innovation, namely: (a) feasibility of the innovative technological solution; (b) viability of the solution, that is the need and potential requirement of the solution; and finally; (c) the success factors, that is acceptability test as in pilot studies and introduction of the new product being an unpredictable product, system or service to the users.

For the requirements of (a) and (c) we have many typical application solutions for adopting the new sensors and verification of their superior features that is throughout five chapters in Chapters 4 to 8 of this book, all enriched with discussions on real and practical applications and case studies enhanced with a variety of experimental results reported by researchers from innovative research laboratories to industrial production lines. For the market requirements and needs of (b) we look into the potential markets under two different sets of categories of the need and service sectors. In the first set we look at 'user need' aspects of distributed sensor systems (DSS) answering the question of 'what sensor systems can do for us'? We then have a closer look at a categorised DSS market for the visualisation of the growing potential markets for emerging sensor-centred intelligent products and services.

1.1.1   User-Based Category

As a response to the question of 'what new sensors can do for us' we identify the following seven generic uses:

Sensor for Monitoring

This feature-based usability is an enhanced use of traditional sensors as an embedded sensing parametric visibility for monitoring critical variables of a system or media of interest. Also a new response to the need for:

- low energy global scale monitoring systems;
- low cost global scale monitoring uses;
- regular data collection applications;
- statistical measurement systems;
- monitoring behaviour system applications;
- location identification;
- ad hoc style information gathering;
- monitoring health and well being;
- monitoring for emergency cases and interventions.

Smart Media

Upon the philosophy of intelligent environment or ambient intelligence many industries, that is electrical, physical, chemical or biological, can now be upgraded using new smart intelligent sensors. Typical broad application areas are associated with the following:

- smart stationery;
- smart home;
- smart office;
- interactive communication enabler;
- behavioural and reactionary functionalities;
- provision of service pre processing;
- data processing, manipulation and ubiquitous environment;
- preparation of information or specifically treated signals;
- information selectivity and effective databases.

System Controllability

As sensor systems grow larger, providing more desirable functions, the more complex they become. Then, in many cases without any regular check-ups, no refined adjustments can be identified leading to a poor status of performance and in some cases they may become unstable. To solve such a growing problem some elements of control could always be desirable, for example:

- sensor-actuator-enhanced systems can provide tight control over few critical components of a system;
- a complex system behaves differently with time;
- extending systems' useful life is always desirable;
- systems using time-variant components require regular tuning;
- many practical systems do not behave perfectly per design, some have side effects and some require extra resources to maintain a proper run. Due to the nature of the problem, these imperfect behaviours grow in time affecting the viability of

the system, which may lead to its obsoleteness, if still stable. We can easily extend a system's life and enhance its stability by adopting integrated intelligent sensing using programmable devices.

Remote Sensor-Actuator Agent

With the rise of globalisation the shape of industries is changing rapidly towards two sustainable equilibriums of:

- Small medium enterprise (SME) style small service industries;
- Agile and globally competitive industries.

The second group of industries should characterise the future of our industrial societies. They include new agile manufacturers, system integrators and distributors who could benefit from enhancements for their remote monitoring, remote configuration and remote control using autonomous embedded technologies. To these industries and many global service providers use of integrated sensor-actuator counts as a major advantage for their market competition enabling them through two basic cost cutting competitive edges of, (a) remote sensing to identify the status of a system and, (b) remote actuation to implement a change without excessive costly visits of the experts.

Dependability

Whilst the impact of globalisation is increasing many new disturbing activities such as the number of computer crimes is on the rise at alarming rates. Whereas, a better use of a distributed security integrated solution using a multi-agent system or intelligent sensor system would be reducing the cost and number of surveillance, casualties and enhancing the global trust (Rashvand et al., 2010).

Sustainability

Depletion of earth resources, frequent massive destructive disasters and a continually changing environment worry many intellectuals, which may result in a demoralisation of public views on the governing bodies. This may also change people's view for supporting future technological developments. To change this trend and reduce the casualties may help to ensure sustainability of life on earth for which we need to establish more effective global monitoring systems to help us deal with:

- Habitual life – the uneven distribution of the population is causing deterioration by the disappearance of valid and healthy villages due to lack of governmental support for their basic needs.

- Urbanisation – pressing issues all over the world of an increasingly poor quality of life for an unachievable expected high-life due to numerous pressures imposed on a large population in poorly organised cities.
- Release of waste and uncontrolled poisonous polluting gas and chemicals causing long-term degradation of life.
- Ongoing human casualties and poverty and the impact of uncontrollable natural disasters.
- Depleting earth's scarce natural resources.
- Poor quality of health due to growing age, growing traffic in heavily congested populated areas with maximum effects on the majority of people.
- Risky and unhealthy habitual activities.

1.1.2 Sector-Based Category

Another way of looking at our needs for sensor-based potential intelligent DSS products, upon maturity of the new cost effective, energy efficient advanced sensors, which enables emerging super mass production capabilities in connection with new advances in wireless, Internet and distributed intelligence technologies, we can broadly categorise five groups as follows:

- Environmental Applications – DSS for greener life and sustainable climate and monitoring earth resources.
- Industrial Applications – automation, heavy economy as an infrastructure for improving the quality of life whilst minimising costs and overheads.
- Medical Applications – surgical, physiological, psychological, increasing age and a higher quality of life.
- Security and Surveillance – safety, immunity, trust, dependability.
- Old Age and Well Being Applications – a growing market with significant pressures on most nations and more on those with social security supports.

Extensive discussions for the above-mentioned categorised applications are included throughout the book with many detailed cases studies and application scenarios in the later parts, Chapters 4 to 8.

1.1.3 Primary Objectives

Now we are in a position to introduce *the book's four primary objectives* and to explain why we have adopted our style of presentation and why we believe this is the best way for the reader to acquire the highest degree of required knowledge using a single volume with the minimum expense of their time to adopt intelligent DSS for their further uses: (a) design of a new application; (b) integration of DSS with another distributed system; (c) engaging in an investigation or product development or (d) conduct a project management for a DSS based service deployment.

Objective 1. Generic Smart Sensing

- In order to capture the maximum use and popularity for deploying DSS on a global scale we provide an in-depth description of new, intelligent sensing devices. This includes the architecture of two or more basic core processing units, one for less flexible, extremely low cost programmable generic functions and a few for smaller size mass production for common application specific features.
- Smart media approach to the future technological developments. With new smart sensors we can embed minimum specific intelligence in various parts of the media for a variety of uses and applications including regular monitoring, intelligent response per request, automatic report generation or systems requirement for actions and warnings.
- Additional global market viability features of the applications are where we can trim off the hardware complexity whilst adopting more flexible blocks in the form of off-the-shelf middleware functions.

Objective 2. Intelligent Specific Sensing

Technical productivity and the natural intelligent features of distributed sensing such as ubiquitous networking, clustering, beamforming, sensor fusion and distributed intelligence make DSS applications superior to all existing smart sensor systems. Though all details are not in the scope of this book we highlight a few specific features of the new generation of intelligent sensors, for example:

- Clustering features a very basic superiority of DSS by providing uniquely efficient target proximity for the point-of-interest (PoI) and facilitating superior fusion for the sensing information collected from the media. This feature provides considerable advantages for two basic functions of data and mobility for a wide range of applications over wireless-only, wired-only and mixed wired-wireless distributed sensors.
- Beamforming features a basic superiority of DSS over the wireless media providing unique directivity and selectivity, where cooperative sensors in an array adjust their antennas configurations automatically to coordinate for a combined transmission lobe towards the target or PoI in the media for the two most effective outcomes of (a) maximum use of transmitted power and (b) minimum interference with other channels and other sensors in the system.
- Exploring the distributed intelligence features the DSS to be able to be integrated with other data oriented intelligent systems such as multi-agent system (MAS), enabling new superior integrated solutions for precision, trust and effectiveness.

Objective 3. Innovation Approach

- Analysing the need for investment in a global scale smart sensor product development project one normally starts with initial market research investigating the viability and success factors of the technology. For this we include innovative features of DSS products including their superiority over the two previous generations as in the first two 'objectives' with a better understanding of market segmentation and its distribution statistics. Then the decision for the investment merely becomes as simple as cost for mass production, cost per unit, returns on investment projection, maturity of the technology and new features of soft production.
- Due to the economical sensitivity of DSS for a sustainable global development and better economical progress we encourage industrial nations along with global organisations to help with this unique adoption of DSS based technological development.

Objective 4. Learning by Example

Learning by example is our adopted approach to maximise the reader's fast understanding of the underlying technological aspects of a versatile and complex system like DSS, without going through volumes of details which one may need for a successful deployment of an application or general understanding of its potential uses and services. This method can also inform both educators and practitioners of the availability of a viable technology in one compact volume about a system which is involved with well over half of the future economy dependent technologies.

Based on our previous discussions in this chapter and upon our four objectives we structure the remaining part of the book. To achieve these objectives through brief but effective materials provided in the remaining part of this and its seven following chapters. In the remaining part of this chapter we examine the basic aspect of the DSS including a brief introduction of the new sensors and actuators covering innovation, distributed intelligent and classification of the DSS applications, where we look for key technologies such as smart and intelligent sensors. Chapter 2 looks into device-based smart sensors with some interesting further classification of sensors. Chapter 3 provides insight into selective smart sensor networking, infrastructure and advanced techniques used in the device-based structure for new generations of smart sensors. The rest of the book consists of five chapters, 4 to 8 covering typical scenarios for advanced applications of the DSS. First we look into medical and consumer applications with five novel case studies demonstrating some typical burning potential application areas of the DSS for the next 5 to 10 years. We then examine a few other application areas of the DSS including three case studies and some other typical applications of the DSS as the tip of the iceberg for upcoming market potential.

1.2 Historical Development

The science of sensing and its associated versatile technology today have been man's best friend since living in caves and enjoying an early agricultural lifestyle. Then, sensors were simple additional enhancement gadgets to their basic but effective collection of tools. Today we have superb multidisciplinary intelligent smart sensors offering something largely different, enabling our overwhelmingly complex systems to help build a sustainable global village.

In order to get a better grip of the new technology, before going any further one should know the answer to the question of 'What is a sensor?' Sensors, though often integrated into a bigger and much more complex system than themselves, represent a well-known technology virtually throughout all today's industries. The enormous range of applications they can offer vary from a humble thermometer checking a new born baby's body temperature to a complex system associated with the nucleus measuring devices of an elaborate radiation measurement system in an atomic reactor.

1.2.1 Sensing

In today's literature the word *sensing* has several meanings, but two most relevant ones are 'intelligence' and 'feeling'. Although both have their own specific projections in human life the first one comes with more relevance to the scope of this book. We can break the first one down further into 'rationality' and 'wisdom'. The combination of these two functions can be regarded as 'intelligent visibility', closely related to the survival of an intelligent human at his early stages of evolution enabling him to master the earth and to overcome the difficulties of a harsh life, to stand out against all the odds of nature, and cope with the inhospitable surrounding environment. This is also the case through the evolutionary process, with shortages of food and primitive shelters causing the spread of infectious diseases. Learning from nature, we then extend the human sensory system for use in our living environment through machines, systems and many other artefacts.

Biologists follow the physical processes of human nature to come up with our well known 'five groups of senses', commonly called human sensory systems associated with seeing, hearing, smelling, tasting and touching:

- Seeing – well known, well utilised (example: camera).
- Hearing – known, partly utilised (example: microphone).
- Smell – little known, some utilisation (example: enose).
- Tasting – little known, little utilised (example: chemicals).
- Touch Feeling – some known, some utilised (example: thermometer).

1.2.2 Historical Sensor Generations

Looking back into the staggering development of sensors over centuries of engineering effort it is possible to trace three distinct development trends for sensor applications, namely transducer sensors, smart sensors and our new intelligent device-based distributed sensors.

The first generation of sensors can be traced back to the early stages of civilisation. They include the actuators with sharp rises in development during the first and second technological revolutions. Due to the limited level of information processing and interconnection capabilities, the uses of these sensors are limited to the potential of their transduction function, where the cost, size and physical placement issues of the products dominate their use. Large, mechanical sensors, controlling actuators, physical sensors, chemical sensors, traditional biosensor, and heavy industrial sensors can be categorised under this generation. Due to the nature of development and versatility in their form, shape, media of application and uses we call them *traditional sensor transducers* (TST). It is easy to notice that, due to their need for continual improvements, most of them cannot make use of low cost mass production and therefore many of them are still developing for practicality factors such as shrinking in size, the cost to the user, and other suitability features.

The development of second-generation sensors started in the later part of the twentieth century and is approaching its peak in the latter part of the first decade of the twenty-first century. These sensor systems are commonly characterised by possession of a common architecture to accommodate the enhanced sensing features including embedded signal processing and light sensor specific computing and communication capabilities.

Their large-scale market enhancement becomes feasible when some basic processing features for carrying out common functions, such as networking, communications, basic data manipulations, low energy devices, application of specific selective smart processing adopted through advanced integrated circuit technology commonly processed in the electrical domain could be integrated.

Being called smart, these sensors are accommodated into small and very limited spaces and enhanced with common computing and communication functions. However, due to the fact that their limited Information and Communications Technology (ICT) features use single core architecture, the production technology naturally cannot respond to the sensor's versatile global market requirements. That is, most of these original smart sensors need to include functions such as sensing specific signal pre processing, data gathering, data distribution, filtering, interference processing, and so on, on top of interfaces and communication protocols. In some cases extra processing, such as energy scavenging, clustering and media specific processes would increase the demand on the single core processor solution far beyond its capability or add cost, power, and so on, so that the extra shrinking boundary would push the originally estimated market into much smaller realistic margins. We therefore see the early sharp rise in both the volume and popularity of what are commonly called smart sensors followed rapidly with an early maturity and all successful integration of ASIC and VLSI solutions for small and miniature sensors went into a saturated market status. Having their applications and usability dominated and therefore characterised by their main basic device capabilities we

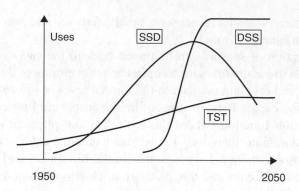

Figure 1.1 Historical developments of three generations of sensor: TST, SSD and DSS. The graphs are their estimated market viability.

therefore call them smart sensor devices (SSD). Some description of these sensors can be found in Chapter 2.

As we learn from the SSD early experiences, in some cases with much simpler media-based processing, some successful applications signify their smart functions to adopt a modularity approach with the capability of being integrated into larger systems should they pave their way up towards the third generation.

The third generation of sensor developments stems from distributed intelligence which makes another evolutionary change from the second generation. The arrival of many mass production oriented new applications under an enhanced architecture for DSS comes under a new cooperative approach for integrating a whole range of essential features into the system. The advanced architecture associated with the distributed approach brings elegant and superior features to sensor developments. As we discuss further details in Chapters 2 and 3, DSS is set to make the most of recent developments and incorporates unique new advantageous features of cooperative sensor and intensive use of distributed computing for intelligent sensor fusion, whilst freeing the production from traditional networking constraints, the limiting factors of previous generations. Figure 1.1 shows a relative estimated potential use of three distinct sensor generations upon their market viability at the global scale.

1.3   Trends and Technology

We address this section under two major views of *market development trends* and *technological developments*, which we have monitored over the last two decades.

1.3.1   Market Development Trends

There have been significant market-enabling social, economical and industrial developments all over the world during the last few decades, triggering many viable new uses of sensor systems for low cost mass production. It is, however, important to mention that a true cost reduction can be achieved only through a proper power integration of both the essential and most popular functions for maximum impact.

These functions would make these devices smart, and in many required cases intelligent enough to suit a much wider range of applications as they stay generic and adaptive enough to perform under variable conditions and changing application scenarios as required.

One should, however, bear in mind that, due to market trends, in many cases we should consider an additional cost for re-engineering and reconfiguration for deployment of the final stage implementation. This extra cost could sometimes grow extensively due to over-generalisation of the core processor, which in turn could push the overall cost to the user far beyond the basic production line.

In general the market boundaries for these systems are expanding rapidly from national or regional scales into global. This trend, however, imposes two new inter-related effects on the market: market size and the overall cost of unit product to the end user. Theoretically, the unit cost goes through a minimum, which strongly depends on two factors of basic mass production and the cost of post deployment service to the user. This minimum can be reduced extensively upon the degree of flexibility embedded in the core processor. That is, early smart sensors using a single core processor are too rigid to be able to find a place for very wide versatility and therefore the market size limitation may not support their position in a very competitive global market. The new generation of smart sensors, also called intelligent sensors using multi-agent technology provide a better market solution. For producing this new breed of sensors, as explained further in Chapter 3, multi-core processing architectures integrating a new common processor with one or more specialised processors is needed. The multi-core processing architecture can provide extensive flexibility at a small extra cost in the mass production, which in turn will enable much greater market opportunity to help us to make use of the many so called sensor cross-road opportunities, see Figure 1.2 (Rashvand et al., 2008).

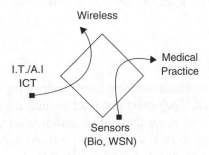

Figure 1.2   The telemedicine cross-road opportunity.

Cross-road opportunities represent techno-economical possibilities that become available under certain social demands in conjunction with a set of complementary technological feasibilities, which provide economical growth possibilities for industrial nations to take advantage of in order to build a better lifestyle whilst benefiting from associated economical success. Such short lived opportunities do not last for a very long period of time, often being lost due to upheavals of negligence and national or global disarray. Thus it is now that we have the best opportunity to deploy telemedicine style services to initiate further developments throughout the industrial nations.

Another new developing market opportunity is the integration of sensor technology with other applications using Internet and wireless technologies. Use of the Internet, these days counts as a unique opportunity for many new distributed services. We have many Internet enabled applications such as remote control systems that could benefit from the DSS. We also have the two fast growing technologies of radio frequency identification (RFID) and Internet of things (IoT) that, when used in cooperation with the new flexible sensors could offer great new market opportunities.

One of the immediate areas of application is an integrated solution with wireless sensor networks (WSN), where networking over the wireless enables a variety of application cases for the cooperation of sensors and RFID devices interconnected throughout the industry for various services at home, office, clinics and for other social activities. The wireless connection provides good mobility, but its use with the Internet makes the service go well beyond the existing borders. The safety and security aspects of RFID can also enhance WSN applications at the lower end of the market bringing a new integrated capability to the services. IoT, on the other hand is still new, but, due to its market potential of integrating various sensors through the Internet in various forms of fixed, moveable, wired and wirelessly connected devices, can bring more application opportunities in the near future.

1.3.2   Technological Developments

Here we scan a few potential technologies that make new smart sensors superior to the earlier generations.

Clustering

Despite their extensive communication and data processing capabilities new sensors are usually light and small. They can be spread around the area of interest in either forms of fixed or mobile to share their information with other sensors as well as the central controller. Due to their ad hoc unstructured networking nature they can form and reform their interconnection with other sensors to achieve their objectives with the minimum use of their scarce resources. For a special objective, or under a certain application scenario, they can easily form a cluster to accomplish their objective for the best result. The cases that make clustering useful depend on the circumstances. For example, wireless mobile sensors can position themselves and approach a target or as close as possible to the point of interest (PoI) in the media for collecting useful data. They can position themselves for a beam forming process to enhance their detection capability.

In the case of fixed sensors they can interact and detect a moving target and communicate in an optimum exchange of information. They can create a smart ambient for detecting activities or provide active interactions with the media or a reliable ubiquitous access service. In general, the clustering service brings a great feature of flexibility to the sensors.

Distributed Sensing

In order to demonstrate the general features of distributed sensing in practice let us consider a simple example where, a simple moisture-sensing device is located in the soil providing a reading for an open agricultural field. Using a single humidity sensor can provide a low cost solution supplying estimated information on the wetness in the soil. However, this sensor, due to the low validity of data, can only provide a poor reading as it can vary significantly upon the location of the sensor under very high natural and practical risks. Usual problems are wind, rain, sun, shade, proximity to the watering pipes and many other natural and man-made variants. Although such a system can almost do the work, if accceptable, it wastes a lot of water, and which, if followed up by processing of the data, can be tolerated, for proper engineering work this solution is far from perfect. However, using collective data from a multiple cooperating sensor system scattered around the field could improve the system's performance significantly.

In general, effectiveness of sensing depends on two inter-related factors of (a) distance between sensing detector and the sensing target, PoI usually referred to as vicinity and (b) accuracy of reading information in relation to the desirable data.

Often the first factor, in practical applications, is a handicap. It could be worse if we are dealing with a moving target or certain obstacles which separate sensors from the PoI. Using distributed sensors, a cluster of sensors or multi-sensor systems, this problem can be resolved. Also, considering the nonlinear signal loss, usually much higher than the order of 2, the problem of accurate reading could become very serious, therefore cooperative sensing can get closer to the target.

Tracking

In some cases a mobile target single sensing device, fixed or mobile, can get distracted by the noise and interfering signals and easily miss out on the target altogether. This feature can be extensively improved using a DSS. There are many advanced algorithms for distributed system, fixed and mobile, to trace a single or a multiple target effectively.

One of the most popular applications is video surveillance for home or plant security where often a mixture of fixed and mobile video sensors equipped with extensive signal and image processing can provide a very reliable surveillance solution.

One very interesting tracking application for the new distributed sensors is in chemical leak localisation. The writing of this book coincides with the BP oil leak in the Gulf of Mexico showing the scale of costs and troubles involved in such an accident. This incident shows how hard it is to detect such a leak and then the seriousness of the problem caused by a delayed detection and the impact to environmental health, sea life and heavy losses to the company.

Chemical leaks, gas or liquid, are mostly harmful and frequently happen in large industrial plants caused by corrosion and other decaying processes. Traditional solutions for detecting such leaks allocate a large number of chemical sensors fixed along the space over high-risk points and weak places mostly close to long pipes and inside processing chambers. Each sensor usually works independently to detect any possible leaks and set an alarm either locally or reported to the plant's central monitoring unit for immediate actions and recovery. Consider that chemical sensors such as odour detectors, which mostly simulate the human nose, are bulky, expensive and wasteful which normally pose a threat to a successful industry. Therefore, under financial pressure many operational managers are forced to compromise and to go along with some risk of not having full coverage. However, instead of a huge number of fixed sensors we can use very few mobile intelligent sensors circulating along some predetermined paths or random roots providing a more reliable sensing system and guarding the whole plant by sharing their findings so that more accurate detection can be activated through a multi-sensing method using smart nanotube gas sensors for a fraction of the cost.

Sensor Fusion

Fundamentally, gathering useful information, also called *data*, is the basic task of a sensor. The accuracy of data depends on the sensing system, stability of the domain and its usefulness. In some cases the collected data is utilised immediately to take an action, but in most cases it is accumulated for a collective decision. In many new applications, collected data is used at a variety of stages. Some are used for pre-processing and some stored in a centralised database for future further processing such as building up reliable statistical data. The usefulness of a sensing system could be compromised due to lack of (a) required accuracy and (b) reliability of data being read by the sensor. Therefore, over sampling, compressed sampling and other pre-processing of the data could become helpful.

The degradation of data cannot be limited to the noise and interference from the sensing signals, but depends on the original reading loss. This usually is caused by some practical factors such as the angle of detection, linearity of the signal and statistical status of the domain at the PoI. The errors and distortions caused by the interfering phenomena are traditionally removed using a filtering system, for example Kalman filtering for sensor fusion, but with new developments these complex processes can be easily compensated or disappear when using multi-sensing by enhancing the original reading of data from the domain.

Energy

Wireless and distributed sensors require a considerable amount of power and usually use local power resources, often batteries. This is often considered a weakness of WSN and other isolated sensing devices.

Though highly improved energy devices have been effective, many new developments have been helping us with this problem in four directions: (a) minimum waste of energy using energy-efficient algorithms, (b) efficient architectures using complementary metal–oxide–semiconductor (CMOS) technology and highly power efficient processors so that no activity wastes can be reduced, (c) highly efficient specially designed architectures and (d) making use of local energy generation such as energy scavenging and solar charging techniques to keep the device always ready for emergency activities.

For example, inefficient traditional style networking and communication activities usually consume a very high portion of the energy. Using cross layer techniques can save a great part of this energy.

Intelligence

The integration of limited intelligence in the third generation of sensors can boost up their applications. We have discussed some of their features under distributed intelligence in Section 1.4 and then throughout the book with applications cases. One of the most interesting applications of these sensors is the creation of smart spaces with clear cases of smart home, smart office under smart ambient. For example, we can look at an established case of using a large number of low cost fixed sensors spread around the area of interest monitoring a moving target. As shown in Figure 1.3 a basic two-dimensional array of fixed ultrasonic sensors fixed on the ceiling of the room can track a vulnerable person whose state of health is at risk and who is not able to call for help when they need it or if they fall. This simple and practical application of distributed sensors is discussed further in Chapter 6 under *smart home* for a better lifestyle.

1.4   Distributed Intelligence

Consider that a successful deployment of the new generation of distributed sensors can be closely related to the use of distributed intelligence, which is regarded as one of the key technologies for a successful global technological development.

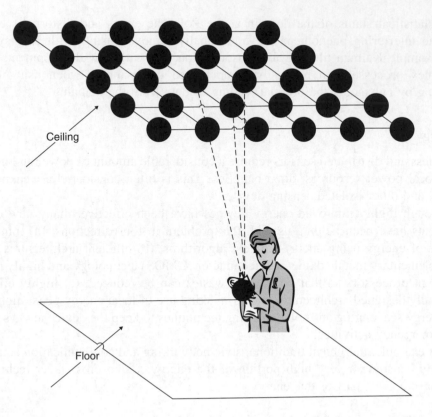

Figure 1.3 Smart home environment application, an example of adopting intelligence at home.

We now examine further developments of distributed intelligence through *process innovation*. Here, we look into the successful development of agent style distributed intelligence through innovation before devising a new method of maintaining the stability in a complex distributed intelligent system.

1.4.1 Innovation

We know by experience that a true *invention* is based on problem solving with a traditional definition of 'a novel idea that has been transformed into reality or given a physical form' is only a starting point towards bigger steps in the chain of invention to innovation to a socio-economical progress. That is, although without invention there is no innovation and therefore no social or economical progress, but only a few selective inventions featuring *feasibility* and fewer feasible inventions can viably enter the domain of innovation and then a bundle of cooperative innovations may trigger further social and economical developments. For turning an invention into an innovation some selective inventions should be tested upon their completion of the follow up steps under the process of innovation upon its traditional definition as 'introducing something new or applying a new idea to meet our needs' indicating two basic success factors of technology and market.

As the time goes on, however, the significant impact requirement of penetration into today's complex, resistive and highly competitive dominating global markets is getting more difficult. We, therefore, need to re examine the process. In order to achieve our extended understanding of innovation we look at two historical views:

'*He who will not apply new remedies must expect new evils, for time is the great innovator*' from Francis Bacon (1561–1626), Philosopher of Science.

'*Many essential human needs can be met only through goods and services provided by industry where industry has the power to enhance or degrade the environment; invariably it does* both' from World Commission on Environment and Development, 1987.

Two immediate deductions:

Our existing survival depends on seeking new solutions for new and old problems because, due to the complex nature of controlling organisations, the industries cannot be trusted blindly, and new and progressive lawsuits are required to control the industries.

The first deduction indicates that we need to maintain the 'problem solving' as a basic process for new developments where researchers, experts and other key intellectuals should recognise 'real, common and pressing problems', understand the most suitable advanced feasible solutions for the most reliable design approach to promote new viable solutions whilst generating minimum or negligible side effects. These side effects or more correctly *innovation side effects* represent a significant factor in the process and therefore require proper attention.

One way to view them is to compare their similarity to the patterns of radiation intensity in a directional antenna delivering its highest power towards a particular direction through the main lobe. But, due to its imperfect design the antenna generates some smaller lobes in different directions causing interference for other transmission systems. In the same way that an innovation brings constructive impacts to society upon its primary lobe, the unavoidable, unwanted side lobes impose undesirable impacts onto society.

For two practical aspects of innovation side effects one needs to examine two factors: (a) *controllability* and (b) the management of the side effects.

When the pace of deployment is slow then both factors are manageable. That is, diffusion of innovation can be controlled and side effects can be gradually removed or converted into progressive and constructive developments, a case true for around 200 years.

The long awaited use of the steam engine made use of Leonardo De Vinci's superior inventive designs. The socio-economical impacts of this major event then triggered a series of effective industrial developments with wider implications in all aspects of Western European societies: for a new industrial life, also referred to as *the first technological revolution*. This is when the progressive technological developments encouraged industrial nations to take on newer challenges, saving European life from the dark ages, a period which is often called *the great transformation* also known as the *Renaissance*. In that period, however, we experienced the innovation as a simple process and we had full control over all aspects of technological, economical and social developments.

As a century of industrial development continued, the impacts spread all over the globe, but at the same time all systems began to become more and more complex. However, then we were still dealing with non-intelligent and controllable systems within two main flagship industries of chemistry and electricity. Much harder, but just as feasible, further developments that continued for another century meant that we succeeded with *the second technological revolution*, enhancing many nations' quality of life all over the globe. Chemistry was then offering new materials previously unknown, such as plastic and fibre, and electricity enabled new ways of generation and distribution of power, light (Edison) and signals to carry information (Morse, Bell, Marconi) all gathering the new essence of a change towards *decentralisation*.

Since then another century has passed with many more impressive innovative ideas and large scale developments triggering many philosophers, economists and technocrats to look for the signs to mark the *third technological revolution*. To some degree, this expectation has been explained by the theory of *Kondratiev long waves* (KLW) featuring, a periodical half-century cycle, superimposed phasing out of innovation with economical booms and busts. Figure 1.4 shows the main factors of the theory. In order to justify their expectations the experts are looking for some key technologies with clear leading flagships.

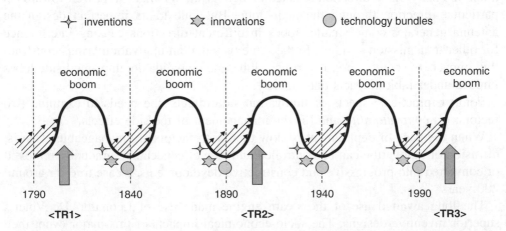

Figure 1.4 Kondratiev's long waves are accompanied by three technological revolutions. The symbols indicate inventions, innovations and bundling new technologies.

With the rapid growth of the telecom industry and continuous developments of micro computing under the flagship of the Internet, in the last few decades of the twentieth century many gathered their hopes upon ICT as a key enabler to deliver another technological revolution. However, as the impacts so far are not convincing enough we are still looking for more effective technologies and expect the new generation of distributed systems such as distributed sensors and distributed intelligence to be the answer. To support this claim we can explain a few important processes of dis-invention, intelligent agent and overlay networking before our proposal of a new quality control management algorithm (Open University, 1996).

1.4.2 Dis-Invention

Following our earlier lobe model of innovation, it is quite common that the excitement of the useful primary lobe could overshadow its side effects at the time of diffusion and will then be forgotten, to remain in society as an interfering process. Three possible cases of innovation side effects:

- passive and controllable, dissolve in time;
- active but controllable, need to be removed sometimes;
- proactive and uncontrollable, may grow in time.

If the problem of controllability is directly related to a system's complexity, then keeping our systems as simple as possible will always help. So, for our progressive technological developments we can remove the side effects by adopting one or both of the following tasks as required.

Keep complex systems under control by breaking down large systems into smaller ones whilst continuously looking for any significant side effects of the previous innovations and remove the harmful side effects through regular stages of dis-inventions.

1.4.3 Intelligent Agent

The most important aspect of using system 'intelligence' for DSS is accountability and therefore systematic manipulation of the data being collected for usability, decision-making and actions. In this form we can promote the implementation of AI in a new approach to integrate a kind of *distributed intelligent system* (DIS) for the use of *artificial intelligence* (AI) to build desirable DSS solutions. This can also help the development of smart sensors (and actuators) into the extreme engineering arts of building new microelectronic devices offering lower cost mass production of new autonomous sensors featuring independence, reliability, intelligence, embedded data possessing, cluster-based data fusion, ubiquitous connectivity and wireless energy scavenging features. That is, use of DIS can add an effective step towards nomadic device capability at system and middleware levels. This, in turn, could imply radical changes to the traditional architecture into new agile intelligent autonomous devices to form the DIS, using fewer simple, but highly effective interactive, loosely connected *distributed intelligent agent* (DIA) systems, which is an interesting development of AI in the distributed style use of agent technology. This follows the last four or five decades of AI being remerged into other forms of its associated technologies, interesting recent developments of AI are smart systems, intelligent agent (IA), in the form of multi-agent systems (MAS) and other agent-based intelligent systems for a wide range of applications such as embedded sensors, smart environments (Rashvand et al., 2010).

1.4.4 Deployment Factor

Deployment of a new system or a new service has always been a complex process. Now, with the advanced systems making more and more use of intelligent devices, the deployment process is becoming more unreliable and often risky. Therefore, implementing a viable system using any form of AI such as intelligent sensors, DAI, DIS and MAS or their combination with other systems should follow a controllable quality control procedure; otherwise, we should expect bitter experiences of non-productive trial and error and associated unavoidable loses.

That is, for a systematic approach to deployment, we should examine the following four system-level key requirements:

Trust – without trust and associated dependability, no proper use can be ensured and no natural global diffusion can be guaranteed.

Objectivity – overall effectiveness of an operation or a service heavily depends on the achievement of its fundamental goals.

Security – the ever-growing public fear of information insecurity due to poorly designed systems has created an unreliable infrastructure, so a new open system service needs its own independent information security control.

Stability – system intelligence inherits a natural factor of instability; therefore, control over critically important system's parameters is essential.

The above four system-level requirements, TOSS for short, should enable the developer to follow various aspects of a system's behaviour during its initial phase of deployment by opening a window of visibility for control, monitoring and management purposes during critical periods. For example, for implementing an intelligent sensor-based DSS application. The embedded intelligence nature of these devices' system behaviour is complex in its nature; therefore, we can claim a quality delivery if our implementation includes all factors of TOSS in its procedure.

In addition, for implementing a DSS service, the deployment configuration could vary significantly from one scenario to another. But the service provider could benefit if he keeps and provides a visually integrated trace of all these TOSS factors as a top-level service, using a simplified design infrastructure to the clients. As we explain the details later, an unstructured networking infrastructure interconnects all intelligent components to a central controller using a reliable communication overlay, where each intelligent sensor is responsible for implementing some objective tasks whilst it makes the best use of its own autonomous capability for a maximum operational efficiency, effectiveness and survival.

To demonstrate this, as a guideline, we propose adoption of a deployment-based algorithm, where the ultimate goals and objectives, TOSS, are acquired through a linear progressive accumulation and distribution using an unstructured overlay system.

1.4.5 Overlay Network

The continuous monitoring of all actions and detailed operations of intelligent sensors in a distributed system put heavy demands on the scarce resources at the expense of wasteful activities to the system as a whole, which could be also disturbing for those sensors that can do most of their tasks independently. That is, the distributed sensors of an intelligent system operate normally in atomic style then they begin a fully controllable unit of an objective task under their own full responsibility. Therefore, a complementary, unstructured, overlay network is sufficient to compliment the atomic operations for coordinating the overall communications within the system. The concept of overlay is not new as it has been around for well over a century and is being extensively used in the telecom industry as part of network design and service delivery management with two different roles: (a) updating and adjustment of network resources as a response to constantly changing demand in operational networks and (b) security, quality control and monitoring of the provision of service upon faults, failures and measuring usage of the components and critical resource utilisation. The Internet is used as, or a part, of the overlay concept for updating information delivered to the distributed sensors: collection of information from the sensors for reporting their status or passing critical information such as hash security keys under classic service location protocol (SLP). More recently, the use of an overlay self-organising network shows a better maintained objective of complex networks under resilient overlay networks (RON). In some distributed sensor applications, distributed programming can benefit from swarm intelligence overlay-enabled cooperative networking, but practical simplicity of these networks resides in their uniqueness and application-based, data-centric, limited functionality used in their protocols such as distributed hash tables and sharing a database in sensor networking and spectrum sharing applications.

1.4.6 Deployment Algorithm

Having discussed the requirement for the adoption of distributed intelligence within the infrastructure of the DSS deployment process, here we introduce a simple but practical mechanism to help us with deployment of distributed sensor applications in all phases of design, test, manufacturing and implementation for both large and moderate complexity systems whilst maintaining our TOSS requirements. The new solutions make use of any well-established and dependable component such as smart sensors, intelligent sensors agents and an overlay mechanism.

Considering almost all active devices used in the third generation, sensors possess some degree of intelligence; for the sake of simplicity, in this algorithm we use the term 'agent'. This phrasing also ensures that all sensor devices that are autonomous and possess intelligence should be included in the algorithm.

For DSS applications, some of these components have been discussed earlier in this chapter but mostly in Chapter 3. Whilst they commonly make use of a P2P protocol in the overlay network, this enables the agents to communicate with each other and with the central controller for maintaining a high-level cooperation. At the physical environment, each agent enjoys some degree of freedom for an efficient operation because of the fact that each member is required to operate reliably and securely in order to provide two sets of predefined complementary functions that are interactive and self-controlled. Provision of a simple debatable or distributed lookup-table (DLUT) translates the system's objectives into an individual agent's task table, which can vary extensively from one application to another and can be defined based on the particularity of the application.

The agents, being either individuals, independent sensors, or representing a cluster of sensors, should face their own, usually unique and dynamic, operational environment and use their own limited but fully controlled capability to achieve their own part of the goal as listed on their task table. Equally, any errors or miscalculations being injected in the design phase, in the translation, such as human error, or due to the drastic changes in the operational environment is converted into a malfunctioning for the agents, especially if the agents are empowered with extensive autonomy or operating under advanced natural algorithms such as evolutionary or swarms are converted into corrective operational objectives and then adjusted and redistributed by the central controller (Weynes 2010).

Process Convergence, Divergence Ubiquitous Access (CDUA)

Although using the integration of distributed intelligence in a MAS style DSS application well behaved agents opens up whole new desirable features which can enable the provisioning of superior solutions for many new applications, in practice, under the realistic circumstances of human errors, unpredictable operational environment and risky behaviour of the agents we need to safeguard the whole process without missing the main target of maintaining any critically important requirements in TOSS criteria.

We then propose an effective superior method for removing unpredictable destabilising interferences through a top-level system stabiliser to maintain a smooth overall system operation. This method based on a high-level quality control system can be implemented using three main complementary processes of convergence, divergence and ubiquitous access (CDUA) control mechanism.

Here we explain CDUA method and its use for implementing a complex distributed intelligent agent-based application to provide a reliable infrastructure to stabilise a complex system to operate within minimum risk margins. In this system the central controller works with many agents and is set to achieve a set of defined system's objectives. In the same way that a distributed agent can be adopted to work under different environments and carry out different tasks, some unique and some common, only its behaviour changes upon the system's objectives. Though all agents are autonomous in their detailed operation, these decisions are shared through the central controller and their tasks, usually a subset of system's objectives are allocated centrally. The overall control and deployment of tasks and objectives, however, is shared between the central controller and the agents. In other words this process of implementation follows a sequence of three complementary phases: (a) convergence of objectives and resources, (b) distribution of objectives translated into agent tasks and (c) provision of ubiquitous access through an unstructured network enabling ad hoc any required connections for monitoring, exchange of information and mutual decisions.

Convergence

This phase is normally divided into two parts of a generation-compilation of a long objectives list (LOL) and the estimation of availability and use of resources and associated optimisation processes.

Long objectives listing (LOL)

Having the ultimate set of goals provisionally defined for an application scenario, the centralised controller compiles the system's objectives accumulated from all agents. These are then combined and integrated into one feasible LOL to be managed by a central controller. This action of convergence removes the heavy burden of stability from agents' lists. LOL usually varies with time, application type, the environment and other variables such as a moving target or agent's mobility. Due to the natural flexibility embedded in the process, the convergence plays an important role in the system's performance.

Resource optimisation

At its top-level optimisation LOL requires the collection of all available agents' resources and their capabilities. Deployment of a mature technology, for example, can be listed more accurately and allocated to slower agents for lesser control overheads whereas new, immature and less known methods can work better with faster agents. In general, we need to consider taking the following actions prior to an optimisation process:

- a measure of processing media;
- update LOL regularly (if adaptive mission);
- build a working agent task table (ATT);
- convert objectives into agent-based tasks (*objective task converter*, OTC);
- identify risks, critical bottlenecks and urgent tasks (RCU).

Task distribution

This phase is the divergence part of the CDUA algorithm. Considering that all tasks should be carried out by agents working under different conditions, the converting system objectives LOL into individual agent objective tasks (AOT) and associated delegation processes normally requires a good application specific distribution algorithm. To ensure timely outcomes under minimum use of resources we need to adopt an optimised distribution mechanism for converting LOL into agents AOT. For this use of simple and mature methods is recommended whereas complex methods could lead to risky behaviour and reduced viability. Best advice for this stage is to avoid the use of any immature methodologies and avoid employing any risky high performance methods.

Ubiquitous access infrastructure

A real time deployment of advanced applications using distributed intelligence requires a highly reliable cluster based on an ad hoc networking infrastructure. Communication between intelligent agents to carry out the tasks in cooperation with the central controller is vital to achieve the system's objectives whilst maintaining the system's stability. Gathering reliable information on progressing tasks and on the behaviour of individual agents, essential requirements of the system should be examined regularly.

In order to be able to maintain a minimum, but always available connectivity a ubiquitous access infrastructure is required to enable both direct and indirect dialogues for exchange of data between all active components, for example overlay. This connectivity therefore is considered an essential part requirement for a CDUA implementation.

Due to the superior capability embedded in this infrastructure any unpredictable faults, interferences and shortcomings are systematically converted into delays in the process as the system is trying alternative routes. That is, as long as the system is in operation and a minimum ubiquitous access of a by-pass mechanism is working the embedded strategy should sub-optimally use some alternative paths for delivery

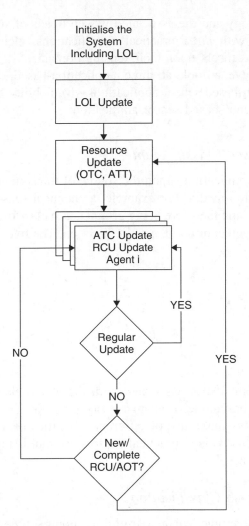

Figure 1.5 The operational flowchart showing the main process involved in Convergence, Divergence Ubiquitous Access (CDUA) algorithm.

of data. The main functions of the main processes of convergence, divergence and ubiquitous access of the proposed CDUA method are shown in the flowchart of Figure 1.5. Due to generality of CDUA algorithm further details are provided in various typical uses of distributed intelligent sensors in appropriate case studies and applications scenarios in Chapters 4 to 8.

1.5   Classifying Application Areas

With a common understanding of sensors and their applications being *'too many sensors too many applications'* one appreciates the need for a better understanding of this versatile technology and the associated woven treads of sustainability through-out other industries with *three* practical classifications, including our traditional grouping of dividing sensors upon their sensing domain.

Therefore, let's have a look at three application-based classification groups: (a) traditional domain-based sensor applications; (b) mobility-based sensor applications and (c) intelligence-based sensor applications.

1.5.1 Domain-Based Classification

Classification of the applications upon the source of transduction in many naturally involves the application media. For example, a chemical sensor normally involves the chemical industry and therefore being grouped as a class of chemical application it provides a helpful perception in the right direction. The five common overlapping domains are:

- physical sensors;
- mechanical sensors;
- chemical sensors;
- electrical sensors;
- biosensors.

There are three general issues associated with the above classification. First, this classification cannot be unique or change every time there is a new development, secondly, many sensors make use of different properties of a domain so we have many overlaps and finally many practical sensors use multi-stage sensing to change their detection domain.

1.5.2 Mobility-Based Classification

One important/effective/practical and useful grouping of new sensor systems is based upon their mobility feature. This simple but very effective classification can help to divide sensor applications into four groups, whether upon the sensor or the target, as in the sensing media, and relatively moving or fixed. See Table 1.1.

Table 1.1 Four mobility based classes of sensor applications

Target	Sensor	Notes
Fixed	Fixed	Relatively fixed, large and complex applications
Moving	Fixed	Embedded solution, smart media applications
Fixed	Moving	Cooperative sensor solution, industrial and monitoring applications
Moving	Moving	Target tracking applications using intelligent sensors

1.5.3 Intelligence-Based Classification

This classification of sensor applications makes use of the grouping discussed in Section 1.2, the historical sensors development with three classes of applications for TST, SSD and DSS.

This classification can be regarded as intelligence-based grouping. That is TST applications, like traditional sensor solutions come with minimum basic intelligence and normally operate in isolation with their full capability composed upon their transduction sensing and/or actuating process.

TST Based Applications

Upon their definition the TST sensors' main objective is the conversion of the phenomena of interest of the target media into a useable form of information with the minimum errors and disparity from the original phenomena's measureable data. Any extra enhancements will then be regarded as a bonus and helpful to the process, but normally not expected, because with the system level of the application a later process is expected to apply a further process before making use of the sensor's information.

This class of sensor application covers the widest and most versatile group of sensing devices including well-established heavy duty industrial, automation traditional, medical sensors and specialised sensors and actuators where the complexity of transduction inhibits their viability for cost effective mass production. Some require work under enormous constraints and operational limitations during their use and cost effectiveness for large-scale markets.

For example, for measuring the temperature we have many sensing options of low cost electrical and electronic devices. These are easy to use and being made abundantly available at a cost of next to nothing. But, as a standalone application there is no potential market requirement unless being integrated with other phenomena, whilst most chemical sensors and biosensors can provide potential desirable applications and come with serious practical constraints restricting their use to only a few costly application cases. In general, viable TST applications suffer from many practical limitations including:

- Weak Data collection – Immature technology and poor transduction process reading due to lack of proper control over the media limits the application's capability.
- Physically awkward placement of the device make it hard to get close to the object or PoI in the media.
- Space limitations often restrict sensors to get to an effective place (e.g. human body).
- Restrictive signalling issues for the device or media can potentially limit application's capability.

- Operational complexity may affect usability of a sensing application.
- Power consumption and provision of energy could reduce an application's lifetime.
- Maintenance and access during the use could increase the cost and limit usability of these sensors.

Chapter 2 provides some description for new TST devices with potential integrated applications.

SSD Based Applications

Use of the generic term of 'smart' in 'smart sensors', commonly used for sensor device technology does not automatically mean 'device intelligence'. Smart could mean simple enhancement to the information provided in the electrical domain within the sensing device, further classic routine information processing in this domain, or it comes with self-configurability of programmability capabilities using a modular structure. Classic smart sensors provide a modular structure to include the following three main sensing functions of (a) interfacing the transduction process with an analogue source for gathering reliable information in an analogue electrical signal, (b) conversion of an analogue signal into digital and (c) provision of a processor bus structure enabling the exchange of information between various modules or units. We explain further details in Chapters 2 and 3, due to the fact that a single core processor is used to provide the overall control of the system as well as programmable capabilities for the remaining modular interfaces and other functions of the application, the degree of intelligence in these devices is quite restricted, which can get directly reflected onto the range of SSD-based applications which in turn get reflected by the mass production of the core device and therefore cost per unit and thus onto any SSD-based deployment venture project. The following points may be regarded as influencing factors for limiting SSD-based applications:

- Single core architecture cannot be versatile enough for the required variations of applications in practice.
- SSD design optimisation can be easily achieved using conventional processing architectures. This, however, comes at the expense of loss in performance.
- Conventional peripherals can be easily adopted to handle the sensor's common functions such as sampling, data conversions, information coding, protocols and basic communication functions. These deceptive cost cuttings can increase power consumption, reduce flexibility and therefore affect the application's performance significantly.
- Cost effectiveness of a mass production could be easily jeopardised if the final applications deviate considerably from the core process.

The early generation of smart sensors originally aimed to develop an optimised 'single' core device to serve virtually all sensor applications, forcing down the production cost to an ideal approaching zero can capture the maximum share of the global market. This idea, however, has proved to be too idealistic due to:

- Single core architecture cannot be versatile enough to cope with the required variations we need for applications in practice.
- Widening the scales of the market imposes higher generalisation to the core architecture which in turn leads to higher customisation costs to the final product.
- As the market expands emerging new applications demand new features bringing new variations.
- Emerging demands change the market behaviour with which a single common core cannot cope without frequent anti productive upgrades.

DSS Based Applications

In general, as mentioned earlier, the market boundaries for many new technologies like sensors are spreading rapidly from national or regional into global scales, creating new market forces of specialisation in a highly competitive market domain where quality and low cost (mass) production become winning factors. Therefore, SSD type single core mass production shows weaknesses, whilst a multi-core architecture approach supports intelligent DSS applications and can play a significant role for cost effective production of new applications upon their potential for significant reductions in cost, power consumption and size whilst enhancing right functionalities and features such as versatility, flexibility, programmability, configurability, packaging, distribution, and provision of the service during their term of service. New intelligent DSS applications using double or in some cases multi-core processing technology make maximum use of agent style smart sensors which are built around a structured integrated device with a common core processor for generic and common functions and one or two specialised modules most adequate to a specific application, ideally from an off the shelf set of modularly compatible complementary processors. For further descriptions of typical examples for these devices and associated architectures see Chapter 3.

2

Smart Sensing Devices

New smart devices, either in the form of sensing, actuating or combined, represent one of very basic core processes of understanding, manipulating and interacting with the environment, both natural and artificial. They are becoming indispensable components of our modern life, industries and developing intelligent systems. A brief description of these devices shows their tendency, capability and potential when deploying their significant roles for new applications to change our lifestyle once again in the near future.

2.1 Specification and Classification

Considering their long sporadic historical development spreading over a wide range of evolving technologies, sensing devices have been accumulating vast volumes of information, knowledge and data, which one can easily locate in both academia and industries. We therefore see that it is meaningless to cover classic and categorised device details in this volume but instead, we shed light on new and upcoming smart and more intelligent sensing devices and their potential capabilities to construct superior applications using a better understanding of upcoming core technologies through new integration-based grouping structure and associated specification-based classification. That is, in order to make it easy to understand the new approach we provide a device-based grouping using our definition of the three natural generations of TST, SSD and DSS, discussed in Chapter 1, Figure 1.1. We expect this approach would also help us to implement this pragmatic development in order to adopt a new family of flexible and programmable devices to acquire higher flexibility in new smart sensing devices using a new multi-core integration approach. That is, using this method should also enable us to extend our processes through the better use of advanced tools such as computer aided design (CAD) and associated rich software tools for design optimisation and error free fabrication processes, which

Distributed Sensor Systems: Practice and Applications, First Edition.
Habib F. Rashvand and Jose M. Alcaraz Calero.
© 2012 John Wiley & Sons, Ltd. Published 2012 by John Wiley & Sons, Ltd.

in turn enable new hardware and circuit level integration design to shorten the distance between our research laboratories and the production lines and then onto the application scenarios.

Based on their cost-complexity factor, viable sensing devices show their unique dominating centre upon their main functioning capability. For this we have a set of three levels of functionality:

Level 1: domain-sensing transduction functions, common but highly versatile in complexity depending on the domains involved.

Level 2: common core for interfacing and processing basic ICT functions, classic smart sensor technology.

Level 3: application-based intelligence, common modular functions to build new devices through an integrated multi core production approach.

That is, though a device may offer some functions at each level as required by the application, in order to ensure the global market criteria for minimum cost under maximum coverage are met, we need to apply two fundamental design rules of maximum performance at minimal cost-complexity.

These rules are simple and natural as a low cost mass production of versatile sensor devices is only feasible with significant flexibility for the most desired functions, which in turn would enable system integrators and industrial implementers to provide a viable final service covering deployment for the highest possible market, a complex but feasible diverse application scenario. This can be achieved through complying with a linear increased complexity of three basic functions of sensing, communications and intelligence in the right order. That is, level 2 functions should be added only after a satisfactory inclusion of all fundamental level 1 functions. This can be extended to provision of level 3 functions on top of all essential level 2 functions, which of course require all basic level 1 functions. Figure 2.1 shows the adoption of a linear regressive development of all sensor devices using these three levels. The non-masked area of the device represents the functional parts that should be included in the core device. This naturally separates optimum TST device requirement as their sensing process is most complex and therefore they can justify devices with a much less complex integrated solution. However, TST applications, also in need of level 2 and level 3 functions can only adopt these functions at the application phase and not at the production phase per scenario.

The mass product viability of these devices depends on: (a) TST functions, as these devices are dominated by the sensing process where higher level functions are not structured; (b) SSD functions, as these devices are dominated by their common core structured ICT required for interconnecting and basic processing nodes through Ad Hoc form of networking and clustering, whilst their level 3 functions are designed and adopted individually and separately; (c) DSS functions, the high-level flexibility of these devices supported by the multi core processes dominate the application-friendly structure of the device.

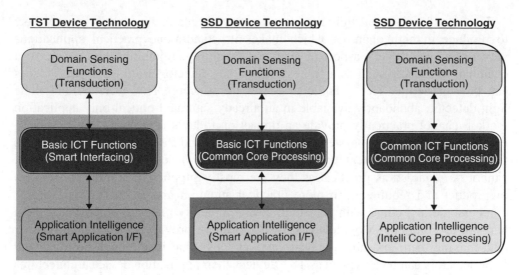

TST Device Technology   **SSD Device Technology**   **SSD Device Technology**

Domain Sensing Functions (Transduction)

Domain Sensing Functions (Transduction)

Domain Sensing Functions (Transduction)

Basic ICT Functions (Smart Interfacing)

Basic ICT Functions (Common Core Processing)

Common ICT Functions (Common Core Processing)

Application Intelligence (Smart Application I/F)

Application Intelligence (Smart Application I/F)

Application Intelligence (Intelli Core Processing)

Figure 2.1 Three-level device structure design associated with three sensor generations of Chapter 1.

In general, level 1, commonly used in all of these devices, plays the fundamental and therefore the most significant role in all sensing and actuating applications. However, as the viability of the product in the market heavily depends on the usable sensing data collected from the domain of interest, an inefficient or non-effective technique for gathering significant valid data could easily become a barrier to the use of the sensor device and therefore threaten the existence of higher-level functions for which no viability of the device production can be predicted. The complex nature of obtaining such valid data potentially depends on the feasibility of accessing the PoI in the domain of interest (DoI), where in many practical cases this access to the domains is not always allowed or could be very complex, even if feasible.

Then, we have the problem of integration of the sensing device with the higher level processing to be used to provide a service for the user. Under today's process integration capability we have full control over a wide range of mass production in silicon materials using many low cost technologies. Our limitations, in cost and other practical issues are therefore mainly associated with integrating them with non-silicon or non-compatible sensing domains. However, if there is full compatibility when integrating electric signals and most circuits and devices into the silicon leading to a hugely cost effective mass production at level 2 and level 3 (core) processes using electric and silicon domains. But if level 1 functions are not naturally compatible or cannot be integrated effectively or interfaced easily with the silicon then an integrated framework cannot be established and the lack of a properly integrated solution of connecting non compatible interworking level 1 devices as attachments to level 2 solutions would make the overall system complex, costly and flow of data loose and unreliable.

For example, a typical light detector is a very simple device and extremely easy to produce in high quantity, which makes its stand-alone practical applications very different, in every aspect of size, cost, mobility, reliability, and so on, to a chemical transducer such as an e-nose poison gas detector. Both these two devices can be regarded as TST (traditional sensor transducers). However, a stand-alone light detector, abundantly available in an already saturated optical-only application market, cannot practically provide an impact on today's user needs, but an e-nose based toxic gas detection system is in very high demand for solving many hazardous gas health and environmental problems. The classic e-nose type solutions, however, cannot be regarded as practical as they are mostly supplied as a rather poor form of product and require much more practical attention and care. They are bulky, cumbersome and come with very high cost-complexity margins which make them hard-to-get solutions. Most low cost devices cannot provide valid reliable reading data making them unusable for many applications. That is, although both gas and light detectors can be classified as TST devices their application device architectures are hugely different. So is their viable deployment. The e-nose has a valid market under a huge price margin whilst the light detector is totally useless for use as a TST device, but it can potentially contribute if adopted into a much bigger system for many useful practical applications such as smart home, remote control, imaging systems integrated with levels 2 and 3 in SSD and DSS device configurations.

The family tree classification of sensor systems shown in Figure 2.2 is the basis of an overall system architecture discussed further in Chapter 3. Its relevance here is associated with the core device structure, indicated in the device grouping provided in Figure 2.1. This further classification of devices extends our application-based classification approach by visualising the natural clear-cut separation of our three generations for capturing their own stable market sectors as they develop further in the future.

The TST class of devices, as shown in Figure 2.2, can take any form of transduction process. They basically convert a domain variable from its originally unusable form into usable and valid form information. This in turn is commonly presented in the electric domain and normally pre-processed before being passed onto level 2. Today, for non-electric and non-silicon domains most of the transformations take place out of the device. However, under emerging technologies domains such as piezoelectric, micro and nanotechnologies are providing alternative or integrated solutions as they become capable of processing non-electric signals. Any further processing and integration with different domains would help the development of smaller, lighter and more effective devices using less complex multi-domain production for sensing various domains in the near future. Further discussions under new developments of new processing domains are provided in the rest of this chapter from Section 2.3.

The devices associated with the second class of Figure 2.2, commonly called smart sensors, come with classic interfaces to the sensing devices and application

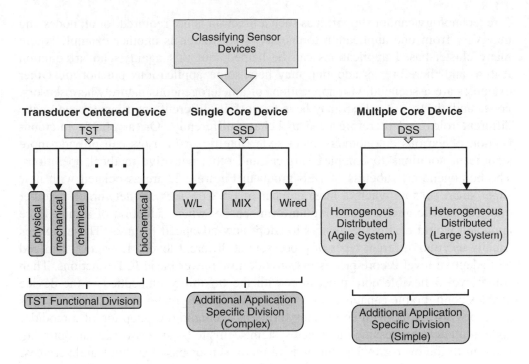

Figure 2.2   Basic application-based classification of sensor devices.

systems whilst providing a basic but highly flexible ICT set of functions. These sensors are normally viewed as small, adaptive and programmable devices so that they can easily interact with each other and form a network. They usually work as a group or clusters, where the cluster head is an active node of a structured or unstructured network. Further descriptions of these devices are provided under typical basic applications provided in the remaining sections of this chapter.

The third class of devices in Figure 2.2, also called DSS devices, are designed based on deployment of the multi core production technology. One may consider this as an evolutionary development from the SSD group as recent discoveries under the WSN and Internet enabled mixed distribution ubiquitous access environment, CDUA as in Section 1.4. That is, recent advanced media, communications and data processing have broken new ground through solutions, which due to practical limitations and market-cost factors cannot be included in a single core power processing device within the classic second class of sensors integrated with the basic level 2 functions in the same device. To mention a few we know the cross-layer technique can be adopted to reduce use of a network's scarce resources at various parts of the nodes of the network. But as decisions for reducing the power consumption through a cross-layer mechanism in a node is set at the application level it cannot be adopted in the same device of the level 2 core processing device, and a single core technology cannot support it as such a function is not required for all nodes and may vary from one application to another. Data fusion is another example where many cluster-based applications can be helped, but this again is not a common feature and therefore its adoption may help some applications but not all. Other examples are associated with applications of a heterogeneous nature where sensors, nodes and clusters of a DSS may have to work in different environments and have different roles and therefore need to behave differently. Or, a reliable interconnection of various autonomous agent style (intelligent) sensors can provide more significant solutions to complex surgical and other sensitive medical operations. The homogeneous subclass of DSS group in Figure 2.2, are associated with new super smart sensors where a large number of DSS devices are networked together under a specific application using uniform sensors where adoption of a multi core approach would enable the devices to adopt new advanced features. These devices usually use two different types of processors at different levels. Low cost, firm and less adaptive level 2 core processors provide low power basic ICT functions. Then some level 3 flexible core processors with the flexibility embedded in the device enables configurable capabilities for wider coverage of greater application scenarios per processor. Now, with continuous shrinking technologies adoption of a modular style multi core processing integrated system seems to be an answer for an optimum solution to the ever-growing demands in many sensor-based systems and services.

2.2 Elementary Sensing Circuits and Devices

Apart from a few other potential and emerging domains such as piezoelectric and nanotechnology, today's dominant processing domain remains electrical and associated silicon the preferred processing domains. That is, in practice most of processing and manipulation of signals and data are usually deployed after the transduction, called post conversion. We have already established very strong foundations for delivering ultimate quality data upon the raw sensing signals obtained to form a domain. We can divide these basic electrical circuit functions into two groups, that is, of elementary functions and advanced functions. Considering that virtually all elementary functions like signal conditioning, signal shaping, sampling, and digital conversions integrated solutions are already made abundantly available throughout the world and that one can easily obtain basic electric and electronic books or look on the Internet for a variety of definitions, and basic knowledge on popular methods. We therefore omit any detailed discussions of the elementary functions from this book into commonly used circuits and interfaces. Instead, however, we explore recent developments of more impressive devices that would help the viewer to appreciate the potential applications and new uses of the sensing devices as they are becoming smart, agile, light, cost effective and more flexible whilst they are less wasteful and perform under reduced energy consumption. One of the areas that one should appreciate is that with recent advances in new materials for integration, new discoveries of wireless communications and new developments in sensor fusion, network coding, clustering, beam forming and cooperative techniques the quality of original sensing signals being delivered at the receiving sinks of the networked distributed devices make very little use of traditional methods. Due to a computerised solution and analysis of the systems, simulation packages for the behaviour and analysis of systems and devices, acquiring knowledge for integrated production, devices programmability, and cost effectiveness of the product and user satisfaction of the systems play more significant roles in managing deployment of the technology and provision of services.

2.2.1   Elementary Electrical Sensors

Considering their natural domain, their relationships with today's analogue and digital worlds of electronic and computing, the three elementary electric components: (a) resistors; (b) capacitors; and (c) inductors can be regarded as basic sensors devices. Under any extended domain-related features they should be able to play a significant inter-mediatory role in representing non-electric sensors and making them usable. The use of these elementary sensors also helps us in a large or complex sensor application for signal integration, signal conversion and signal transformation and in practice can play significant roles in a system optimisation process. Practical examples of the systems are their use in the lab-on-a-chip (LOC) and micro total analysis systems (μTAS).

As a simple sensor, resistors can commonly be used as simple transducers with resistive physical parameters of the element such as changing its resistance with the temperature, as a thermometer device. Equally, its expansion could be used as a simple actuator switching on and off a circuit controlled by a signal. Also, converting light in a photo-resistant device to change the ambient temperature into a resistance is a thermometer, a form of transducer.

A capacitor can be considered as a transducer producing a signal in the capacitor circuit in relation to a phenomenon in the domain affecting the capacitance of the device or equally converting some charging effective parameters of the medium. Many physical, chemical and biochemical domains have been already experienced for the use and there are many application cases throughout the book of capacitors used in the core process of the transduction. Figure 2.3 shows the basic relationship between various adjustable parameters of a capacitor to vary as a sensing device: C is the capacitance, S is the effective area of the plates' surface, d is the distance between two plates and ε_r is the relative permittivity of the material occupying the space between the plates.

Designing a circuit capable of functioning over a large range can be accomplished using the parameters, which show potential linear flexibility in a dynamic configuration. In practice however, a programmable gain circuit should enable the achievement of a wide match in the dynamic range. This means that regardless of

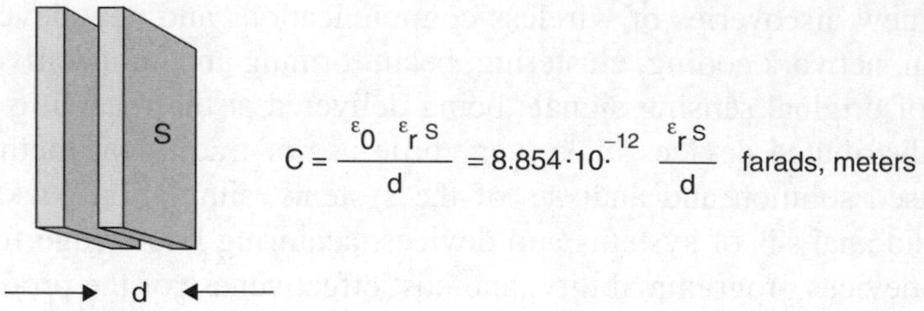

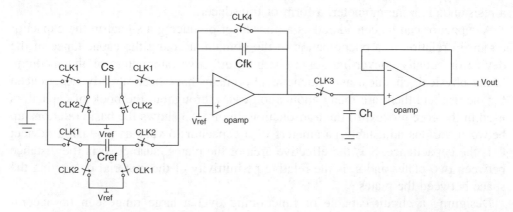

$$C = \frac{\varepsilon_0\ \varepsilon_r\ S}{d} = 8.854\cdot 10^{-12}\ \frac{\varepsilon_r\ S}{d}\quad\text{farads, meters}$$

Figure 2.3 Using a capacitor as a sensing device, where the formula shows the relationship between surface, distance and capacity.

the capacitance change for the experiment range that specific range will be converted to full output voltage by tuning the gain of the system. (Song et al., 2011) In more advanced applications such a dynamic system is implemented using a switched capacitor controlled mechanism. In the simplest format we use two capacitors in the circuit, the sensing capacitor C_S and a reference capacitor C_{ref}. These two capacitors should be laid down in the identical domain and under very similar conditions for producing exactly the same initial capacitance values using the same reference voltage V_{ref}. Then, the sensing capacitor is exposed to the sensing object appearing within its measuring domain without changing the condition in the reference capacitor. As shown in Figure 2.4, in the process of switching all the charge on the sensing capacitor is used to charge a large feedback capacitor C_{fk} before the reference capacitor C_{ref} discharges it. Then, the capacitance difference as a measure of change due to the presence in the sensing domain appears as the residual ΔQ charge,

$$\Delta Q = V_{ref}(C_S - C_{ref})$$

Figure 2.4 A typical switched capacitor and implementation of the capacitance sensing. © 2011 IEEE. Reprinted, with permission, from Song et al (2011) A fully-adaptable dynamic range capacitance sensing circuit in a 0.15mm 3D SOI process.

If the charging and discharging can be repeated, for example N times then under ideal conditions and linearity in the process the resolution of the sensing device can be increased by a factor of N due to a larger total charge accumulated on the feedback capacitor, $\Delta Q = N \cdot V_{ref}(C_S - C_{ref})$. This process also behaves like a low pass filter reducing effects noise and all other independent interfering signals.

An inductive sensor is best used as an electronic proximity transducer detecting metallic objects without touching them. The sensor consists of an induction loop. Electric current generates a magnetic field, which collapses generating a current that falls toward zero from an initial level. The inductance of the loop changes according to the material inside it and since metals are much more effective inductors than other materials the presence of metal increases the current flowing through the loop.

2.2.2 Low Energy Integration

One of most interesting features of multi core sensor devices is their integration capability at level 3 supported at levels 2 and 1. This idea can be approached under two important factors of ubiquity and low energy integration, being a multiplexing or a low level scheduling facilitated by an enabling advanced ubiquitous sensing environment design at the integrated circuit level.

(Seok et al., 2010): In order to achieve such a critically important design require-ment we look into some special cases of an ultra low power core processor, a pW voltage reference circuit and a highly efficient DC-DC voltage convertor for a load in the region of sub-μA. This would enable us to achieve true mm^3 low-cost sensing nodes with hybrid power sources and integrated features such as embedded scaveng-ing subsystems and a micro battery, providing long lifetime and reliable operational applications. For this we can use today's advanced semiconductor technology com-ponents so that the integrated systems are acceptable for mass production for the battery or alternative scarce energy solutions. For example, 1 mm^2 zinc/silver bat-tery with a capacity of 100 μAh/cm^2 and output of 1.55 V a power consumption of 177 pW guarantees a year of battery life. For this we examine a few typical ultra-low power common devices such as power gating processor, memory, reference voltage and DC-DC switched capacitor converter.

For reducing the processor's energy consumption we can reduce the supply volt-age (V_{dd}) of the digital logic. That is, CMOS circuits continue to function with a V_{dd} well below their nominal threshold voltage (V_{th}) down into the sub threshold regime using a specific design approach. The main power saving functions in the processor are shown in Figure 2.5, where a power gating switching approach is used. Where several unrequired modules, in certain periods such as CPU, IROM, clock generator, and IO the block can be completely turned off.

There are some modules that cannot be gated due to their data retention require-ments, but the low leakage custom design modules can help. The processor shows a power consumption 297 nW in active mode and 29.6 pW in standby mode at

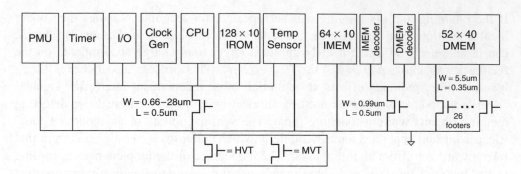

Figure 2.5 Diagram of the different modules provided by a power gating processor (Seok et al., 2010). © 2010 IEEE. Reprinted, with permission, from Mingoo et al (2010) Circuit Design Advances to Enable Ubiquitous Sensing Environments.

$V_{dd} = 0.5\,V$, which, due to its extremely low power consumption makes a good candidate as a core processor for sensing devices.

Another frequently used circuit that can help us with energy is voltage reference. Although several low power voltage references are already known, in many practical cases using an embedded multiple voltage reference system comes with clear benefits. Figure 2.6 shows a 2T voltage reference with virtually no power overhead and comes with good temperature and voltage insensitivity. The zero-voltage-threshold (ZVT) device is identical to a normal MOSFET with a near-zero

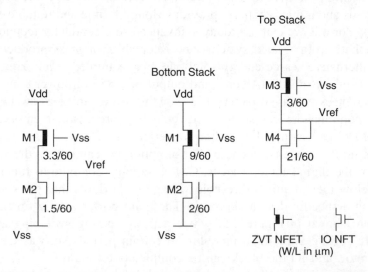

Figure 2.6 The 2T (on the left) and 4T (on the right) voltage references. © 2010 IEEE. Reprinted, with permission, from Mingoo et al (2010) Circuit Design Advances to Enable Ubiquitous Sensing Environments.

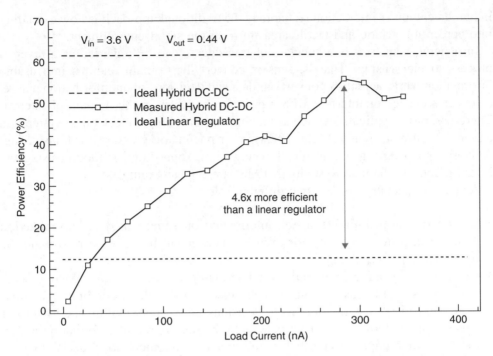

Figure 2.7 Power efficiency of the SCN DC-DC converter. © 2010 IEEE. Reprinted, with permission, from Mingoo S et al (2010) Circuit Design Advances to Enable Ubiquitous Sensing Environments.

threshold voltage. The 4T voltage reference cascades two 2T voltage references to generate a higher output.

Most devices use constant supply voltages to work with them and in practice DC-DC converters provide the voltage. When considering size and integration, Lithium-based thin-film batteries are particularly attractive. However, the relatively high output voltages from most chemical batteries, typically above 3 V, makes the conversion down to for example 0.5 V more difficult. In an experiment using a hybrid approach consisting of a 5:1 Fibonacci switched capacitor network (SCN) manufactured in a 0.13 μm CMOS process in an area of 0.262 mm^2 and providing an efficiency behaviour shown in Figure 2.7 as a function of load current, its efficiency peaks to 56% with a load of 285 nA at 444 mV.

2.3 Actuator Interface Structures

Actuators, or actors, represent a complementary class of devices to the sensors. Being traditionally used in heavy industries for automation or remote control, they are normally visualised in a mechanical format though in effect they can be considered as a realistic approach to fanatical telekinesis, which under recent developments in human computer interaction technology the traditional approach is now extending into perceptible haptic and tactile areas to form the smart environment.

In general, actuators as a subclass of transducers can be regarded as a reversal process to the sensing. That is, sensors convert the domain features into usable information whilst actuators convert our data and decisions of an information nature, to inject a change or interfere with a process or feature in the domain of interest. Therefore, basic applications such as moving or controlling a mechanical system via an electric current, hydraulic fluid pressure or pneumatic pressure for transferring, delivery or converting a controllable energy into some kind of motion using the Micro-Electromechanical Systems (MEMS) are typical examples.

Actuators generally possess two different roles:

- To convert an informative signal, information or a piece of data into an action or physical form of activity for example mechanical in isolation provoking the media or causing a change.
- Agent style dynamic intervention or interfacing a media for a proactive sensing process. In the case of cluster style fixed or moveable interconnected smart devices as in DSS with some local decisions being made within the system or clusters the system can achieve potentially higher degree of information and/or more effective impact using their mobility and autonomous local decisions.

Haptic, as the technical part of the generic term 'haptic', however, is a tactile feedback technology that takes advantage of a user's sense of touch by applying forces, vibrations and motions to the user. This physical stimulation can assist in the creation of virtual objects or augmented reality for controlling objects as an enhancement to the classic remote control mechanism leading to the next-generation actuator technologies offering a wider range of effects. Promising haptic actuator technologies include MEMS, electro-active polymers, Piezoelectric and electro-static surface actuation. To conclude our basic understanding of actuators' technology we may consider them as part of our intelligent sensing devices where these autonomous featured devices can proactively adjust their own working environment for more effective sensing results.

(Jo et al., 2010): One good example is the use of sensors and actuators in emerging applications of smart space. Using reliable and trustworthy systems we can develop ideal smart space, where for more effective use of sensors they can be interactive under the ubiquitous smart space approach using some controllable actuators.

As we discuss later in Chapter 6, smart home applications use a control mechanism to activate the devices. However, most of the existing devices do not comply with the new system and we therefore provide interim solutions using a heterogeneous network. As an example, the integration between heterogeneous protocols using a middleware technique used for the following techniques provides a practical solution:

- Universal Plug and Play (UPnP) Protocol.
- Digital Living Network Alliance (DLNA).
- Ultra Mobile Broadband (UMB).

The only limitations that remain are function-level service detailed description and network dependent device compatibility. The solutions are mainly at the design level, such as making a connection of legacy devices using simple external attachable devices and providing services with user-level abstraction. The same with an abstract service description, such as 'manage illumination' or 'control the temperature', which can be provided easily within a context. Figure 2.8 shows a typical ubiquitous core functional block diagram for the controller device and Figure 2.9 shows a service viewer and controller using a smart-phone.

   Although the need to have tactile sensors has been identified for some time, most traditional sensors dealing with touch are in the form of mechanical sensors, which are far from generating meaningful signals replacing the touch feeling. We therefore prefer to call upon some tactile actuators. To understand this new technology we consider a wearable tactile actuator (Koo et al., 2006). This brief is based on an ambitious research work that uses a two dimensional array of soft actuators

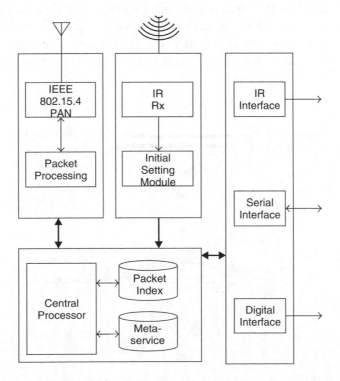

Figure 2.8 Typical ubiquitous core functional block diagram. © 2010 IEEE. Reprinted, with permission, from Kyunam et al (2010) Service-Oriented Actuator for Ubiquitous Smart Space.

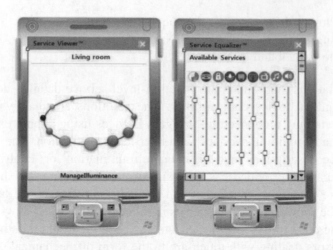

Figure 2.9 Service viewer and controller using a smart-phone. © 2010 IEEE. Reprinted, with permission, from Kyunam et al (2010) Service-Oriented Actuator for Ubiquitous Smart Space.

for stimulating a touch feeling received through new wearable artificial material due to some prepared sets of information. The actuation, in principle, uses some elastomeric material on the electromechanical transduction of a two parallel plate capacitor. A schematic of basic actuation is shown in Figure 2.10. When a voltage is applied across the elastomeric polymer film coated with compliant electrodes on both sides, the material is compressed, reducing its thickness whilst expanded in the

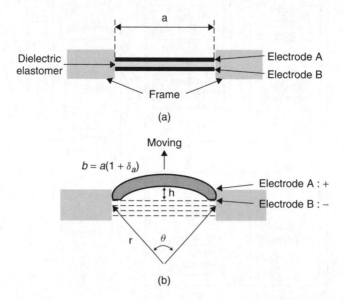

Figure 2.10 Actuation principle of elastomeric material. (a) Without actuation; (b) With actuation © 2006 IEEE. Reprinted, with permission, from Koo et al (2006) Wearable Tactile Display based on Soft Actuator.

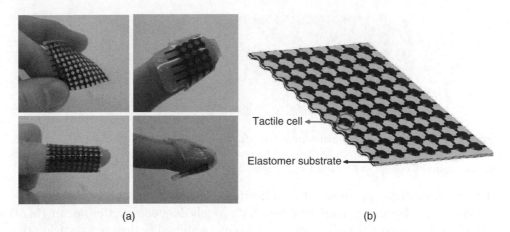

Figure 2.11 Wearable Tactile: (a) Various applications on the finger, (b) Fabrication. © 2006 IEEE. Reprinted, with permission, from Koo et al (2006) Wereable Tactile Display based on Soft Actuator.

lateral directions. This expansion produces a mechanical force within the limited housing provided for the cell causing a concave or convex bending of the elastomer film, also called 'buckling'.

The device contains multiple tactile stimulating cells embedded on a flexible substrate as illustrated in Figure 2.11. However, the device itself is just a single polymer sheet, which has embossed soft actuators polymer structures. As tactile stimulating cells, compliant electrodes with conductive layers are partially printed on the embossed region, which preserves the substrate's flexibility and reduces its cost fabrication. The embossing also secures the direction of bending.

2.4 Physical Phenomena Sensing Devices

2.4.1 Optical Sensors

One of the potential areas of physical sensor development is due to its progress in the optical domain. Their importance is due to the availability of light and their natural compatibility with our daily life activities. Though its wireless version as in free space optics (FSO) has its own features, the most advanced applications are in the form of a guided style, which is complemented by silicon miniaturised integrated products and poses potential developments in a variety of medical and industrial applications.

For many medical and instrumental applications optical sensors are heavily in demand. One good solution uses the silicon-on-insulator (SOI) substrate to design an optical nanowire sensor device. It benefits from the ultra sharp bending capability as well as CMOS integration compatibility. The ultrahigh contrast index and its ultra small cross section of the evanescent field in the nanowire make it strong and

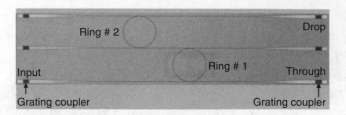

Figure 2.12 Integrated optical nanowire as an optical sensor-on-a-chip solution. © 2011 IEEE. Reprinted, with permission, from Daoxin and Sailing (2011) Ultracompact Silicon Nanowire Circuits for Optical Communication and Optical Sensing.

suitable for practical optical sensing applications. The ultra small arrayed waveguide gratings using photonic crystal reflectors with Mech-Zehnder interferometer (MZI) coupled in the micro rings offers high sensitivity. A digital optical sensor using two cascade rings with different spectral ranges is shown in Figure 2.12. This integrated optical spectrometer can be used as a low cost and portable optical sensor on a chip solution. (Daoxin and Sailing, 2011).

2.4.2   Image Sensing

One of the development areas with an extensive number of applications is the image sensor technology. An image sensor converts an optical image into an electronic signal. It is used mostly in digital cameras and other imaging devices. Early sensors use video camera tubes, but recent devices use a charge-coupled device (CCD) or a CMOS technology, as two of the most popular analogue and digital solutions available today. Figure 2.13 shows a typical CCD image sensor.

Despite industrial pressure to advance the CCD devices for most imaging potential applications CCD cannot match CMOS mainly due to its integration capabilities with other systems. A CMOS device is an active two-dimensional array of CMOS pixel sensors where upon further circuitry next to each photo sensor light energy is converted to a voltage before signals are converted to digital data for further processing or in many cases the light signals can be processed in their original analogue form. One area of concern, however, is their development has been far slower than most others in the last decade.

An analogue image CMOS sensor device commonly comprises of photodiodes, each producing a small photocurrent to be amplified. Here, we briefly examine some of the new features of analogue CMOS sensor circuits and address the existing issues:

- Crosstalk caused by flow of wrong currents in the photo detectors die is high.
- The noise level in CMOS image sensors is too high.
- The existing colour image processing methods such as using colour filter arrays and transverse field detector are far too complex for most applications.

Figure 2.13 Typical CCD image sensor circuit board. This photo is distributed under Creative Common License: GNU Free Documentation License. Http://en.wikipedia.org/wiki/File:Ccd. Sensor.jpg.

- CMOS image sensors are sensitive to artefacts flickering a light source such as incandescent light caused by the AC power signal.
- Most image sensors do not really stimulate the human eye's perception.
- For real time applications such as surveillance, automotive and interactive machine vision the general-purpose architectures cannot perform as effectively as required and also their practical operations such as matrix transformation depends on digital storage at the expense of excessive power and space.

The crosstalk issue of an image sensor can be resolved by adopting a 3D IC fabrication. Here, the photo detector arrays and electronic circuits are free to adopt different processes. For example, in a vertically integrated (VI) CMOS image sensor the VI-CMOS sensor comes in a silicon die of CMOS read-out circuits in a transparent die to an un-patterned array of photo detectors. As shown in Figure 2.14, we can implement this using an operational amplifier with a logarithmic feedback to control the voltage at the input node. The common-drain configuration is found to be the most suitable one for applications involving megapixel arrays from considerations of power consumption. The circuit shown in T is a common-drain transistor. The R_{ph} represents photodiode. (Marzuki et al., 2011).

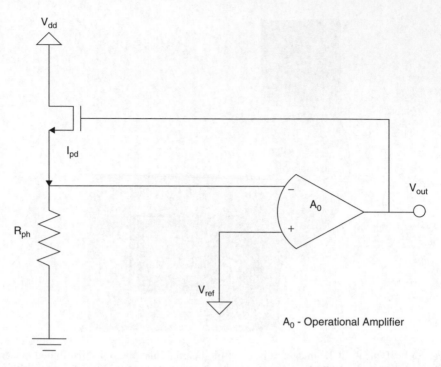

Figure 2.14 Circuit configuration of an operational amplifier using the common-drain VI-CMOS technology.

In order to manipulate the pixel sensors in higher levels in the image processing a photodiode can be used as a current source in a differential pair circuit as shown in Figure 2.15(a) to form a multiplication of photocurrent and input voltages. Rather similar circuit level operation can be applied to the photodiode's sensor voltage to form an improved logarithmic intensity for better visibility, Figure 2.15(b).

2.5   Biological and Chemical Phenomena Sensing Devices

One of the most interesting aspects of biosensors is that most practical biosensors make use of elementary electrics. As mentioned earlier in this chapter, one commonly used element is capacitance. The main feature lies in the fact that it is able to sense a bio phenomenon, or, changes to the physical features of the domain, which can be best identified by capacitors. Having the signal readily in an electrical format rather than needing further processing and manipulation of data forms a suitable media for extensive use of information. That is, micro systems are becoming increasingly an accepted platform for conducting biological sensing and analysis and other health related sensing. Also, measurement activities upon LOC and micro total analysis systems for many applications in clinical diagnosis, cancer research, drug

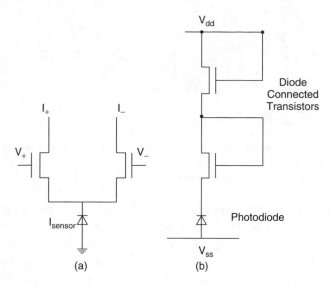

Figure 2.15   Diagrams of CMOS sensor basic operations: (a) Multiplication of two inputs by the photocurrent, (b) Logarithmic intensity compensation.

discovery and screening, stem cell, assays, single cell applications, neuroscience, microbiology, migration and chemo-taxis, intra- and intercellular signalling, cell mechanics, tissue models, assisted reproduction and various applications. (Song et al., 2011) The most interesting part of the LOC system process is to integrate the micro fluidic and biological reagents with CMOS technology. Then, CMOS technology offers integrated circuits, programmability and control and embedded sensors in a single device perform. Furthermore, the shrinking size of CMOS technology allows the circuits to be produced in approximately the same size as the cells and bio-particles being analysed. CMOS technology has been used to create microelectrodes, micro coils, photodiodes, bipolar transistors and ion sensitive field effect transistors (ISFETs) that serve as front end detection devices for a variety of applications ranging from temperature sensing to pH sensing to glucose monitoring and dielectrophoretic manipulation. As mentioned in Section 2.2 the capacitive sensing using the CMOS technology has been developed already. This can be used for antibody antigen recognition, monitoring bacterial growth, DNA detection, toxic gas detection, cell localisation and monitoring. In this system the biological particles and agents of potential targets for sensing vary in size and are in charge of making a universal measurement with a variable range desirable. A typical system would be suitably sensing changes down to the femtofarad range and sensing cell attachment up to the nanofarad range as with cell proliferation measurements. Figure 2.16 shows a capacitance sensor used to detect the presence of a variety of biological materials from protein or antibodies to cells.

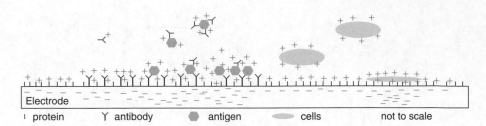

Figure 2.16 Capacitance sensors used to detect the presence of variety of biological materials from protein or antibodies to cells. © 2011 IEEE. Reprinted, with permission, from Song et al (2011) A fully-adaptable dynamic range capacitance sensing circuit in a 0.15mm 3D SOI process.

Taking advantage of new multifunctional transduction processes being made available in many new biochemical transducers, acting as a sensors or an actuators, can be easily integrated into more comprehensive complex devices at far less cost with much less complexity than ever before. One good example of such integration is from Chung et al. for combining three basic health-monitoring functions of *electrocardiogram* (ECG), Accelerometer and oxygen saturation (SpO_2).

This is a wearable ubiquitous healthcare monitoring system using integrated ECG, accelerometer and SpO_2 sensors. This non-intrusive wireless smart multi-sensor can be used by healthcare industry or equally is integrated into smart home development. Upon the cover of wireless sensor network (WSN) its use can be extended to wearable ubiquitous sensor network (USN) with minimum battery power to support RF transmission. The system designed is based on a wide area coverage and can be set up such as node, wearable devices for a wide range of small and large applications on the wrist, chest pulse oximeter sensors with low power ECG, accelerometer and SpO_2 sensors board have been developed using various wearable USN nodes for user's health monitoring. The wearable ubiquitous healthcare monitoring system allows physiological data to be transmitted across a wireless sensor network using IEEE 802.15.4 from on-body wearable devices to a base station connected to a server database (Chung, 2008).

In order for the monitoring system to function per requirement of the ubiquitous healthcare specification it is split into two parts of (1) wearable sensor devices and (2) server as the application central controller. The wearable devices include the wearable USN node, chest sensor belt and wrist pulse oximeter. The physiological measure data collected from the human body is then sent to a base-station for further analysis using IEEE 802.15.4 wireless protocol, where it is passed onto the server to display the output waveforms for ECG, accelerometer and SpO_2 sensors.

The device as a smart sensor node features an ultra low power Texas Instruments MSP430 micro-controller with 10 KB RAM, 48 KB flash memory and 12-bit A/D converter, with some low power operating modes, consuming as low as 5.1 uA in sleep mode and 1.8 mA in the active mode. The CC2420 wireless transceiver node is IEEE 802.15.4 ZigBee compliant. It has programmable output power, maximum

Table 2.1 Specifications of U-Healthcare wearable system

Sensor	Feature	Specification
Accelerometer (MMA7260Q, Freescale)	# Axis	3-axis
ECG (2 electrodes)	Gain	300 (24.8 dB)
	Cut-off Frequency	0.05–123 Hz
A/D Converter (Embedded MSP430F1611)	Sampling Rate	200 Hz
	Frequency Band	2.4 GHz–2.485 GHz
Wireless Transceiver (CC2420, Chipcon)	Sensitivity	−95 dBm
	Transceiver Rate	250 Kps
	Current Draw	Rx: 18.8 mA
		Tx: 17.4 mA
		Sleep: 1 μA
Power	Battery Powered	3.3V

data rate of 250 Kbps and hardware to provide PHY and MAC layer functions. The CC2420 is controlled by the MSP430F1611 through SPI port and a series of digital I/O lines. The node uses M25P80, an 8 Mb serial flash memory with write protection mechanism and can be accessed via a SPI bus. In order to reduce the size of the wearable node a separate module programming board has been adopted, where all application programs are made locally available to a central controller to be downloaded when the node acts as a base-station.

The wearable sensor node comes on a small board integrated with an ECG and a three-axis accelerometer sensor, where the ECG signal is amplified to gain 300 (24.8 dB) and filtered with the cut-off frequencies of 0.05 Hz and 123 Hz. The sensor board also comes with a three-axis accelerometer sensor (MMA7260Q) for measuring the acceleration signals. Table 2.1 shows specifications of the wearable system.

Two wearable u-healthcare sensor systems: a chest belt and a wrist sensor system (Wrist oximeter) have been developed. Figure 2.17 shows the belt wearable chest sensor, which comes with conductive fabric electrodes, an ECG and accelerometer sensor. To obtain an ECG signal from the patient's skin we use a wearable chest sensor, equipped with 8 cm conductive fabric electrode. The wrist sensor comes with a SpO_2 sensor, Nonin OEM Module (Nonin, USA), size 53 mm * 20 mm * 15 mm.

Figure 2.18 shows output waveforms of ECG, accelerometer and SpO_2 sensor when the person was walking, running and resting on a treadmill. The ECG signals provide a record of electrical events occurring within the heart. Acceleration signals provide valuable information about an individual's activity such as walking, running, resting and falling. SpO_2 and heart rate are received from PPG data.

A good example of chemical sensing devices is a practical use of microwave waveguide cavity technology, which can be integrated with micro fluidics sensing for a variety of chemical and biochemical sensor applications. (McKeever et al., 2010).

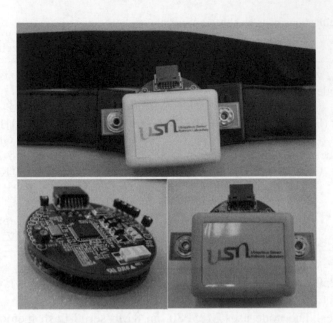

Figure 2.17 Family of wearable chest belt sensor systems. © 2008 IEEE. Reprinted, with permission, from Chung et al (2008) A wireless sensor network compatible wearable u-healthcare monitoring system using integrated ECG, accelerometer and SpO_2.

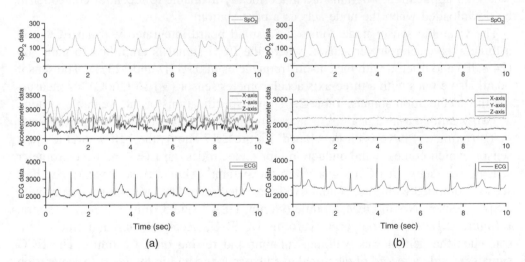

Figure 2.18 Data signals gathered from a person in two cases of (a) Running and (b) Resting on the treadmill. © 2008 IEEE. Reprinted, with permission, from Chung et al (2008) A wireless sensor network compatible wearable u-healthcare monitoring system using integrated ECG, accelerometer and SpO_2.

Micro fluidics micro-scale devices make use of very small processing volumes such as micro litre or smaller spaces to perform the required sensing. These systems use chemical, electrical or optical detection techniques for rapid sensitive analysis of samples. Though higher scale applications are widely available in chemical industry devices, most innovative applications for very small scale devices can be manufactured using micro fabrication technology to produce channels, pumps and valves for a wide range of solutions such as biosensors in the development in micro fluidics. One commercially available biosensor is the glucose monitor, which takes a micro litre of blood and measures the concentration of glucose. These sensors generate a detectable signal by using the chemical reactions in the attachment with a fluorophore or enzymatic label. The use of electromagnetic (EM) waves identifies the dielectric properties of biological substances such as tissues and cell suspensions, where three distinct regions associated with the electrical properties of tissues change with frequency. These are called α-, β-, and γ-dispersions for low frequencies ($<100\,Hz$), intermediate RF frequencies ($100\,Hz$ to $10\,MHz$) and frequencies

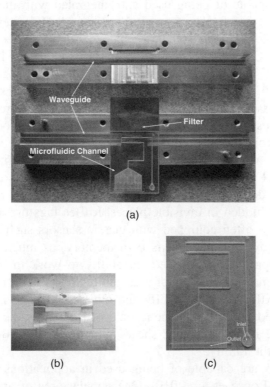

Figure 2.19 A prototype for chemical micro fluidics sensing applications: (a) The prototype, (b) Finite element design filter and (c) The laser machined micro fluidic channels. © 2010 IEEE. Reprinted, with permission, from McKeever et al (2010) Label-free chemical/biochemical sensing device based on an integrated microfluid channel within a waveguide resonator.

above 1 GHz, respectively. The waveguide inserts a filter and a micro fluidic channel, as shown in Figure 2.19(a). A finite element method (FEM) software package from Ansoft, Figure 2.19(b), is used to design the filter with two resonators. The combined feature of these is then used for sensing. The micro fluidic channels in the prototype are engraved 350 μm deep providing an inlet port and two parallel channels align with the resonators and a micro fluidic pump at the outlet, Figure 2.19(c). These sensors are suitable for non-invasive, contactless label-free detection solutions and provide fast and precise measurements.

2.6 Other Sensors and Actuators

The most interesting part of recent sensor-actuator developments is the multi domain integration approach adherent to the miniaturisation process and many domains can be integrated. Therefore, multi functional devices rise to dominate the future of level 1 TST based applications.

Tag style devices, due to their wireless connectivity, can be classified as wireless sensors and capable of being used and integrated with the sensors at various levels. In general we can further classify them into (a) Barcodes, (b) RFID devices and (c) RuBee devices. These standard general-purpose devices have a variety of typical applications in industry. Figure 2.20 shows the core RFID and extended technologies.

RuBee is the IEEE standard 1902.1, a two way, active wireless protocol designed for high security asset visibility applications, as a new low frequency member of wireless networks uses an on-demand peer-to-peer (P2P) protocol. It uses narrowband long wave propagation optimally at 131 kHz signalling at 1,200 baud. Due to the nature of its magnetic radiation and non line-of-sight its indirect propagation loss is minimal as it goes through liquids and buildings so that the rays reach the destination for detecting RuBee tags and communicating with them wherever they are located or hidden in invisible places. RuBee tags use an Internet Protocol address (IP) and are often equipped with classic sensors such as temperature and humidity. A tag may hold data in its own memory, as much as 5 KB is usually enough for their application. RuBee's capability to work in harsh environments, interworking with many thousands of tags and relatively long range like 1 to 30 m make them very different to the RFID and other radio tags. Due to the fact that the energy is carried upon the magnetic field of the waves the antennas are much shorter than classic radio. Table 2.2 shows superiority of RuBee technology over other members of the family.

Some active tags are capable of being used in applications beyond their basic processes. (Benito-Lopez et al., 2010). One popular area of application is various functions of the body. The potential of being used with wearable chemical sensors for monitoring body status during movements, exercising, jogging, running, climbing, and so on would be to check on one or more body fluids such as tears, sweat,

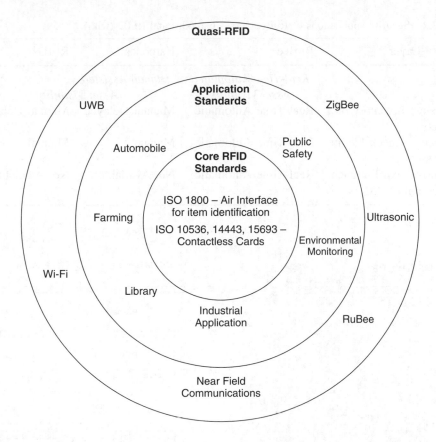

Figure 2.20   Diagram of various wireless standards of RFID and associated technologies.

urine and blood as they occur rather than being sampled, collected and delivered to clinics or hospitals for examination. One popular measurement is to monitor one's body sweat's pH during a particular activity. Here a wearable system built on disposable barcode ionogels can do this measurement using a much simpler process than complicated chemical sensors. The pH of one's sweat can be obtained by looking at the barcode's colour pattern in comparison with a standard chart drawn over a period of time.

As shown in Figure 2.21(a), in this experiment a barcode system with four independent reservoirs uses a polymethyl methacrylate (PMMA) and pressure sensitive adhesive (PSA) material to form a five-layer device using laser cutting tools, Figure 2.21(b). In difficult wearing conditions the device is closed by a two-layer lid, which also has four holes positioned exactly over the polymerised ionogels.

Microphones, for example, are widely used in different applications like cellular phones, electronic devices, hearing aids and a wide variety of measurement devices.

Table 2.2 Security applications of RuBee versus barcodes and RFID (wiki)

Security Layer	RuBee	Barcodes	RFID
	Real-Time Automated Asset Visibility	*Human Assisted Asset Visibility*	
1. Physical Inventory of Assets	Real-Time Automatic	Manual Delayed	Manual Delayed
2. Issuance Check In/Out of Assets	Real-Time Automatic	Manual	Manual
3. Sesntive Exit/Entrance Detection of Assets	Real-Time Automatic	Not Available	Not Available

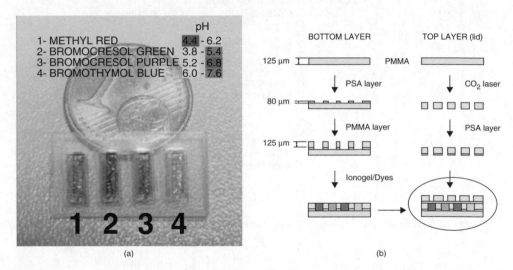

Figure 2.21 Barcode used as a sweat pH reader: (a) Whole system using four indicator dyes and their pH activity range; (b) Barcode fabrication process. © 2010 IEEE. Reprinted, with permission, from Benito-Lopez et al (2010) Simple barcode system based on inonogels for real-time ph-Sweat.

As mentioned earlier the traditional microphones are commonly used from capacitive transduction, but due to their size, weight, energy and other limitations cannot be used for many sensing processes in new applications. These capacitive microphones also show limitations in their sensitivity where use of optical technology enhances the sensing device for its capability due to the capacitance independence feature as a result of much smaller displacement sensing capability.

Here, the integration of optical fibres with micromechanical devices provides a new sensing mechanism with a quality acoustic sensor, which is immune to the external and mechanical noises hampering the quality of traditional devices. Also this technology provides new properties such as smaller devices and out of plane light emission as it consumes much less energy.

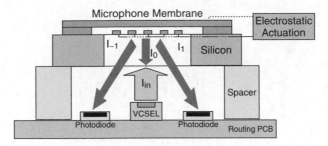

Figure 2.22 Schematic diagram showing the optically enhanced acoustic sensor where a built in phase-sensitive diffraction detector transfers the displacement caused by the acoustic pressure onto the photodiodes. (Qureshi et al., 2010). © 2010 IEEE. Reprinted, with permission, from Qureshi et al (2010) Integrated low voltage and low power CMOS circuits for optical sensing of diffraction based micromachined micrphone.

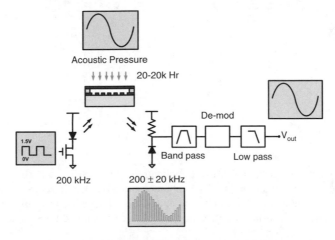

Figure 2.23 For a hearing aid application scenario the transceiver circuit consists of an optical pulse generator and the process of combining a pair of two sensors, an optical and a mechanical (MEMS) coupled together. © 2010 IEEE. Reprinted, with permission, from Qureshi et al (2010) Integrated low voltage and low power CMOS circuits for optical sensing of diffraction based micromachined micrphone.

The schematic in Figure 2.22 illustrates the process of detection, which measures the movements of a high quality membrane being transferred through the diffraction of an optical reflector caused by an optical displacement. Considering the flexibility of the precision of the optical techniques used for this device it can be deployed for both continuous wave and pulse modulating laser signals coming from the base. Use of vertical-cavity surface-emitting laser (VCSEL) can help to reduce the power consumption.

For example, for a hearing aid application our transceiver circuit is like that in Figure 2.23 where we can consider the system as a pair of two sensors, optical

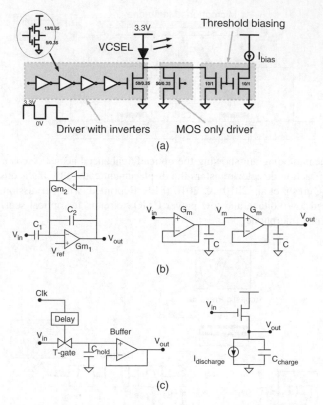

Figure 2.24 Components of optical MEMS application: (a) VCSEL pulse source; (b) The band-pass filter and the low-pass filter; and (c) The coherent and envelope detectors, repeatedly. © 2010 IEEE. Reprinted, with permission, from Qureshi et al (2010) Integrated low voltage and low power CMOS circuits for optical sensing of diffraction based micromachined micrphone.

and mechanical (MEMS) coupled together. A typical incident 200 kHz input pulse train generated by VCSEL is modulated with the audio band acoustic pressure onto the MEMS sensor. At the receiving end of the modulated light, a photodiode along with a passive on-chip resistor enables the front-end detection. The size of the photodiode was designed to completely capture the laser beam. Using a band-pass filter cantered at 200 kHz removes all low frequency content as well as any undesirable 1/f noise. Then, demodulation extracts the desirable 20 kHz band limited audio signal after being processed by a 20 KHz low pass filter removing all possible ripples and unwanted high frequency contents.

All components of Figure 2.23 are shown in Figure 2.24 in further details. The circuit in Figure 2.24(a) is a typically VCSEL pulse source, which usually requires only a few milli-amperes to operate. The two circuits in Figure 2.24(b) are band-pass and low-pass filters. The two circuits in Figure 2.24(c) are the coherent and envelope detectors.

3

Smart Sensing Architectures

There is rising hope for a firm and well-structured networking infrastructure interconnecting the ever shrinking smart sensors for adopting new and low cost integrated miniaturised technologies in conjunction with sensor friendly wireless technology and Internet infrastructure.

To this effect we celebrated the last decade as an exciting period for global push towards deployment of homogeneous sensors using the wireless as an enabler.

Despite heavy global investments and dedication of research communities, however, we feel whilst traditional communications are pushing for higher bandwidths and wider coverage, the promotion of energy critical scarce resource networking such as WSN gives the wrong idea to younger researchers. For this reason we see more research than deployment of smart sensor systems than expected.

To this end the content of this chapter stays brief but objective. We therefore look at the uses and applications of DSS at the architectural level to show the need for a set of simple but adaptive and flexible infrastructures that can use them as a single core or multi-core processing smart sensor and actuator devices to support new viable applications upon an understanding of:

- immediate deployment of ready to use potential applications;
- modularity in the design for a wider deployment coverage;
- appreciation of global scale market factors;
- deployment under agile production for reduced cost per unit;
- inclusion of higher intelligence and autonomy in smart sensing devices;
- further feasibility in networking infrastructure for less networking and more unstructured peer-to-peer data exchange;
- energy and other scarce resources efficient objectives dominate all others.

We have therefore structured this chapter to start with the architectural aspects of sensor nodes, being regarded as smart, flexible and intelligent devices that continue

Distributed Sensor Systems: Practice and Applications, First Edition.
Habib F. Rashvand and Jose M. Alcaraz Calero.
© 2012 John Wiley & Sons, Ltd. Published 2012 by John Wiley & Sons, Ltd.

to play the central role in constructing versatile and adaptive applications upon any unstructured network infrastructure. Operating systems as the basis for all micro-controller based systems provide a basic appreciation of smart devices. We then turn into network formation to discuss through typical examples how a networking infrastructure can support and enable nodes to interact with each other with a discussion on 'node placement'. Some experts view sensor networking as nothing but node placement with few hierarchies of overlay networking. The chapter then considers cross-layer aspects of DSS as a growing technique to bypass top heavy classic layering structures for obvious reasons of saving energy and other scarce resources. With the data centric approach to the future of DSS a brief, partially mathematical, discussion on inference and aggregation would favour the new tendency towards a bottoms up approach to the architecture of clustering and networking the data as the main objective. Then this chapter concludes with two popular case studies of smart camera and collaborative beamforming to exemplify the extreme diversities that DSS needs to handle.

3.1 Smart Sensor Nodes

In meeting the requirements of smart sensor applications the sensor nodes, often simplified as 'the nodes', play a very basic but at the same time most important role in constructing sensor applications. With the basic digital capability integrated in them they can operate with minimum processing power and provide some autonomous capabilities under control of an application, a system or a central controller. In advanced professional and versatile applications such as medical operation systems they, however, can possess more intelligence and upon predefined restrictive rules they can possess higher degrees of autonomy and self-control functions.

Although these extra features may not be required for all applications, in general *clustering* counts as one of the nodes' basic functions. This may give a hierarchical structure for multi-level clustering, but in principle they need to be able to form a cluster under certain joint objectives to carry out some common tasks. Other functions for which they are responsible include collaborative deployment of task sharing, smoothing and local processing of sensing signals, communication in wireless form or when wired with other nodes, or as a sink role for delivery of the network to final destination, or a relaying role, other networking functions, and so on.

Nodes cluster in the form of networkable elements of a homogeneous network and count as a node in an inhomogeneous interactive set of autonomous components. At the cluster level, with a large number of sensors, a homogeneous clustering works very well, but if the system requires the sensor nodes to take different tasks upon different system objectives it indicates very low commonality. In order to provide a stronger system it then encourages less clustering actions, which results in a more independent style operation.

That is, though the clustering feature enables them to benefit from their inherent similarity to other systems, their ability to act independently, but differently as a node, when required, should enable them to count as superior in a networked environment. In principle during an operation a node can take one of the following clustering status:

- unclustered, not being part of any clustering system;
- clustered, already assigned to be a member of an immediate grouping system with clear tasks and objectives;
- cluster-head, usually with membership functionality, it takes responsibility for managing one or more cluster operations.

From the activity point of the view and in order to minimise consumption of scarce resources and reduce interfering with other parts of the system their operations are equipped with a mode for sleep.

In this mode, a node, as an un-clustered sensing device or as a member in a cluster, stays in the state of hibernation with minimal energy consumption whilst waiting for a wake up signal.

In the case of longer sleeping periods with minimum false reactions and quicker processes the sleeping and waking up is essential for higher performing, long life-time applications.

The basic architecture of the nodes and central controllers, however, may not be affected significantly upon the form of interconnection being used. That is, ad hoc style peer-to-peer unstructured connectivity is mostly favoured but under certain circumstances a much firmer interconnection using a traditional networking topology, often in form of clustering, could provide a preferred solution.

Now, in support of our discussions in the previous two chapters, we continue to claim that we need development of more intelligent autonomous smart sensors to respond to the ever-growing demands, whereas one implementation approach is to use modular multi core processors. In order to overcome sensor deployment problems we need to take a new approach to the production of smart devices, as follows:

- further reduction to the device complexity at level 1, limited to very basic functions such as low energy sensing, data enhancement and low power interfacing;
- further reduction to the device complexity at level 2, limited to very basic but flexible ICT functions;
- modular style application core processing at level 3 using off the shelf application-based agile modules to respond to (a) changing market environment, (b) changing application intelligence, and (c) self optimisation under varying conditions, coverage, and so on.

3.1.1 Hardware

Alternative solutions are commercially available under smart sensors. The core processor can make use of a common-purpose low power microprocessor. In special cases they can use application specific multiprocessors available at low cost to play the central role in a smart sensor node. For wireless sensor nodes, usually inclusion of a large onboard memory low power radio module with an onboard or external antenna for wireless communication over the allocated wireless band, offers connectivity over the radio for specific maximum transfer speed as required for the application. In most commercial core-processing platforms (e.g. IMote2) usually there is no ADC onboard allowing it to interface with a user-selected sensor unit or board over its basic connectors. That is, the ADC usually comes with the sensor device. Some sensor devices use compressed sensing for higher sensing efficiency. Using the VLSI for less complex sensing nodes we can easily produce all these modules in one tiny device. Due to the wide variety of hardware configurations we refer the reader to the examples provided in the later parts of this section.

As for most single core and often multi core DSS deployment, the use of commonly mass-produced devices when cost is important and availability in large numbers is critical to deployment for large-scale DSS applications, the use of Mote style wireless devices is highly recommended. For example, Mica family comes with a range of platforms including Mica2, Mica2Dot, Mica2Z using an 8-bit ATmega128l processor. TMote Sky uses a 16-bit MSP430F processor; IMote uses a 32-bit ARM7 processor; XYZ uses a 32-bit OKI ML67Q5002 processor; and Stargate uses a 32-bit Intel XScale processor. For direct utilisation of these platforms one should consider their practical limitations. For example, for a video application the following could limit the product: (a) due to their limited computational capability and memory they cannot perform video acquisition and computing and do not have enough processing bandwidth to maintain quality streaming; (b) mostly come with balanced capability – that is, one with higher computational power comes with lower communication bandwidth radio or cannot support large distances required for video applications; (c) use of existing PDA type platform radios within the IEEE 802.11 family is acceptable for real-time video streaming, but, due to their generic function design approach, they do not optimise for DSS applications and their performance is limited.

3.1.2 Software

The software, usually embedded, is an essential component of the node for real-time communications and data acquisition. Here, we look briefly at four basic components of the software for application development: the operating system, software architecture, time synchronisation and sensing module.

Operating System – There is a variety of sensor node operating systems available worldwide, but in many cases the popular embedded wireless sensor networks, TinyOS, is used (on the IMote2). TinyOS (www.tinyos.net) is a component-based operating system written in the embedded systems version of C, NesC. It usually minimises the memory requirements. The software supports an event-driven concurrency model executing the tasks in an interruptable first-in-first-out (FIFO) fashion, where using asynchronous interrupts allows the system to interact with the hardware.

Software Architecture – A modular style service-oriented architecture with a framework that usually consists of three main elements: foundation services, domain-specific application services, and tools and utilities. A typical application would combine several foundation and application services. Several of the key foundation services support real-time sensing and include reliable communication and synchronised sensing.

Time Synchronisation – Precise time synchronisation serves two key purposes in real-time sensing: (a) providing consistent global timestamps for synchronising the data acquired from different sensor nodes, and (b) scheduling communication. It is quite common to make use of a custom time synchronisation protocol for a particular application.

Sensing Module – In general, the sensing interactions between the core processor and the sensing module are accomplished through driver commands.

In the case of IMote the user first specifies the desired channels, sampling rate and number of samples. The application relays this information to the driver when posting a sensing task. When the driver is initialised, sensing begins and the data is passed to the application through a buffer. Sensing continues until the desired amount of data has been acquired.

Further detailed structured features of smart sensor nodes showing their significant variations from above a generic structure can be identified through the following examples.

One example is of using a monitoring wearable sensor system via the environmental sensor network of a patient in his own home. The device collects a set of physiological health indicators transmitted via some low energy wireless communication to mobile computing devices. Three application scenarios are implemented using the proposed network architecture. The self organised wireless physiological-monitoring system monitors and reports certain patient postures for convenient and timely intervention by a healthcare supervisory system. One use of sensing devices is the cardiac ECG sensor connected directly or indirectly to the Internet at all times. In connection with other sensing devices some wireless communication motes have been integrated with medical sensors. Wearable sensor devices are designed for physical contact with the substance or object being measured to record physiology such as blood pressure, heart rate, ECG, weight, body temperature, and so on. For example, the CodeBlue project developed a range of small and wearable wireless vital sign sensors based on the Mica2, MicaZ, and Telos commonly-used sensor platforms. The developed devices include a wireless pulse oximeter, wireless two-lead EKG, and some specialised sensor motes.

In order to run over one or two dry batteries the system uses micro-sensing chips at its physical layer featuring low-cost computing, communication functions for its sensing objectives over a long period of time. The hierarchical networks are separated so that the communication and control transmission paths use reconfigurable mapping and pipelines for efficiency and a reduction in the power consumption.

The device incorporates Bluetooth technology. The transmission rate of 1 Mbps over 10 or 100 meters allows seamless wireless connections with most wireless with other Bluetooth-enable devices. Under so called Piconet protocol it is able to create small networking groups in which one master device operates up to seven active slave devices. Further details on the physical layer and medium access control for low-rate wireless personal area networks can be found in the IEEE 802.15.4 and ZigBee standards. A typical structure of this sensing system as a smart node is shown in Figure 3.1 (Huang et al., 2009).

Here, the wireless sensor system (WSS) technology generates digital signal measurements from five components of biomedical sensors (for the heart beat and body temperature), analogue to digital converter (ADC), data processing and communication modules. The Bluetooth controller facilitates data transmissions over short distances.

In this system four sensing nodes measure body temperature and ECG. The ADC can sample the changes in an electrical signal, and perform analogue-to-digital conversion, digital filtering, digital amplification and down sampling. The data processing module performs data encoding and encryption. Figure 3.2 shows the fabric belt, the controller box and a lithium battery organised using silver coating textile. It provides 5 MB capacity memory for temporary storage, biomedical digital signal management and transmission to the MCD through a Bluetooth connection using MICAz motes supporting 2.4 GHz ISM band, IEEE 802.15.4 RF transceiver for 250 data rate. The MICAz mote sensors provide physical parameters such as light, temperature and relative humidity.

A featured example for the low end of the market, based on a general study of wireless smart sensor nodes, is from Pei et al., 2008. They look into various general-purpose wireless smart sensors suitable for many DSS applications. The devices being investigated are equipped with various standard wireless connectivity functions designed to support a flexible operation so that the sensor can be programmed for a wide variety of sensing applications. The main stress is on the wireless communication capability of devices to operate at low power at low cost in short distances. Part of the study is related to the use of ZigBee device, where a

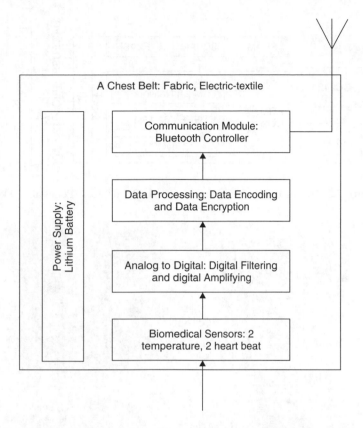

Figure 3.1 Block diagram of a typical chest belt sensor device (Huang et al., 2009).

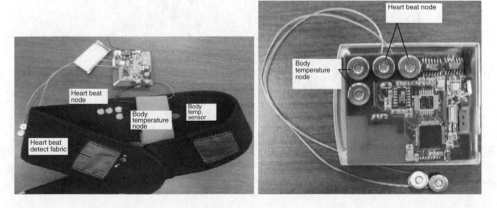

Figure 3.2 Sensor placement in a prototype of a fabric belt with sensing capabilities (Huang et al., 2009). © 2009 IEEE. Reprinted, with permission, from Huang et al. (2009). Pervasible, Secure Access to a Hierarchical Sensor-based Hearlthcare Monitoring Architecture in Wireless Heterogeneous Networks.

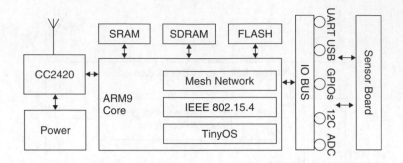

Figure 3.3 Block diagram structure of Cicada 3.0 devices (Pei et al., 2008).

Figure 3.4 Some hardware parts integrated in Cicada 1.0 WSN nodes (Pei et al., 2008). © 2008 IEEE. Reprinted, with permission, from Zhongmin et al. (2008). Application-oriented wireless sensor network communication protocol and hardware platform: a survey.

self-organised WPAN protocol can be set up using IEEE 802.15.4-2003 MAC and physical layers. The variations are ZigBee Alliance, ZigBee 1.0, ZigBee-2006 and ZigBee-Pro. The ZigBee device enables various networking topologies including star, tree and mesh at the device level.

At the device level the node uses an ARM core processor with attached radio and sensor interfaces on an integrated device as in Figure 3.3.

The device developed in Tsinghua University, uses a MC9S08GT60 processor and MC13193 RF devices. It comes with a USB interface for the sink. Sensing functions included the node having an embedded wireless module (EWM) and integration of a temperature and humidity sensor board, a smoke sensor board, an infrared sensor board and a gas sensor board. Figure 3.4 shows some parts of the node used in an experiment.

In this experiment, the home security robot system, sensing functions include monitoring temperature and humidity, gas leak, fire detection disaster and house-breaking. The nodes use point-to-point communications for up to 100 meters in the open environment.

3.2 Embedded Operating Systems

Traditionally, the sensor platforms are composed of a set of interconnected devices. Each device is usually controlled by a microcontroller. The microcontroller has attached sensors, which are used for monitoring the environment, to an external memory and to a communication interface either wired or wireless to transmit the sensed data.

In this scenario, the key resides in the correct programming of the microcontroller, which is the heart that governs the rest of the components. This microcontroller needs to be programmed with a set of tasks, which need to be carried out by the device.

- To mediate the communication microcontroller-sensor. It includes the ability to perform sensing tasks at specific rates, to process the information and to configure the sensors when required. It is commonly referred to as the sensor drivers.
- To mediate the communication microcontroller-communication interface. It included the set of methods for sending and receiving messages using the attached communication interface. Note the incredible amount of implementation of this basic set of methods, which include a wide range of MAC, Routing and Transportation protocols. It is commonly referred to as the networking drivers and protocols.
- To facilitate the microcontroller-sensor communication. This is also especially relevant in devices in which multi-core capabilities are available.
- To perform smart network features such as congestion avoidance, in-network data recovery, multi-path routing for reliability, and so on.

- To collect, process and aggregate locally the information sensed. It includes smart and adaptive data sensing, inference algorithms over the data sensed, artificial intelligent algorithms, smart device-device coordination, and so on.
- To control the access to external memory to store the data.
- To enable an easy reprogramming of the device in order to be flexible enough to be used in a wide set of scenarios and applications.

Without the usage of an operating system, a developer may need the manual programming of all the previous tasks per each different development of distributed sensing platforms. The usage of an operating system into the microcontrollers builds a common structure in which all the previous tasks are modular, extendable and flexible and can be configured easily. Hence, the developer can reduce significantly the efforts needed in order to develop a new distributed sensing platform just focusing on the new functionalities, which are specific to the problem being tackled or those not directly covered in the current implementation of the operating system.

There is a set of features, which define diverse aspects of the operating system. Regarding *architecture*, there are several architectures in the well-known operating systems: monolithic, micro-kernel, virtual machine, layered and modular architectures. Monolithic is based on OS, which does not have any structure. Services provided by an OS are implemented separately and each service provides an interface for other services.

These services are selected, configured and collected statically at compilation time generating a customised OS image which is loaded into the device, *TinyOS* (Hill et al., 2000), *Nano-RK* (Eswaran et al., 2005) and *MagnetOS* (Barr et al., 2002) are some monolithic examples. Micro-kernel architecture using a minimal set of functions loaded in the kernel enables a significant reduction of the kernel size. Then, the rest of the functionalities like filing systems, timing, and so on. are provided at user-level with the associated degradation of performance. Another alternative is the virtual machine architecture in which virtual machines are exported to user programs, which resemble hardware. A virtual machine has all the needed hardware features. The key advantage is its portability and a main disadvantage is typically a poor system performance. *Maté* (Levis and Culler, 2002) and *VMSTAR*[1] are some operating systems which this architecture uses. Layered is another type of OS architecture; it implements services in the form of layers. *MantisOS* (Bhatti et al., 2005) is an example of a layered operating system for sensing platforms. Finally, a modular architecture enables the dynamic configuration of the different modules at run-time stage. *Contiki* (Dunkels et al., 2004) and *LiteOS* (Cao et al., 2008), *CORMOS* (Yannakopoulos and Bilas, 2005) and *SOS* (Han et al., 2005) are some examples of OS with this architecture.

Other feature associated with the OS is the *execution model*. There are different execution models used in the current variety of OS for distributed sensing

---
[1] VMSTAR is available at http://senses.ucdavis.edu.

platforms: event-based, thread-based. The earlier versions of *TinyOS*, and other operating systems like *SOS, CORMOS, EYES* (Dulman and Havinga, 2002) and *PEEROS* (Mulder et al., 2003) do not provide any multi-threading support and application development strictly follows the event driven programming model. They work very well in uni-processor scenarios, but are not very suitable for the next generation of multi-threaded sensing devices. Other operating systems like *MastisOS* and *Nano-RK* offer a programming model purely based on the development and implementation of threads. Moreover, these programming models are not incompatible and some hybrid approaches are available. These approaches are primarily event driven models, but support multi-threading as an optional application level library. This is the case of *TinyOS 2.1* or later, *Contiki* and *LiteOS*.

One of the main important features required in the OSs for distributed sensing platforms is the re-programmability of the sensors in order to enable the adaptation, reusability and flexibility of the platform. Note that efficient reprogramming of devices reduces significantly the cost associated with upgrade processes in the distributed sensing system, since it may avoid the necessity to un-deploy and redeploy the sensors again with the new programmable functionality. An efficient reprogramming enables the remote programming of the device and makes this process easy. Usually, this functionality is associated with code distribution protocols. There are different granularities in which the reprogramming of devices can be done: application, modular, instruction and variable levels. *Application* level reprogramming replaces the entire application image. For example, *TinyOS* and *SenOS* (Koshy and Pandey, 2005) operating systems provide this application level reprogramming. *Modular* level reprogramming replaces or updates the module of an application. There is a wide set of operating systems providing this level of reprogramming: *SOS, MantisOS*, *Contiki* and *CORMOS*, among others. *Instruction* and *variable* levels give the flexibility in changing instructions and tuning parameters of the application respectively. Usually, these fine grain changes in the applications are supported by operating systems based on virtual machine architectures. Maté is a good example of an operating system which enables reprogramming at instruction level.

Scheduling is another aspect of the operating system. It determines the order in which tasks are executed on the microcontroller. In traditional computer systems, the goal of a scheduler is to minimise latency, to maximise throughput and resource utilisation, and to ensure fairness. However, this definition also needs to have in mind power when scheduling is applied into distributed sensing platforms. Having it in mind, the selection of an appropriate scheduling algorithm depends on the nature of the application. For applications with real-time requirements, a real-time scheduling algorithm must be used whereas non-real-time scheduling algorithms are sufficient for other applications. Only a few sets of operating systems have been designed with real-time scheduling algorithms. Some good examples are *Nano-RK*, *CORMOS* and *PEEROS*. However, there is a wide range of operating systems, which are directly designed to be suitable in non real-time environments. Good examples are *TinyOS*, *SOS, Contriki, MagnetOS*, *MantisOS*, *and SenOS*.

Another feature considered in the design of operating systems for distributed sensing platforms is *power management*. In this sense, different approaches can be adopted in order to perform an efficient power management of both micro-controller and attached devices. For example, *TinyOS* provides an API in order to conserve and manage the power of both radio and processor properly. This enables the processor to sleep whenever possible upon the next clock. *MantisOS* achieves energy efficiency using a power-efficient scheduler, which makes the processor sleep after all active threads. Power-aware scheduling is implemented by the idle thread, which may detect patterns in CPU utilisation and adjust kernel parameters to conserve energy. Another OSs like *EYE* or *PEEROS* support a low power mode. This mode started at the lowest possible layer that is hardware. Thus, a sensor node provides several power-saving modes, which are enabled, and different layers and modules in order to reduce energy consumption.

Memory Management is another aspect, which defines the operating systems for distributed sensing. It refers to the strategy used to allocate and de-allocate memory for different processes and threads. There are two main memory manage-ment techniques used: static and dynamic memory management. The *static memory management* is simple and useful when dealing with scarce memory resources, but it is inflexible because run-time memory allocation cannot occur. *TinyOS* and *Nano-RK* are a good example of static memory management. The *dynamic memory management* yields a more flexible system because memory can be allocated and de-allocated at run-time. *Contiki, MantisOS* and *LiteOS* are also good examples of OSs with dynamic memory management support.

Note that the memory management is related to *memory protection*. It is referred to the protection of one process address space from another. This protection can be available in any operating system regardless of the memory management approach implemented. Thus, *TinyOS* offers memory protection in a static memory manage-ment approach whereas *Nano-RK* does not provide protection in the same approach.

3.3   Network Formation

There are many reasons to believe that sensor systems benefit from interconnection, clustering and communications with each other as well as with one or more cen-tralised controllers and databases. In nature, however, they tend not to be committed to following any particular formation. To this effect, they generally benefit from some flexibility to suit sensor applications, being in the form of classic WSN, wired or mix and match configurations, which stem from their task style data seeking and collection behaviour rather than a telecom style service requirement.

Due to unpredictable resource hungry behavioural and other complexities of wire-less networking the idea of ad hoc networking has been proposed so that they

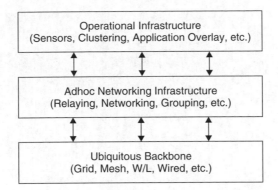

Figure 3.5   Infrastructure to help the deployment of sensing devices.

break away from structured traditional networking into a resource conservative style relaying form of connecting the sensors and clusters together. Then, in order to make use of well-established dominating networking technology we rather use 'unstructured networking' than 'networking' whenever possible. Therefore, DSS applications under unstructured networking approach take immensely versatile networking topologies.

Due to vast common resources available for traditional networking we cannot include a significant discussion, but in a limited space we suggest a designer should be looking at various classic topologies without spending any significant time or energy in drawing up a basic infrastructure, mostly already available and provided by the suppliers before considering a working infrastructure for DSS based applications. That is, a basic topological objective review feature would reveal a subset out of the large number of topologies to limited desirable architectures before designing a typical architecture would help with building up a practical networking solution for the smart sensor nodes. For example, the overlay approach, as shown in Figure 3.5 would help with more direct and efficient structure in little design time.

Following the above infrastructure and CDUA deployment algorithm in Chapter 1 we consider the use of agile and dynamic overlay approach to DSS applications optimum for medium and large systems. The overlay networking in here should be built upon ad hoc networking principles due to their short life span, particularly if some changes are required to the dynamics of the application scenario, if simplified processes using light protocols are desirable. For cluster-based DSS applications clustering and network reconfiguration should happen at a different and lower tier of the network whilst there are overlays for system application, user interactions and network management in the upper tier of the system.

Using a two-tier, in sensor network architecture as shown Figure 3.6, helps to split sensors into sub groups such as clusters or so called micro-sensor nodes

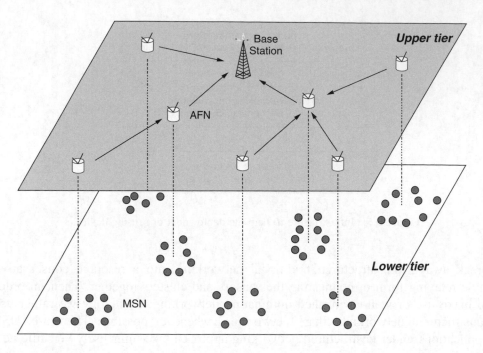

Figure 3.6 Example of a two-tier network topology. Reprinted from: Younis et al (2008). *Strategies and Techniques for node placement in wireless sensor networks: a survey*, with permission from Elsevier.

(MSN) to act as an aggregation-and-forwarding node (AFN) for simplifying the process for reducing processing and energy direct costs to the sensors. AFNs, located in the upper tier possess relaying functions in this layering structure and can adopt different networking topologies, which could vary extensively from one application scenario to another. AFN overlay plays a critical role in overall system performance and therefore all AFNs are required to be energy efficient and prolong an acceptable lifetime.

The most dynamic developments of DSS originated from wireless technology under the common name of WSN. Working at the lower end of the networking hierarchy WSN technology provides unique contactless connections between small and miniature components at extremely low power for most effective smart sensor applications. Standardisation of WSN plays a most significant role in their deployment in DSS for which the generation of unified structure and inter-operational protocols count as the most useful part of this process. Some useful details under application-oriented wireless sensor network (AOWSN) come from Pei et al., 2008.

As discussed in a previous section ZigBee provides a self-organised WPAN protocol. The protocol is an agile and cut-down version of common IEEE wireless networks with the main function of peer-to-peer coordination being starting up a

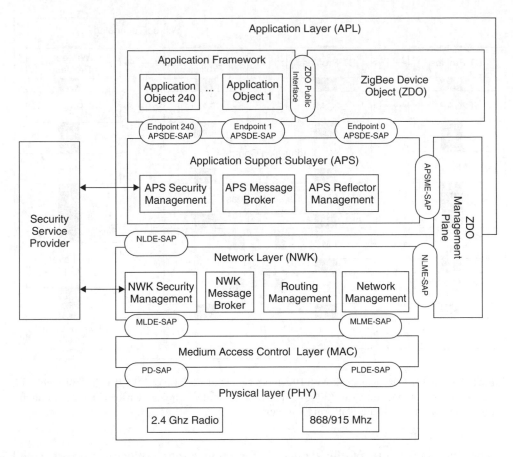

Figure 3.7 Protocol stack for ZigBee protocol (Pei et al., 2008). © 2011 IEEE. Reprinted, with permission, from Zhongmin et al. (2008). Application-oriented wireless sensor network communication protocol and hardware platform: a survey.

network and assignment of main network parameters using IEEE 64 bit address for all variations of addressing such as direct, indirect, group and broadcast addressing mechanisms. From an application point of view ZigBee, as shown in Figure 3.7, provides a suitable mechanism for accessing the network and application layers to manage two service entities of data and management information. The application support sub-layer provides foundations to ease up two main application level objects of networking and application oriented functions.

Another variant of short distance wireless technology is Nokia's Wibree protocol, as shown in Figure 3.8. This open industry standard extends wireless connectivity for much shorter distances of up to 10m for very small devices of under 1 Mbps. It comes with two alternative structures of (a) a stand-alone and (b) dual-mode use with Bluetooth.

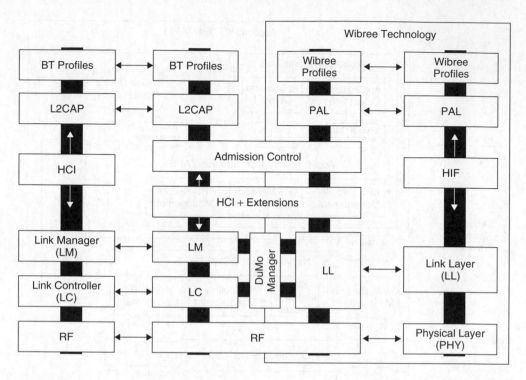

Figure 3.8 Wibree protocol stack with Bluetooth (Pei et al., 2008). © 2011 IEEE. Reprinted, with permission, from Zhongmin et al. (2008). Application-oriented wireless sensor network communication protocol and hardware platform: a survey.

The stand-alone use of the device is for applications collecting a very small amount of data and dual-mode for Wibree used in conjunction with Bluetooth devices. That is, when integrated with a Bluetooth device Wibree enables utilisation of Bluetooth components such as Bluetooth RF more effectively for connecting to tiny battery-powered devices at a small incremental cost.

Other short distance wireless protocols worth mentioning are, Z-Wave, RuBee, UWB, Bluetooth, RFID, IrDA, IEEE 802.15.4a and Wi-Fi. The short range wireless standards of Z-Wave is for two-way communication using a 908 MHz radio system for very low data rates up to 40 Kbps. Whilst RuBee standard is another tag style system for bidirectional, on-demand peer-to-peer network protocol, designed to work with thousands of tags within a range of 3 to 15m which is most efficient with energy using frequency band of 450 kHz. The RuBee tag system is supported by many industries and also has the support of IEEE P1902.1 standard. IEEE P1902.1 enables real-time tag-search using IPv4 addresses, where tags can be located or monitored over the World Wide Web using common search engines.

One good example is the medical application health monitoring case of using a WSS system, with an architecture shown in Figure 3.9.

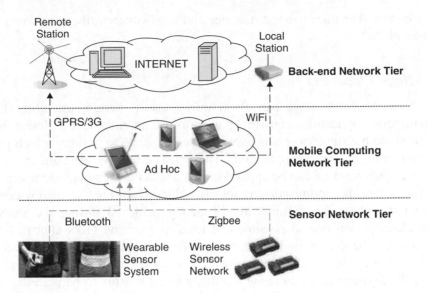

Figure 3.9 Hierarchical network architecture for monitoring health using a wearable sensing system (Huang et al., 2009). © 2009 IEEE. Reprinted, with permission, from Huang et al. (2009). Pervasible, Secure Access to a Hierarchical Sensor-based Hearlthcare Monitoring Architecture in Wireless Heterogeneous Networks.

This application uses three networking tiers: (a) Sensor networking tier, (b) Mobile computing tier, and (c) Back-end networking tier. In the sensor networking tier two types of sensor systems are used, one to capture the individuals' vital signals and one for the environmental physical parameters.

For the mobile computing tier the architecture uses a number of mobile computing devices (MCD), such as the PDA for ad hoc networking to route through either multiple hops or over an infrastructure-based network to connect for a fixed remote or a local station. Each MCD has enough computation capability to capture and analyse physical records so that it does not transmit massive signals to the main data storage over a long period of time, typically a few months or years. However, a significantly high level of data collection storage is required in the back-end network database, but usually through an infrastructure-based operation mode. The ad hoc mode occurs in this tier when one medical person in motion must deliver monitored WSS or WSN data to another staff person on the move, where a mobile-to-mobile text-based alarm message can show real-time abnormal findings via cellular or satellite networks.

For the back-end networking tier the Internet is used at an application-level service provision, exchanging and bringing a vast amount of data from numerous MCD devices. Security of data is an essential part of this application arrangement for a third party set up on the Internet required to open access areas such as hospitals or nursing homes. The third party certificates and keys cover all the way down to the device level.

3.3.1 Node Placement

Under the ad hoc networking part of the infrastructure pseudo structuring of DSS for a particular application scenario we need to use an embedded efficient mechanism for smooth change over a network from one shape to another which plays a significant role in system operation. The most effective mechanism is *node placement*. This function can be applied in two distinct design placement strategies: (a) static, where the optimisation is mostly achieved during the deployment and (b) dynamic, where the optimisation is mainly achieved during the operation. Further classification is also possible and usually upon the tasks allocated to the sensor nodes. The design formation using static node placement can be achieved based on the way the location of the node is being selected. For convenience and speed of deployment when there are too many smart sensors to be placed a simple spreading process like seed plantation in the factory place or by hand would satisfy the requirement. This method is called random placement where there are no rules applied, but it needs to be as uniform as possible. On the other hand, the placement could be more regulated following certain rules like equally separated or a more intelligent form of being denser in some places for more sensing requirement than others to serve the design purpose better. Further controlled node distribution may come from more objective placement. This subclass of static placement is application-oriented such as maximised overall sensing data collection or maximum network life. There is also another static group of node placement where the method makes the best use of communication infrastructure. For wireless, for example, it could be different, using sensor centred design or cluster-based design. In some cases the relay centred design optimisation would be the best if making most use of relay devices counts as the best strategy.

Static node placement strategies provide the network with adequate flexibility for high performance under the design environment as they enjoy high reliability and feature natural stability during the operation. From the performance optimisation, however, their tolerance to change is quite limited. That is, once the position of nodes in the network is decided principally they cannot be changed which can cause performance bottlenecks if intended design objectives deviate from the original proposal.

Alternative design strategy is a network that extends the design procedure into two stages of pre-design networking infrastructure complemented with a complementary node placement during the operation. For example, for a moving target tracking application the networks need to be highly flexible. The infrastructure dictates to the sensors to follow the target or spread all the way from the target to the sink, but their location is decided upon a dynamic strategy whilst managing their critical

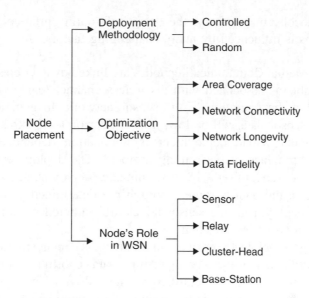

Figure 3.10 Common node placement classes (Younis and Akkaya, 2008). Reprinted from: Younis et al (2008). Strategies and Techniques for node placement in wireless sensor networks: a survey, with permission from Elsevier.

performance with sensors unnecessarily duplicating data and regular replacement of dead sensors to improve the network lifetime upon a more advanced strategy. For example, nodes closer to the target and the sink require much greater energy consuming communications than others and therefore they could die much more quickly. Helpful strategies like regular opportunistic displacement could enhance the overall network's lifetime upon longer than average sensor lifetime. Figure 3.10 shows the main family classes of node placement design.

The dynamic node placement strategies usually go up to ad hoc level of the DSS networking infrastructure, whilst static strategies firm up all the way up to include most of the operational infrastructure in Figure 3.5.

3.4 Networking Protocols

Wireless networks have several restrictions, for example limited energy supply, limited computing power, and limited bandwidth of the wireless links connecting sensor nodes. One of the main design goals is to carry out data communication while trying to prolong the lifetime of the network and prevent connectivity degradation by employing aggressive energy management techniques. (Al-Karaki and Kamal, 2004a) provide a complete overview of the challenges associated with networking protocols in distributed sensing platforms and also provide a survey about different networking protocols. In summary, the design of routing protocols in distributed sensing platforms is influenced by many challenging factors.

- Transmission Media: Communicating nodes are linked by a wireless medium. The traditional problems associated with a wireless channel (e.g. fading, high error rate) may affect the operation of the sensor network. In general, the required bandwidth of sensor data will be low, in the order of 1–100 kb/s.
- Node deployment: Node deployment is application dependent and directly affects the performance of the routing protocol. The deployment can be either deterministic or randomised. In deterministic deployment, the sensors are manually placed and data is routed through pre-determined paths. However, in random node deployment, the sensor nodes are scattered randomly creating an infrastructure in an ad hoc manner.
- Data Reporting Model: Data sensing and reporting is dependent on the application and the time criticality of the data reporting. Data reporting can be classified as: 1) time-driven (continuous), event-driven, query-driven, and hybrid. The time-driven delivery model is suitable for periodic data monitoring; 2) in event-driven and; 3) query-driven models, sensor nodes react immediately to sudden and drastic changes in the value of a sensed attribute due to the occurrence of a certain event or if a query is generated by the base station; 4) a combination of the previous models is also possible. The routing protocol is highly influenced by the data reporting model with regard to energy consumption and route stability.
- Energy consumption: Sensor nodes can use up their limited supply of energy performing computations and transmitting information in a wireless environment. Energy-conserving forms of communication and computation are essential. Sensor node lifetime shows a strong dependence on the battery lifetime. In a multihop network, each node plays a dual role as data sender and data router.
- Node/Link Heterogeneity: Often all sensor nodes were assumed to be homogeneous, that is, having equal capacity in terms of computation, communication and power. However, depending on the application a sensor node can have a different role or capability. Even data reading and reporting can be generated from these sensors at different rates, subject to a diverse quality of service constraints, and can follow multiple data reporting models and this is a decisive factor for the networking protocol.
- Data Aggregation: Since sensor nodes may generate significant redundant data, similar packets from multiple nodes can be aggregated so that the number of transmissions is reduced. This technique has been used to achieve energy efficiency and data transfer optimisation in a number of routing protocols.
- Scalability: The number of nodes deployed in the sensing area may be in the order of hundreds or thousands. Any routing scheme must be able to work with this huge number of sensor nodes. In addition, routing protocols should be scalable enough to respond to events in the environment.

Table 3.1 Types of routing protocols and representative contributions

Category	Representative Protocols
Location-based protocols	MECN, SMECN,GAF, Span, TBF, BVGF
Data-centric protocols	SPIN, Directed Fusion, Roumor Routing, COUGAR, Gradient-based Routing, Energy-aware Routing, Information-Directed Routing
Hierarchical protocols	LEACH, PEGASIS, HEED, TEEN, APTEEN
Mobility-based protocols	SEAD, TDD, Joint Mobility and Routing, Data Mules
Multipath-based protocols	Sensor-Disjoint Multipath, Braided Multipath
Heterogeneity-based protocols	IDSQ, CADR, CHR
QoS-based protocols	SAR, SPEED, Energy-aware Routing

- Network Mobility: Routing messages from or to moving nodes is more challenging since route stability becomes an important issue to take into account when selecting a routing protocol.
- Coverage: Each sensor node obtains a certain view of the environment. A given sensor's view of the environment is limited both in range and in accuracy; it can only cover a limited physical area of the environment. Hence, area coverage is also an important design parameter in sensing platforms.
- Quality of Service: Data should be delivered within a certain period of time in some critical applications. Otherwise the data is useless. Latency for data delivery is another condition for time-constrained applications. Hence, energy-aware routing protocols are essential to capture this requirement.

According to the previous indicated challenges, a wide set of routing protocols have appeared in the communication in order to cope with them. Singh et al. (Singh et al., 2010a) provide an interesting classification of networking protocols for a distributed sensing platform as seen in Table 3.1 This table shows the different categories of networking protocols and the representative protocols of each category and this classification will drive the layout of this section.

3.4.1 Location-Based Protocols

Most of the routing protocols for distributed sensing networks require location information for sensor nodes. In most cases location information is needed in order to calculate the distance between two particular nodes so that energy consumption can be estimated. Location information can be utilised in routing data in an energy efficient way. For example, if the region to be sensed is known, using the location of sensors, the query can be diffused only to that particular region which will eliminate the number of transmissions significantly.

This kind of protocol, in turn, can be divided into two different approaches: those which are designed primarily for mobile ad hoc networks and consider the mobility of nodes during the design and those in which sensor networks are not so mobile.

A good example of location-based protocol is Minimum Energy Communication Network (MECN) (Rodoplu and Ming, 1999). It maintains a minimum energy network for wireless networks by utilising low power GPS. Although the protocol assumes a mobile network, it works better in sensor networks, which are not mobile. The idea is to calculate a minimum-power topology for stationary nodes, including a master node. MECN assumes that the master node is the information sink or base station.

MECN identifies a relay region for every node. The relay region consists of nodes in a surrounding area where transmitting through those nodes is more energy efficient than direct transmission. The relay region for a node pair (i, r) is depicted in Figure 3.11.

The enclosure of a node i is created by taking the union of all relay regions that node i can reach. The idea is to find a sub-network, which has a lower number of nodes and requires less power for transmission between any two particular nodes. Thus, global minimum power paths are found without considering all the nodes in the network.

This path discovery is implemented using a two-steps localised search algorithm for each node considering its relay region. Firstly, the node takes the position of its neighbourhood in a two-dimensional plane and constructs an enclosure graph. This construction requires local computations in the nodes. The enclosed graph contains globally optimal links in terms of energy consumption. This table is maintained periodically in order to keep the routing table up-to-date. After that, it is found that the optimal links on the enclosure graph use a Belmann-Ford shortest path algorithm with power consumption as the cost metric. In case of mobility the position coordinates are updated using GPS.

The small minimum energy communication network (SMECN) (Li and Halpern, 2001) is an extension to MECN considering possible obstacles between any pair of nodes. Thus, the subnetwork constructed by SMECN for minimum energy relaying is smaller (in terms of number of edges) than the one constructed in MECN. As a result, the number of hops for transmission will decrease using less energy than MECN.

Another good example of location-based routing protocol for distributed sensing networks is the geographic Adaptive Fidelity (GAF) routing protocol (Xu et al., 2001). It is an energy-aware location-based routing algorithm designed primarily for mobile ad hoc networks, but applicable also to stationary sensor networks.

The network area is divided into fixed zones forming a virtual grid. Nodes collaborate with each other to play different roles inside each zone. This interaction,

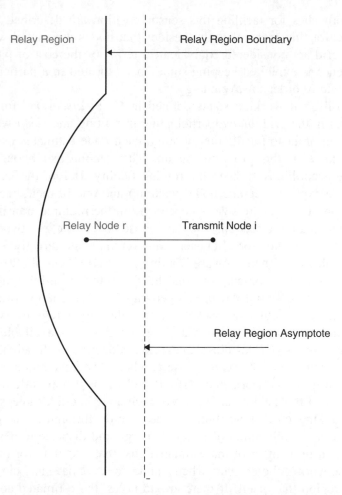

Figure 3.11 Relay Region defined in the routing protocol MECN for delimiting a power efficient area for packet transmission (Rodoplu and Ming, 1999).

for example, includes the election of a node to be awake for a certain time period following a similar approach to the clustered-head node in cluster-based routing protocols, previously described. This node is responsible for monitoring and reporting data to the base station on behalf of the nodes in the zone. Thus, although GAF is a location-based protocol, it may also be considered as a hierarchical protocol, where the clusters are based on geographic location.

GAF conserves energy by turning off unnecessary nodes in the network without affecting the level of routing fidelity. Originally, each node uses its GPS in order to indicate its location and to associate itself with a point in the virtual grid. However, the GPS device could be replaced by a priori insertion of the node location in stationary networks or any alternative approach to know the sensor location.

Different approaches for tackling this sensor position are described in this book. The main idea of this protocol is to consider that nodes associated with the same point on the grid are considered equivalent in terms of the cost of packet routing. This equivalence is exploited keeping some nodes located in a particular grid area in sleeping state in order to save energy.

There are different working states defined in GAF: *discovery*, for determining the neighbours in the grid, *active* participating in routing and *sleep* when the radio is turned off. In order to handle the mobility, each node estimates its leaving time of the grid and sends this to its neighbours. The sleeping neighbours adjust their sleeping time accordingly to keep the routing fidelity. Before the leaving time of the active node expires, sleeping nodes wake up and one of them becomes active.

The selection of the square size is dependent on the required transmitting power and the communication direction. They are chosen such that any two sensor nodes in adjacent vertical or horizontal clusters can communicate directly.

The Geographic and Energy Aware Routing (GEAR) (Yu et al., 2001) is another good example of location-based protocol. The idea is to forward the packets towards a target region. To do so, GEAR uses a geographical and energy aware neighbour selection heuristic to route the packet towards the target region. Two cases can be considered. The first one is when a closer neighbour is available to reach the destination; in this case GEAR picks a next-hop node among all neighbours that are closer to the destination. The second case is when all neighbours are further away (there is a hole in the network) then GEAR picks a next-hop node that minimises some cost value of this neighbour. To do so, each node in GEAR keeps an estimated cost and a learning cost of reaching the destination through its neighbours. The estimated cost is a combination of residual energy and distance to destination. The learned cost is a refinement of the estimated cost that accounts for routing around holes in the network. A hole occurs when a node does not have any closer neighbour to the target region than itself. If there are no holes, the estimated cost is equal to the learned cost. The learned cost is propagated one hop back every time a packet reaches the destination so that route setup for the next packet will be adjusted.

There are two phases: (1) Forwarding packets towards the target region: Upon receiving a packet, a node checks its neighbours to see if there is one neighbour which is closer to the target region than itself. If there is more than one, the nearest neighbour to the target region is selected as the next hop. If they are all further than the node itself, this means there is a hole. In this case, one of the neighbours is picked to forward the packet based on the learning cost function. This choice can then be updated according to the convergence of the learned cost during the delivery of packets, and (2) Forwarding the packets within the region: If the packet has reached the region, it can be diffused in that region by either recursive geographic forwarding or restricted flooding. The former is more efficient than the latter in high-density networks and vice versa. This splitting and forwarding process continues until the regions with only one node are left.

In case the reader is most interested in location-based routing protocols, (Al-Karaki and Kamal, 2004b) provide a complete and comprehensible survey in the field.

3.4.2 Data-Centric Protocols

In many applications of sensor networks, it is not suitable to assign a global address scheme to identify each node due to the inherited sheer number of nodes deployed. Such lack of global identification along with random deployment of sensor nodes makes it hard to select a specific set of sensor nodes to be queried. Therefore, data is usually transmitted from every sensor node within the deployment region with significant redundancy. Since this is very inefficient in terms of energy consumption, routing protocols that are able to select a set of sensor nodes and utilise data aggregation during the relaying of data have been considered. This consideration has led to data-centric routing, which is different from traditional address-based routing where routes are created between addressable nodes managed in the network layer of the communication stack.

In data-centric routing, the sink sends queries to certain regions and waits for data from the sensors located in the selected regions. Since data is being requested through queries, attribute-based naming is necessary to specify the properties of data. One of the most represented examples of data-centric protocol is SPIN (Sensor Protocols for Information via Negotiation) (Kulik et al., 2002), which is based on data negotiation between nodes in order to eliminate redundant data and save energy.

In essence, the idea behind SPIN is to name the data using high-level descriptors or meta-data. Before transmission, meta-data are exchanged among sensors via a data advertisement mechanism, which is the differentiating key feature of SPIN. Each node upon receiving new data, advertises it to its neighbours. Then, the interested neighbours retrieve the data by sending a request message. SPIN's meta-data negotiation solves the classic problems of flooding such as redundant information passing, overlapping of sensing areas and resource blindness, thus achieving energy efficiency. There is no standard meta-data format and it is assumed to be application specific. There are three messages defined in SPIN to exchange data between nodes. These are: ADV message to allow a sensor to advertise a particular meta-data, REQ message to request the specific data and DATA message that carry the actual data.

Figure 3.12 shows an overview of the processes involved in the SPIN protocol. In essence, Node A starts by advertising its data to node B (a). Node B responds by sending a request to node A (b). After receiving the requested data (c), node B then sends out advertisements to its neighbours (d), who in turn send requests back to B (e-f).

Directed Diffusion (Intanagonwiwat et al., 2003) is a very good and interesting data-centric routing protocol based on diffusing data through sensor nodes by using a naming scheme for the data. All data generated by the sensor nodes are

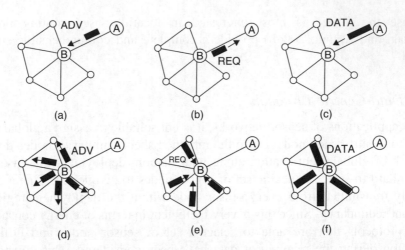

(a)   (b)   (c)

(d)   (e)   (f)

Figure 3.12 Steps involved in the SPIN routing protocol and the associated type of propagation (Kulik et al., 2002). Reprinted from: Akkaya and Younis (2005). A survery of routing protocols for wireless sensor networks, with permission from Elsevier.

named as attribute-value pairs. The main idea is to combine the data coming from different sources in the path (in-network aggregation) by eliminating redundancy, minimising the number of transmissions; thus saving network energy and prolonging its lifetime.

The base station requests data by broadcasting interests. An interest query package uses the list of well-known attribute-value pairs such as name of objects, interval, duration, geographical area, and so on. for describing the data in which the sender is interested. Such query package diffuses through the network hop-by-hop, and is broadcast by each node to its neighbours.

Each node receiving the interest can do caching for later use. The nodes also have the ability to do in-network data aggregation. The interests in the caches are then used to compare the received data with the values in its own interests. Each query also contains several gradient fields. Each sensor receiving the interest set up a gradient toward the sensor nodes from which it receives the interest. This process continues until gradients are set up from the sources back to the base station. In this sense, a gradient is a reply link to a neighbour from which the interest was received. It is characterised by the data rate, duration and expiration time derived from the received interest's fields. Hence, by utilising interest and gradients, paths are established between sink and sources. Note that several paths can be established so that one of them is selected by reinforcement. Figure 3.13 shows the basic protocol steps involved in directed fusion protocol, previously described.

Rumour routing (Braginsky and Estrin, 2002) is another data-centric routing protocol which consists of a variation of directed diffusion and is mainly intended for applications where geographic routing is not feasible. Directed diffusion uses

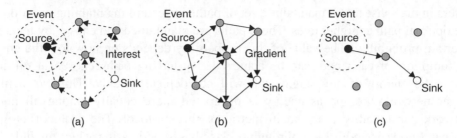

Figure 3.13 Phases of the Directed Fusion protocol (Intanagonwiwat et al., 2003). (a) Interest propagation; (b) Initial gradients setup; (c) Data delivery along reinforced. Reprinted from: Akkaya and Younis (2005). A survery of routing protocols for wireless sensor networks, with permission from Elsevier.

flooding to inject the query to the entire network when there is no geographic criterion, however, in some cases there is only a small amount of data requested from the nodes and thus the use of flooding is unnecessary. An alternative approach is to flood the events if the number of events is small and the number of queries is large. The key idea is to route the queries to the nodes that have observed a particular event rather than flooding the entire network to retrieve information about the occurring events. Thus, it proposes a logical compromise between query flooding and event flooding schemes.

In order to flood events through the network, the rumour routing algorithm employs long-lived packets, called agents. When a node detects an event, it adds such an event to its local table, called an events table, and generates an agent. Agents travel the network in order to propagate information about local events to distant nodes.

Agents traverse a network and inform each sensor it encounters the events that it has learned about during its network traverse. An agent travels the network for a certain number of hops and then dies. Each sensor maintains an event list where every entry in the list contains the event and the actual distance in the number of hops to that event from the currently visited sensor. Therefore, when the agent encounters a sensor on its path, it synchronises its event list with that of the sensor it has encountered. The sensors that hear the agent update their event lists according to that of the agent in order to maintain the shortest paths to the events that occur in the network.

When a node generates a query for an event, the nodes that know the route may respond to the query by inspecting its event table. Hence, there is no need to flood the whole network, which reduces the communication cost. Note that this routing scheme will perform well only when the number of events is small. For a large number of events, the cost of maintaining agents and event-tables in each node becomes infeasible.

Other interesting data-centric routing protocol is Energy Aware Routing protocol (Shah and Rabaey, 2002). Although this protocol is similar to directed diffusion, it differs in the sense that it maintains a set of paths instead of maintaining or enforcing one optimal path at higher rates. These paths are maintained and chosen by means of a certain probability. The value of this probability depends on how low the energy consumption of each path can be achieved. By having paths chosen at different times, the energy of any single path will not deplete quickly. This can achieve longer network lifetime as energy is dissipated more equally among all nodes. Network survivability is the main metric of this protocol. The protocol assumes that each node is addressable through a class-based addressing which includes the location and types of the nodes. The protocol initiates a connection through localised flooding, which is used to discover all routes between source/destination pair and their costs; thus building up the routing tables. The high-cost paths are discarded and a forwarding table is built by choosing neighbouring nodes in a manner that is proportional to their cost. Then, forwarding tables are used to send data to the destination with a probability that is inversely proportional to the node cost. Localised flooding is performed by the destination node, which is responsible for keeping the paths alive. The main disadvantage is that it requires gathering the location information and setting up the addressing mechanism for the nodes, which complicates route setup compared to the directed diffusion.

In case the reader would like to have a more complete overview about other data-centric routing protocols, (Akkaya and Younis, 2005) provide a complete and understanding review of the field.

3.4.3   Hierarchical Routing

A single-tier network can cause the gateway to overload with the increase in sensors density. Such an overload might cause latency in communication and inadequate tracking of events. Moreover, the single-gateway architecture is not scalable for a larger set of sensors covering a wider area of interest since the sensors are typically not capable of long-haul communication. Networking clustering has been pursued in some routing approaches to efficiently maintain the energy consumption of sensor nodes by involving them in multi-hop communication within a particular cluster and by performing data aggregation and fusion in order to decrease the number of transmitted messages to the sink. Cluster formation is typically based on the energy reserve of sensors and sensor's proximity to the cluster head.

In this kind of protocol, a network is composed of several clumps (or clusters) of sensors. Each clump is managed by a special node, called a cluster head, which is responsible for coordinating the data transmission activities of all sensors in its clump. Figure 3.14 shows a basic layout of a cluster. It is composed of a base station, which only communicates directly to head clusters. Each cluster is composed of a head cluster and a set of simple nodes. All of them communicate to the base station or to nodes in other clusters by means of the respective cluster head nodes.

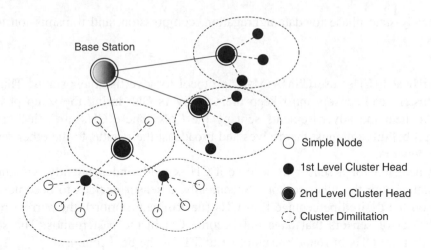

Base Station

○ Simple Node

● 1st Level Cluster Head

◉ 2nd Level Cluster Head

◌ Cluster Dimilitation

Figure 3.14 Layout of a basic cluster, where there are two level of cluster heads. Reprinted from: Akkaya and Younis (2005). A survery of routing protocols for wireless sensor networks, with permission from Elsevier.

Low-energy adaptive clustering hierarchy (LEACH) (Heinzelman et al., 2000) is one of the first and most popular energy-efficient hierarchical clustering algorithms for distributed sensing platforms. In LEACH, the cluster head (CH) role is rotated among the nodes, based on duration. Direct communication is used by each CH to forward the data to the base station (BS). It is an application-specific data dissemination protocol that uses clusters to prolong the life of the wireless sensor network. LEACH is based on an aggregation technique that combines the original data into a smaller size of data that carry only meaningful information to all individual sensors.

Given that energy dissipation of the sensor depends on the distance and the data size to be transmitted, LEACH attempts to transmit data over short distances and reduce the number of transmission and reception operations. The key features of LEACH are:

- randomised rotation of the CH and corresponding clusters;
- local compression to reduce global communication;
- localised coordinaion and control for cluster set-up and operation.

LEACH uses a randomise rotation of high-energy CH position rather than selecting in a static manner, to give a chance for all sensors to act as CHs and avoid the battery depletion of an individual sensor. The operation of LEACH is divided into rounds, each of which has mainly two phases:

- a setup phase to organise the network into clusters, CH advertisement, and transmission schedule creation;

- a steady-state phase for data aggregation, compression, and transmission to the sink.

Cluster heads (CHs) use CSMA MAC protocol to advertise their status. Thus, all non-cluster head sensors must keep their receivers ON during the setup phase in order to hear the advertisements sent by the CHs. These CHs are selected with some probability amongst themselves and broadcast their status to the other sensors in the network.

The decision for a sensor to become a CH is made independently without any negotiation with the other sensors. Specifically, a sensor decides to become a CH based on the desired percentage P of CHs (determined a priori), the current round, and the set of sensors that have not become CH in the past rounds. The sensor nodes that are CHs in round 0 cannot be a CH for the next round.

Once the network is divided into clusters, a CH computes a TDMA schedule for its sensors specifying when a sensor in the cluster is allowed to send its data. Thus, a sensor will turn its radio ON only when it is authorised to transmit according to the schedule established by its cluster head, therefore yielding significant energy savings.

Power-Efficient Gathering in Sensor Information Systems (PEGASIS) (Lindsey and Raghavendra, 2002) is another good example of hierarchical routing protocol. It can be considered as an extension of LEACH. The basic idea of the protocol is that in order to extend network lifetime, nodes need only communicate with their closest neighbours, creating a communication chain and they take turns in communicating with the base station. When one round of all nodes communicating with the base-station ends, a new round starts and so on. This reduces the power required to transmit data per round as the power draining is spread uniformly over all nodes. The chain construction is performed in a greedy way.

The aggregated form of the data will be sent to the base-station by any node in the chain and the nodes in the chain will take turns in sending to the base-station. To locate the closest neighbour node in PEGASIS, each node uses the signal strength to measure the distance to all neighbouring nodes and then adjusts the signal strength so that only one node can be heard. Note that PEGASIS avoids cluster formation and uses only one node in a chain to transmit to the BS (sink) instead of using multiple nodes.

In the PEGASIS routing protocol, the construction phase assumes that all the sensors have global knowledge about the network, particularly the positions of the sensors, and use a greedy approach. When a sensor fails or dies due to low battery power, the chain is constructed using the same greedy approach by bypassing the failed sensor. In each round, a randomly chosen sensor node from the chain will transmit the aggregated data to the base station, thus reducing the per round energy expenditure compared to LEACH.

HEED (Younis and Fahmy, 2004) extends the basic scheme of LEACH by using residual energy and node degree or density as a metric for cluster selection to achieve power balancing. It operates in multi-hop networks, using an adaptive transmission power in the inter-clustering communication. HEED is proposed with four goals:

- prolonging network lifetime by distributing energy consumption;
- terminating the clustering process within a constant number of iterations;
- minimising control overhead;
- producing well-distributed CHs and compact clusters.

HEED periodically selects CHs according to a combination of two clustering parameters: residual energy of each sensor node (used in calculating probability of becoming a CH) and intra-cluster communication cost as a function of cluster density or node degree (i.e. number of neighbours). The residual energy is used to probabilistically select an initial set of CHs while the intra-cluster communication cost is used for breaking ties. The HEED clustering improves network lifetime over LEACH clustering because LEACH randomly selects CHs (and hence cluster size), which may result in faster death of some nodes. The final CHs selected in HEED are well distributed across the network and the communication cost is minimised. However, the cluster selection deals with only a subset of parameters, which can possibly impose constraints on the system. These methods are suitable for prolonging the network lifetime rather than for the entire needs of WSN.

An energy-efficient routing paradigm is proposed that utilises data aggregation and in-network processing to maximise the network lifetime as proposed under Virtual Grid Architecture routing (VGA) (Al-Karaki et al., 2004). Due to the node stationary and extremely low mobility, a reasonable approach is to arrange nodes in a fixed topology. A GPS-free approach is used to build clusters that are fixed, equal, adjacent and non-overlapping with symmetric shapes. For example, square clusters can be used to obtain a fixed rectilinear virtual topology.

Inside each zone, a node is optimally selected to act as cluster head. Data aggregation is performed at two levels: local and then global. The set of cluster-heads, also called Local Aggregators (LAs), perform the local aggregation, while a subset of these LAs is used to perform global aggregation. However, the determination of an optimal selection of global aggregation points, called Master Aggregators (MAs), is a NP-hard problem. Figure 3.15 illustrates an example of fixed zoning and the resulting virtual grid architecture (VGA) used to perform two level data aggregation. Note that the location of the base station is not necessarily at the extreme corner of the grid, rather it can be located at any arbitrary place.

The objective of selecting a number of MAs out of the LAs that maximise the network lifetime is the purpose. A way of addressing that is that LA nodes form, possibly overlapping, groups. Members of each group are sensing the same phenomenon, and hence their readings are correlated. However, each LA node that

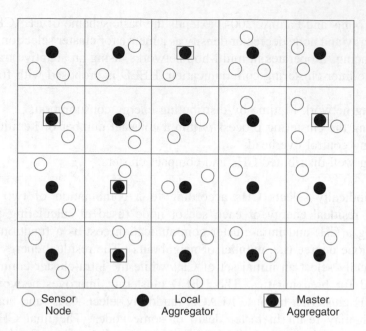

Figure 3.15 Example scenario where a Virtual Grid is built with both local and master aggregators.

exists in the overlapping region will be sending data to its associated MA for each of the groups it belongs to. This can be implemented using a similar approach to a bin-packing problem providing an approximate algorithm which produces results which are not far from the optimal solution.

For a complete survey about hierarchical protocols for wireless distributed sensing platforms, (Singh et al., 2010b) provides a complete state-of-the-art in the topic.

3.4.4 Mobility-Based Routing Protocols

Mobility brings new challenges to routing protocols in distributed sensing platforms. Sink mobility requires energy-efficient protocols to guarantee data delivery originated from source sensors toward mobile sinks. A network with a static sink suffers from a severe problem, called an energy sink-hole problem, where the sensors located around the static sink are heavily used for forwarding data to the sink on behalf of other sensors. As a result, those heavily loaded sensors close to the sink deplete their battery power more quickly, thus disconnecting the network.

This problem exists even when the static sink is located at its optimum position corresponding to the centre of the sensor field. To address this problem, a mobile sink for gathering sensed data from source sensors is suggested. In this case, the sensors surrounding the sink change over time, giving the chance to all sensors in the network to act as data relays to the mobile sink and thus balancing the load of data routing on all the sensors.

One important contribution in the mobility based routing protocols is the data MULES approach (Shah et al., 2003). It is based on an architecture in which mobile entities are moving in the sensor field and collect sensed data from the source sensors. After that, when mobility entities are in proximity to deliver the information to an 'access point', which is any node acting as a base station, this will change the data into the network. Figure 3.16 shows an example scenario in which a bus acts as a mule collecting the data available in the different sensing devices when they are in proximity and then sending it to the base station when it come back home. Note that MULE architecture helps the sensors save their energy as much as possible and thus extend their lifetime. Since the sensors directly communicate with the MULEs through short-range paths, they deplete their energy slowly and uniformly. This protocol is explained in detail in Chapter 7 since it is intensively used in many outdoor application scenarios. Moreover, a practical case study is also presented in Chapter 7 in which a case study for sink mobility is fully addressed.

Scalable Energy-Efficient Asynchronous Dissemination (SEAD) (Karp and Kung, 2000) is a trade-off between minimising the forwarding delay to a mobile sink and energy savings. It considers data dissemination in which a source sensor reports its sensed data to multiple mobile sinks and consists of three main components:

- dissemination tree (d-tree) construction;
- data dissemination;
- maintaining linkages to mobile sinks.

It assumes that the sensors are aware of their own geographic locations. Every source sensor builds its data dissemination tree rooted at itself and all the dissemination trees for all the source sensors are constructed separately. SEAD can be viewed as an overlay network that sits on top of a location-aware routing protocol.

In essence, it assumes that one source generates the sensory update traffic possibly on behalf of a group of local sensors. Then, the data updates are disseminated along a tree to the mobile sinks in an asynchronous manner in which each branch of the tree may have its own update rate depending on the desired refresh rate of the downstream observers.

When a mobile sink wants to join the d-tree, it selects one of its neighbouring sensor nodes to send a joint query to the source of the tree, the selected node is called the sink's access node and it is used to represent the moving sink when the optimal d-tree is built, amortising the overhead in the presence of mobility.

The tree delivers data to the fixed access node. In turn, the access node delivers the data to the sink without exporting the sink's location information to the rest of the tree. In this context, the tree is updated only when the access node changes, in fact, as the sink moves, no new access node is chosen until the hop count between

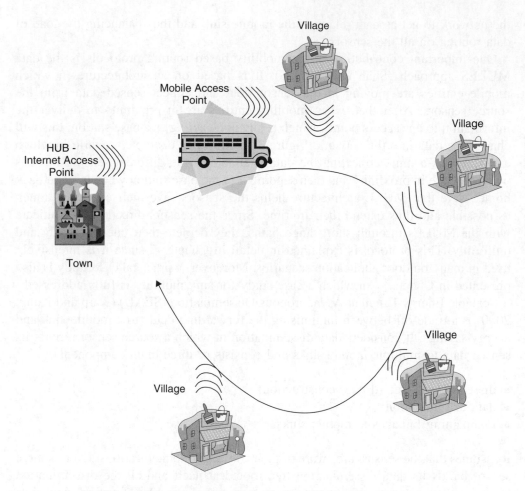

Figure 3.16   Example Scenario for Data Mule routing protocol.

the access node and the sink exceeds a threshold. It achieves a trade-off between path delay and energy spent on reconstructing the tree.

Source data is replicated at selected nodes between source and the sinks defined as replicas store a copy of the source data. They temporarily store the latest data incoming from the source and asynchronous disseminate it to others along the tree.

Another interesting mobile-based routing protocols is Dynamic Proxy Tree-Based Data Dissemination (Chang et al., 2004). It is a dynamic proxy tree-based data dissemination framework for maintaining a tree connecting a source sensor to multiple sinks that are interested in the source. This helps the source disseminate its data directly to those mobile sinks.

In this protocol, a network is composed of stationary sensors and several mobile hosts, called sinks. The sensors are used to detect and continuously monitor some mobile targets, while the mobile sinks are used to collect data from specific sensors, called sources, which may detect the target and periodically generate detected data or aggregate detected data from a subset of sensors.

Because of target mobility, a source may change and a new sensor closer to the target may become a source. Each source is represented by a stationary source proxy and each sink is represented by a stationary sink proxy. The source and sink proxies are temporary in the sense that they change as the source sensors change and the sinks move. A source will have a new source proxy only when the distance between the source and its current proxy exceeds a certain threshold. Likewise, a sink will have a new sink proxy only when the distance between the sink and its current proxy exceeds a certain threshold. The design of such proxies reduces the cost of pushing data to and querying data from the source and sinks proxies.

3.4.5   Other Routing Protocols

In addition to minimising energy consumption, it is also important to consider quality of service (QoS) requirements in terms of delay, reliability and fault tolerance when routing in distributed sensing platforms. Some routing protocols provide QoS, finding a balance between energy consumption and QoS requirements.

Sequential Assignment Routing (SAR) (Sohrabi et al., 2000) is a good example of this kind of protocol. It is based on a table-driven multi-path approach striving to achieve energy efficiency and fault tolerance. Routing decision in SAR is dependent on three factors: energy resources, QoS on each path and the priority level of each packet.

The SAR protocol creates trees rooted at one-hop neighbours of the sink by taking the previous metric into consideration. By using created trees, multiple paths from sink to sensors are formed. One of these paths is selected according to the energy resources and QoS on the path. Failure recovery can be achieved by enforcing routing table consistency between upstream and downstream nodes on each path. Any local failure causes an automatic path restoration procedure locally. The objective of SAR algorithm is to minimise the average weighted QoS metric throughout the lifetime of the network. A handshake procedure based on a local path restoration scheme between neighbouring nodes is used to recover from a failure.

Other routing protocol facing QoS requirements in distributed sensing platforms is SPEED (Tian et al., 2003). The protocol requires each node to maintain information about its neighbours and uses geographic forwarding to find the paths. SPEED strive to ensure a certain speed for each packet in the network so that each application can estimate the end-to-end delay for the packets by dividing the distance to the sink by the speed of the packet before making the admission decision. Moreover, SPEED can provide congestion avoidance when the network is congested.

The beacon exchange mechanism collects information about the nodes and their location. Delay estimation at each node is basically made by calculating the elapsed time when an ACK is received from a neighbour as a response to a transmitted data packet. By looking at the delay values, it selects the node which meets the speed requirement. If such a node cannot be found, the relay ratio of the node is checked.

The relay ratio is calculated by looking at the miss ratios of the neighbours of a node (the nodes which could not provide the desired speed). If the relay ratio is less than a randomly generated number between 0 and 1, the packet is dropped. This is combined with a backpressure-rerouting method used to prevent voids, when a node fails to find a next hop node, and to eliminate congestion by sending messages back to the source nodes so that they will pursue new routes.

In heterogeneity sensor network architecture, there are two types of sensors: line-powered sensors, which have no energy constraint, and battery-powered sensors having limited lifetime. For these scenarios their available energy efficientcy can be explored maximising their potential of data communication and computation. In this section we discuss uses of heterogeneity for WSN applications.

A good example of routing protocol coping with heterogeneity is Information-Driven Sensor Query (IDSQ) (Chu et al., 2002). It tries to maximise information gain and minimise detection latency and energy consumption for target localisation and tracking through dynamic sensor querying and data routing. To improve tracking accuracy and reduce detection latency, communication between sensors is necessary and consumes significant energy. In order to conserve power, only a subset of sensors needs to be active when there are interesting events to report in some parts of the network.

The choice of a subset of active sensors that have the most useful information is balanced by the communication cost needed between those sensors. Useful information can be sought based on predicting the space and time interesting events would take place. In IDSQ protocol, the first step is to select a sensor as leader from the cluster of sensors. This leader will be responsible for selecting optimal sensors based on some information utility measure.

There are two routing paradigms: single-path routing and multipath routing. In single-path routing, each source sensor sends its data to the sink via the shortest path. In multipath routing, each source sensor finds the first k shortest paths to the sink and divides its load evenly among these paths. A good example of routing protocols, which addresses the multipath is Disjoint Paths (Lindsey et al., 2002). This routing protocol helps find a small number of alternate paths that have no sensor in common with each other and with the primary path. The primary path is best available whereas the alternate paths are less desirable as they have longer latency.

The disjoint makes those alternate paths independent of the primary path. Thus, if a failure occurs on the primary path, it remains local and does not affect any of those alternate paths. The sink can determine which of its neighbours can provide it with the highest quality data characterised by the lowest loss or lowest delay after the network has been flooded with some low-rate samples. Although disjoint paths are more resilient to sensor failures, they can be potentially longer than the primary path and thus less energy efficient.

3.5 Cross-Layer Optimisation

Cross-layer optimisation is an escape from the pure concept of the OSI communications model with virtually strict boundaries between layers. The cross layer approach transports feedback dynamically via the layer boundaries to enable the compensation for example overload, latency or other mismatch of requirements and resources by any control input to another layer but that layer is directly affected by the detected deficiency. At least two or more layers have to be involved in a cross-layer optimisation. The last target is to improve the communication by means of collaborative working among such layers. Thus, several research works are available in the literature using different sets of layers to be optimised. In case the reader is keen to learn more, Melodia et al., 2006 provides a complete state of the art in different cross-layer optimisations for WSN.

A clear example is the cross-layer optimisation between MAC and Routing OSI layers. This optimisation is usually focused on improving the performance and reducing power consumption in end-to-end communications. For example, avoiding congestion in wireless sensor networks not only causes packet loss but also leads to excessive energy consumption. Congestion needs to be controlled in order to prolong a system's lifetime and to increase network performance. The Priority-based Congestion Control Protocol (PCCP) (Wang et al., 2007) is a power-efficient traffic control congestion protocol which is trying to mitigate both node and link level congestions. This congestion control protocol is based on the cross-layer optimisation of the medium access control (MAC) layer and the routing layer.

Let us assume a scenario in which sensor nodes are sending continuous upload traffic to the sink node since this is one of the most typical scenarios for evaluating congestion avoidance protocols. Moreover, it assumes a CSMA-like MAC protocol. Each sensor node could have two types of traffic: source and transit. The former is locally generated at each sensor node, while the latter is from other nodes. Note that a given sensor can be an intermediate node for a number of other sensors since it relays the traffic to all of them. Then, when a node is source node and intermediate node at the same time, both transit and source traffic converges at network layer and has to be relayed to the sink node. Henceforth, the intermediate node could be referred to as parent and the source nodes as children.

The output traffic rate of the node is determined by the time invested in the MAC layer itself in processing such messages. Moreover, with the assumption of CSMA-like protocol, the number of active sensor nodes as well as their traffic density influences such performance.

The congestion control approach is based on flexible and distributed rate adjustment in each senor node as shown in Figure 3.17. It introduces a scheduler between

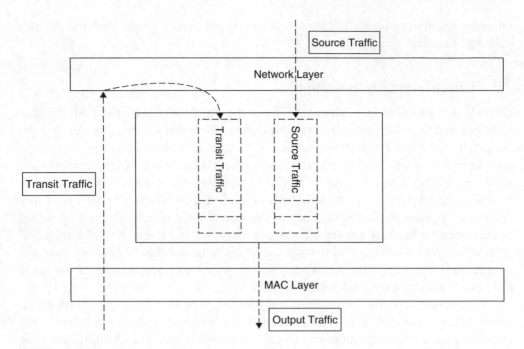

Figure 3.17 Common Congestion Control mechanism using two queues between MAC and Network layers (Wang et al., 2007).

the network layer and the MAC layer, which maintains two queues: one for source traffic and another for transit traffic. Then this protocol can adjust the source traffic generated reducing locally the sampling (or reporting) frequency and also can adjust the transit traffic through rate adjustment at the previous adjacent node.

The two queues can be used practically to guarantee fairness between source and transit traffic, as well as among all the nodes. To this end, the queues can use effective queuing protocols such as weighted fair queuing (WFQ) or weighted round robin (WRR). Moreover, a priority index of source traffic and transit traffic is used as the weight for establishing a level of priority for the different traffic flows.

PCCP consists of three components: intelligent congestion detection (ICD), implicit congestion notification (ICN) and priority-based rate adjustment (PRA).

The ICD component is in charge of the detecting congestion situation. To do so, it relies on packet inter-arrival time and packet service time. The first one is defined as the time interval between two sequential arriving packets from either source or for the transit traffic. The second one is defined as the time interval between when a packet arrives at the MAC layer and when its last bit is successfully transmitted. Note how the join of these values provides helpful and rich congestion information.

A congestion degree (CD) index is defined as the ratio of average packet service time over average packet inter-arrival time over a pre-specified time interval. Note that CD > 1 indicates that inter-arrival time is smaller than the service time and vice versa which gives CD < 1.

The ICN consist of piggybacking the congestion information in the header of data packets. The congestion notification is triggered either when the number of forwarded packets by a node exceeds a threshold or when the node overhears a congestion notification from its parent node. The piggybacked information includes: mean packet service time, mean packet inter-arrival time, global priority and the number of intermediate nodes.

A sensor node then computes its global priority index by summating its source traffic priority index and all the global priority index of its child nodes, which is piggybacked in the received data packets.

The PRA is the one used to manage scenarios in which sensors have different priorities. Each sensor node is given a priority index. Hence, this component is designed to guarantee that the node gets more bandwidth proportionally to the priority index. Note that the global priority index previously introduced in the ICN composed is calculated using such a value. In summary, this component performs the exact adjustment of the scheduling rate in order to mitigate/reduce the congestion in the network using the information provided by the CD and priority indices.

PRA needs to adjust the scheduling rate and the source rate after overhearing congestion notification from its parent node, in order to control both link-level congestion and node-level congestion. To calculate the scheduling rate, the following cases are proposed to calculate such adjustment:

Case 1. If one of the child nodes is turned idle, packet inter-arrival time of the parent is increased and congestion degree is decreased. This fact means that a parent node could scale-up its scheduling rate to improve link utilisation.

Case 2. If the child node becomes active, the reverse actions are taken. Note that doing this action; the parent node sets its scheduling rate to the maximum allowable rate in order to guarantee fairness and high link utilisation.

Case 3. When the number of intermediate nodes remains constant but some nodes don't have enough traffic, the congestion degree will become smaller, and therefore such a node can scale-up its scheduling rate to improve link utilisation.

Case 4. When any node previously producing small traffic now produces more traffic, congestion degree is increased and may possibly be larger than 1. Then, the parent node i resets its scheduling rate back to the allowable rate as in case 2 so that it can mitigate congestion while maintaining high link utilisation.

Regarding the source rate, it is updated based on its source traffic priority index and global priority index. Note that the global priority index is used to update the scheduling rate while the source traffic priority index is used to calculate source rate.

Another example of cross-layer optimisation is between physical (PHY) and MAC OSI layers. This optimisation is done usually to reduce the power consumption.

For example, the spatial correlation in the observed physical phenomenon can be exploited for medium access control. A sensor node can act as a representative node for several other sensor nodes and thus, a distributed, spatial correlation-based collaborative medium access control (CC-MAC) (Yu and Yuan, 2005) can use such information in order to improve significantly the performance in terms of energy, packet drop rate and latency. Another alternative is the analysis of how various carrier-sensing schemes affect the exposed-node (EN) and hidden-node (HN) phenomena. These phenomena can be alleviated using a cross-layer optimisation in which it used a MAC-address-based physical carrier sensing to determine if the medium is busy. The idea is that the addresses of transmitter and receiver of a packet are incorporated into the PHY header. Making use of this address information for its carrier-sensing operation, a node can drastically reduce the detrimental effects of EN and HN. In fact, it has been demonstrated by means of simulations that this optimisation yields superior throughput and fairness performance to traditional methods (Chan and Chang-Liew, 2006).

Other examples of cross-layer optimisation involve more than two layers. This is the case of the cross-layer optimisation in which PHY, MAC and routing layers are involved. The idea behind this is usually related to the maximisation of the lifetime of energy-constrained wireless sensor networks. The problem of computing a lifetime-optimal routing flow, link schedule and link transmission powers can be formulated as a non-linear optimisation problem (Madan et al., 2005). Then link schedules can be restricted to the class of interference-free time division multiple access (TDMA) schedules. In this special case, the optimisation problem can be formulated as a mixed integer-convex program, which can be solved using standard techniques. The idea is to provide an adaptive link scheduling and computation of optimal link rates and transmission powers for a fixed link schedule.

The recently emerged cross-layer approach still necessitates a unified cross-layer communication protocol for efficient and reliable event communication that considers transport, routing and medium access functionalities with physical layer effects for WSN. Several architectures have been recently provided in order to provide an efficient architectural structure to cope with cross-layer optimisations. A common understanding is the inclusion of a cross-layer module (XLM), which may extend or replace the entire traditional layered protocol architecture that has been used so far. The basis of communication in the XLM module is based on the intrinsic functionalities required for successful communication. The idea is to tweak the XLM module in order to optimise all the layers according to the specific particularities associated with the distributed sensing platform, which has to be deployed. In fact, note how the previously described works in a cross-layer when optimised fit perfectly in this architectural proposal.

3.6   Inference and Aggregation

With the rise of smart sensors for adopting the distributed intelligent sensing the old idea of processing information for more decisive deduction takes a new momentum. The move, under the term *inference* or more precisely 'the science of deduction', attracted many scientists and engineers to adopt new methods for making use of mathematical methods for maximising the information being collected from clustered sensors and other sensing systems.

Avoiding the usual unnecessary jargons on the top of classic data and signal processing methods such as Kalman filtering, we see two distinct approaches to addressing the problem of enhancing our certainty of knowledge/information as *Bayesian inference* and *Dempster–Shafer theory* (DST). Both approaches make systematic use of new information, called evidence, to upgrade our certainty or knowledge. Whilst Bayesian method looks into the source of collecting the information DST provides subjective features to the parameters of source and belief.

Bayesian Inference

The word inference, which comes from the verb infer with a meaning of deduce, means deduction upon evidence. Bayesian inference is a mathematical method for updating the certainty parameters upon receiving new evidence using statistical methods. That is, given a set of decisions [D] being made upon a set of hypothesis [H] using a certainty matrix [C] we have [D]=[C][H]. Then, arrival of new evidence [E] can change the decision to [D']=[C'][H]. Using the classical Bayes theorem:

$$\mathbb{P}(H/E) = \frac{\mathbb{P}(E/H)\mathbb{P}(H)}{\mathbb{P}(E)}$$

Bayes theorem quantifies the probability of hypothesis H, given that an event E has occurred. P(H) is the a priori probability of hypothesis H, P(H|E) states the a posteriori probability of hypothesis H. P(E|H) is the probability that event E is observed given that H is true.

Given multiple hypotheses, Bayesian inference can be used for classification problems. Then, Bayes' rule produces a probability for each hypotheses H_i. Each H_i corresponds to a particular class:

$$\mathbb{P}(H_i/E) = \frac{\mathbb{P}(E/H_i)\mathbb{P}(H_i)}{\sum i \mathbb{P}(E/H_i)\mathbb{P}(H_i)}$$

Dempster–Shafer Theory

The Dempster–Shafer theory is a generalisation of the Bayesian theory, where a degree of belief (mass) is represented as a belief function instead of probability distribution of a Bayesian.

Often used as a method of sensor fusion, the Dempster–Shafer theory is based on two ideas: obtaining degrees of belief from subjective probabilities and a rule for combining degrees of belief. In practice this could be applied using extra sensing or information gathering mechanism to enhance the sensors' data.

In the course of design and engineering of data-efficient smart sensors, however, many other development approaches have been adopted for wireless enabled high performance WSN application systems. These approaches usually come under terms of *aggregation, data fusion* and *sensor fusion*. Due to their similarity term information fusion will be represented by data fusion.

Aggregation

Considering successful deployment of the DSS, particularly when it uses wireless networking (WSN) due to restrictions in the practical systems such as battery operated devices or radiation interference and so on, the network needs to run under several constraints whilst using minimum scarce resource conditions. One most dominant constraint affecting an application's lifetime operation is the energy.

By definition classic smart sensors are required to carry out three basic functions of sensing, computation and communication. As in the original designs, both sensor nodes and whole application are well aware of the resources available for sensing and computation so surprises may come from the resource consumptions associated with communication. For fixed applications and established wired media using reliable channels variations are not surprising, but in the case of ubiquitous access with wireless links the variations can be very severe.

A simplified energy formula for a signal transmitted over a distance of d is $E_d = k.d^n$, where k is a constant but n varies extensively upon the path being taken for the transmission. For a usual ad hoc connection the path could be direct, indirect via some relaying pass-on nodes and may encounter reflections, refractions with or without unpredictable noise and interference. To appreciate the extent of energy requirements a guess of 10 m transmission costing as much energy as required in the execution of 1000 instructions is reasonable.

Aggregation is a technique commonly used for energy conservation. At the aggregation point the node collects readings from a subset of nodes, combining the values for a single message forwarded to the destination node. That is, when multiple copies of the same data generated independently for a base station via long paths the intermediate node upon aggregation process suppresses the duplicated transmission copies and thus saves energy consumption, bandwidth and other resources whilst reducing the overall level of interference in the network. Most aggregators provide basic extra computations, often as simple as producing an average of the collected values.

Aggregation could also be combined with other functions in the network, such as routing. Due to the nature of the sensor networks they use resource-constraints routing. If as much routing in sensor networks is data-centric then combining data aggregation with the routing can also save extra activities essential to fault-tolerant reliable routing with the ability to discover alternative routes and self-organise when an existing route fails.

For example, an energy-efficient routing that makes use of aggregation in conjunction with network processing for a maximised network lifetime for low mobility nodes uses a fixed topology routing and divides the region into clusters of sub region square zones. Inside each zone, a node is optimally selected as a cluster head. The aggregation process is performed at two sequential stages by: (a) local aggregators and (b) master aggregators. Figure 3.15 illustrates an example for fixed zoning for virtual grid architecture.

As the energy consumption increases exponentially with distance selection of sensors far away from the centre of the region as a cluster-head can be anti-productive meaning that a more sophisticated aggregation algorithm based on sensor node position is required. As shown in Figure 3.18 the energy efficiency of the above method can be improved, for energy and interference, if any sensor located in the corners of the region with a radius of 'r' beyond 'a' is not selected as a cluster-head.

Aggregation can also be used in higher levels of networking such as supervisory overlay for packet and information processing.

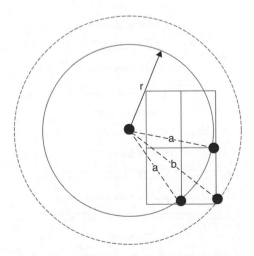

Figure 3.18   Scheme, in which a two level routing aggregation process is shown for improving energy and interference.

Data Fusion

Data fusion encompasses theory, techniques and tools conceived and employed for exploiting the synergy in the information acquired from multiple sources of information such that the resulting decision is in some form qualitatively or quantitatively and in terms of accuracy, robustness, and so on. better than these sources if used individually without such extra synergy exploitation.

Sensor Fusion (Elmenreich, 2002)

In order to produce enhanced data in a more usable format we adopt a sensor fusion process of combining the sensory signals or data derived from sensory systems to enhance its usefulness for various objectives being robustness, extended spatial awareness, temporal coverage, increased confidence, reduced ambiguity, reduced uncertainty and improved resolution. Due to the fact that sensor fusion is a versatile mechanism, which heavily depends on the application we have not commonly agreed to sensor fusion, but in general the process of manipulation has many, say four different stages of processing and refinement converting the raw individual sensor signal into its final format. One generalised process is waterfall sensor fusion shown in Figure 3.19.

The waterfall model is simple and fast, but due to the lack of any feedback data flow it suffers from limited precision in many practical applications. For enhancement some feedback loops such as observe, orient, decide and act (OODA) loop as shown in Figure 3.20.

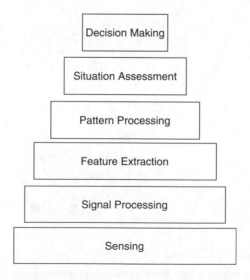

Figure 3.19   Waterfall sensor fusion process.

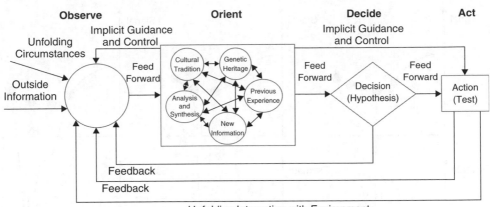

Figure 3.20 Basic loop of observe, orient, decide and act in which it can be seen how data is interconnected along the different steps of the loop.

The concept of sensor fusion, however, could be categorised under some architecture, but very little commercial benefits could be initiated from the modelling. However, there are many interactive distributed sensing applications such as intelligent robots which benefit from a simplified model like omnibus with a long loop of four sequenced processes of (a) hard decision triggering some actuators attached to the sensor nodes; (b) sensing process to observe new information or evidence; (c) the fusion processes such as pattern processing and feature extraction and finally (d) processing the data and making decisions.

Signal processing methods commonly used for reconstructing the samples in connection with the sensor fusion are of three types: (a) smoothing; (b) filtering and (c) prediction. The smoothing is normally applied in real-time, where a sample data is enhanced upon values of some previous and some of the past sampled values. The filtering process adjusts the existing sample according to many of the previous samples. The prediction, however, is usually more complex by making use of previous samples providing an estimated value for a sample immediately or later in the future. Figure 3.21 compares the delay time behaviour of these three processes.

3.7 Case Study: Smart Camera Networks

A *Smart Camera Network* is intended as a distributed deployment of a set of interconnected cameras, which are governed by means of a distributed intelligence for a given purpose. Several applications such as remote monitoring of locations shapes, tracking objects, video vigilance and surveillances may provide a direct usage of such *smart camera networks*. This case study is focused on the usage of smart camera networks in order to perform remote video surveillance.

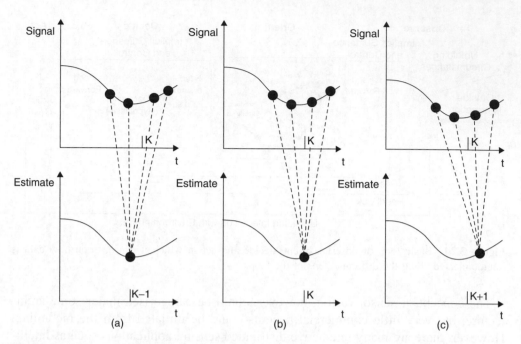

Figure 3.21 Comparison of delay time behaviours of three common signal processing methods for reconstructing samples sensor fusion: (a) Smoothing; (b) Filtering; (c) Prediction.

Surveillance consists of the monitoring of the behaviour, activities, or other changing information, usually of people. This is really important and critical in many scenarios such as outdoor and indoor monitoring, military applications, security systems and public safety systems, amongst others. The usage of a distributed sensing platform for video surveillance can provide benefits over the current camera networks. It acts as a traditional video surveillance system in which there are also many low cost wireless sensors deployed in the monitored given area. (Chen et al., 2008) propose a clear example of this kind of system. They deploy a distributed sensing system to sense and report the information of events back to the sink node. The sink node is connected to a central processing unit. Then, it uses the event received from the sensing platform to control the cameras, changing the rotation angle of the camera to monitor and send video streams back for data analysis of the interested zones. Note that since video streams contain event information, the surveillance system saves significant manpower for data analysis achieving high monitoring quality.

The sensors are spread in the monitor region which is uniformly aware of their position. Moreover, the sensors are equipped with sensing capabilities to detect the objects themselves as phenomena, for example, using presence detection sensors based on light sensibility (photodiode), volumetric levels (microphones), and so on. The sensors are used to track and report locations of objects to a sink using a multi-hop tree-based routing protocol.

When sensors are used for detecting events, they start to collect data for a specific period in order to minimise either false positive and false negative events produced by poorly sensed information of the presence sensors employed. After that, they send such information to the sink, including also the position of the object. Note that the usage of multiple sensors would detect simultaneously the same event in several sensors, causing a large number of reported packets and this high traffic load and collision possibility. To deal with this issue, authors propose to perform in-network event position tasks. When a sensor S_i detects an event, it broadcasts an event-position containing both the Event_ID and sensed data. A higher value of the sensed data means that the sensor's position is closer to the event. The sensors use a time delay trigger from the detection of the event to the transmission of the package. The calculation of this delay is directly proportional to the data of the sensed data. Thus, it prioritises the sensors, which are closer to the object so can send their messages first.

Then, when other sensors receive an event-position packet from its one-hop neighbours, they check if their neighbour also detects the same event (or a different one) and then compares the received packets from its neighbours to determine which sensor is the closest to the event. The way in which sensors determine if multiple events are occurring at the same time is based on the correlation between the sensed data provided by the neighbours and their locations. Then, for the same event, only the sensor with the highest sensed value (i.e. the closest one to the object) sends the event to the sink node.

The communication between the sink nodes and the video camera rotation engines is done using a Wi-Fi wireless communication interface (IEEE 802.11) for receiving the video streams from the cameras and to update the rotation angle when necessary. Figure 3.22 shows a sample scenario in which the proposed DSS video surveillance scenario is depicted. Note the homogeneous distribution of the sensors and the notification of the presence of objects only for those sensors most close to the detected object due to the in-network data processing.

The goal is to use fewer cameras to achieve the same surveillance quality as traditional surveillance systems. A few cameras cannot cover the whole sensing region and many interested events may happen simultaneously. Thus, Chen et al. propose a heuristic method to solve the problems of monitoring regions allocation and inter-task handoff. In essence, the camera knows a priori how many sensors are covered (recorded) on each of the possible positions of the rotation engine (i.e. n_p set of n nodes associated with the rotation position $p = n^1{}_p, n^2{}_p, \ldots n^i_p$. For each position p and per each node in $n(n^i{}_p)$, it calculates a node set composed of such a node and all the one-hop neighbours, that is $S(n^i_p)$. Then, at first, each camera

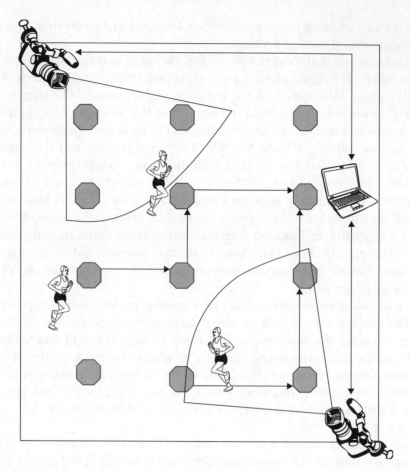

Figure 3.22 Sample scenario where a distributed sensor system for video surveillance is deployed.

is assigned to a position p having the maximum number of different nodes in the node set associated with the covered nodes, that is $S(n_p)$ which is defined as the different sensors of the group: $\sum_{n=0}^{i} S(n_p^i)$ Moreover, the sensors associated to a given camera (n_p^i) are removed from the node sets calculated for the other cameras. Note that this calculation is achieved to cover the widest area possible with the given cameras.

When an event M is now moving in the region belonging to camera X, if the next predicted position is still in the region of the current monitoring camera X, nothing happens. However, if the object enters the predicted covered region of other cameras Y and Z, the system notifies this fact to such cameras to record the associated information. If the prediction failed, the system can still catch up this moving object by an original detecting method and the sink reassigns the right camera to monitor it. Moreover, the set of simultaneous events associated to a given camera determines the better position of such a rotation engine maximising the number of possible simultaneous recorded events.

3.8 Case Study: Collaborative Beamforming

Based on the principles of electromagnetic theory an antenna generates the wave energy, as in isotropic antenna, to the same degree in all directions of three-dimensional space. The radiating energy in practice, however, does not get propagated the same in all directions. The directional power intensity, called directivity gain, defined as the ratio of the power over the average power given in dB, provides a good visual perception of the antenna's directional propagation property. Being also called the gain pattern of the antenna, either as a transmitter or as a receiver, it characterises an antenna and in practice is used as an indicator of an antenna's quality and performance. Due to numerous techniques available for designing antennas in their lengthy life of development we have a great number of gain patterns for fixed antennas. A typical pattern demonstrating main parameters of an antenna, the main lobe, half-power beam width (HPBW) of the main lobe, back lobe, side lobes and nulls, are shown in Figure 3.23.

As a simple example, a straight piece of wire element, or a dipole, generates a uniform gain pattern in all directions on a plane, orthogonal to the element. But when we put two of these elements located vertically side-by-side in parallel at a distance and fed them with a carrier with specific phases then the overall horizontal beam pattern changes from a single wire's uniform/circular to a pointed pattern. As we add more elements to this array it gains sharper main lobe and smaller side lobes. One common use is to apply a saw-tooth style changing signal to the array that would result in gradually moving the direction creating a radar style scan capability to the main lobe. Another use is provision of a feedback to enable finding or locking onto certain items.

The array antennas have a strong presence in multiple-input multiple-output (MIMO) system style beamforming capabilities for many applications in wireless communication systems such as WiMAX and mobile communication networks. Due to their historical development of array style beamforming they are best categorised under smart antennas, but from a beamforming point-of-view we call them traditional beamforming systems. Though they cannot be commonly adopted in DSS applications they may be worth a brief discussion in here.

Smart antennas also known as adaptive arrays, multiple antennas and MIMO systems are made up of an array of antennas and work using a clever signal processing algorithm with specific spatial signal properties of the direction of arrival (DOA) of the signal for a cooperative calculated beam vector to track or locate a complementary antenna system as the target to establish an effective one way or two way transmission link. Smart antenna techniques can be used, such as RADAR, radio

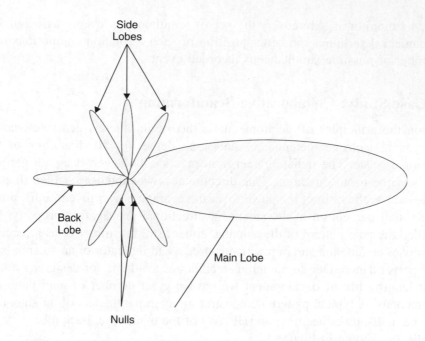

Figure 3.23 A typical antenna gain pattern showing various beam lobes.

astronomy, acoustics, and are most commonly used in advanced wireless networks such as IEEE 802.11n, WiMAX and cellular systems like W-CDMA and UMTS.

Smart antennas' main functions include DOA estimation, beamforming and other spatial information coding such as multiplexing and diversity coding.

Although their cooperative beamforming effect is significant due to the space requirement of multiple antennas MIMO style use of the cooperative transmission in DSS is very limited. In DSS applications of WSN the nodes are small and each node can have only one antenna. Other variations of smart antennas such as multiple-input single-output (MISO) antennas have some uses but are not very popular. We therefore limit using the smart antennas as an extra wideband accessory but in general consider them not part of the beamforming objective for DSS applications.

The beamforming that we need for DSS is expected to help us with many pressing problems hindering WSN applications in association with: (a) agility, (b) mobility, and (c) lifetime, which in turn can be translated into energy, lifetime, distance, beam width, SNR, performance, side-lobe interference, bandwidth and clustering deficiencies.

For the energy issues of wireless smart sensors, for example, the array style beam-forming used suits only specific cases where the sensor node is fixed and connected to the mains. For convenience we call them classic array-based beamforming (ABF) as opposed to *distributed beamforming* (DBF). The DBF techniques suit many practical DSS application scenarios including clusters and other groupings of distributed and moving sensor nodes, which commonly come with a limited source of supply energy so generally use simple antennae.

The main differences between ABF and DBF, from a complexity-point-of-view, is that the geometry of the former is usually known *a priori*, which in dynamically changing ad hoc networked sensor nodes the exact location of the nodes are unknown or should not be acquired. For slow changing and high power sensing systems their relative locations can be estimated using one of many available adaptive algorithms. But for low power devices low SNR locating the sensor nodes accurately may cause serious problems. This plus the effect of other impairments such as local oscillators, phase jitter, and synchronisation among the collaborating nodes simply degrade the overall performance of the system. For related details and practical examples of the geometry the reader is encouraged to look at some of the case studies in Chapter 7, such as sink location in Section 7.5.

For practical uses of DSS, the advent of beamforming can be best achieved through a coordinated propagation within a closed set of smart sensor nodes, where they can coordinate their transceiver signals for a relative right magnitude at the correct phase. This coordinated feeding process applies to a selected set of antennae amongst the whole network, for a collaborative operation. Then with all beams generated from these active antennae in the set towards a common target, a *collaborative beamforming* (CBF), they can help the sensors and clusters to achieve their objectives at a much higher efficiency. This process, if not designed precisely would require considerably more data and signal processing efforts for locating the targets, clusters, sensors and the antennae.

One important use of CBF is in improving overall systems performance under limited signal power embedded in heavy highly unpredictable noise and disturbances. The performance improvement and SNR management can be for both directions of propagation, transmitting the signal or receiving the signal. Fortunately, for wireless communications, all materials, devices and media in the working environment are reciprocal, linear and independent from the direction. That is, exactly the same properties of transmitting signals are applied to the reverse direction of receiving them. Thus, we can make use of beamforming for all aspects of DSS nodes, clusters and sensing devices for collecting sensing activities all the way down to and from any particular PoI/RoI radiation for all directions without any worries about the direction of propagation.

The error performance of a DBF receiver is related to the overall effective signal-to-noise ratio at the receiving detection point according to,

$$\text{SNR} = k_c \sum_{i=1}^{m} w_i \frac{P_i}{\sigma_n^2}$$

Where,

- SNR is the overall CBF signal to ratio indicating the error performance and associated capacity for the speed, bandwidth, and so on;
- kC is the CBF system efficiency, ideally 1 but as in practice largely depends on the efficiency of the algorithm as well as operational application scenario it is smaller than 1;
- m is the total number of collaborating node members, $I = 1.m$, usually fixed for practical algorithms, but it may vary for dynamic systems;
- wi is the weighting factor for the ith collaborating node, which indicates relative effectiveness of the ith collaborating member node's signal source;
- Pi is signal power received from the ith collaborating node member;
- σ_n^2 is the effective noise power at the receiver as usually all collaborating links are exposed to almost the same level of noise, otherwise an i should be included to the index of this component in the formula changing it to $\sigma_{n,i}^2$.

That is, in order to be able to use the above SNR for a collaborative beamforming process at a node upon receiving signal further adjustments in a combination process it is necessary to include the phase and magnitude and other path specific details for the signals travelling from the source.

Localisation or in a simpler term, estimation of distances between the nodes and with other active stations in WSN has been one of the recent issues of adopting CBF in wider scales. Therefore, generalisation of the techniques used for localisation becomes very complex, mainly for making them applicable for both a moving target using a mobile and a system using fixed sensor nodes, usually arranged in specific or random locations or scattered around the RoI.

Furthermore, for smart sensor application use ad hoc style networking inter-connecting the sensor nodes insufficiently, accurate directional knowledge of the destination may not be sufficient if missing the *a priori* information. This is due to the fact that a precise acquisition of knowledge may not be feasible, then the location estimation ambiguity for the collaborating sensors can easily affect the overall performance considerably.

Also, the sensor nodes being either mobile or distributed in places with access difficulties for maintenance, for example on top of buildings, they should operate for years or at least several months without battery replacement. In practice each antenna element requires supportive hardware and electronic devices, each sensor node is likely to be equipped with a single antenna which prohibits the use of autonomous beamforming systems for most DSS applications. On the contrary, sensors may share some *a priori* information and collaboratively transmit the data synchronously then it is possible to perform a beamforming in a distributed manner.

Another potential application of CBF in DSS is the production of stronger unified signal power. That is, for more spread applications, the nodes make a dynamic combined effort to send or receive information over longer distances for much higher power efficiency. Also, as sensors can send or receive in some specified directions without any significant interference with the neighbouring sensors located in other directions, the intended receiver needs to receive the data with much higher signal power. This also gives a chance to low power nodes to transmit combined power to communicate more effectively with a significantly increased channel capacity.

All DSS applications have the aptitude to benefit from CBF, but the most significant achievements of this technique are in distributed sensing applications, whose sensors are hidden with either no access or may have very little access to the sensing object. Applications like scanning the brain for abnormalities, injuries inside the body, or communicating with remote places such as the stars as in astronomy, that is, a far-field space communications for new discoveries through distributed antenna system.

In general using CBF provides significant improvement capabilities to save wasteful propagation energy, reducing interferences, and reducing possible health hazards of high power radio waves.

Further applications of CBF in smart sensors are enormous, which include sensor-integrated systems like radar, sonar, seismology, radio astronomy, speech, acoustics and biomedicine. Below are more typical examples for uses of CBF in DSS.

This is one of the uses of CBF, where a group of acoustic sensors adopt an adaptive beamforming process. Here the array processing attempts to influence the signals in its spatial as well as its temporal aperture. In accordance with temporal sampling, which leads to the discrete time domain, spatial sampling by sensor arrays forms the discrete space domain. The processor, a beam former, combines temporal and spatial filtering signals for beamforming based on finite spatial apertures.

Some FIR filters sitting in the sensor channels help the beam former to achieve its required properties. The design may be one of two class categories: (a) data-independent beam formation, where filters are designed statistically independent from sensor data; (b) data-dependent beam formation, where filters use known estimated statistics of the data to optimise the response.

The arrays provide: (a) detection of the source of a signal, (b) estimate the value of temporal waveforms or their spectral content, (c) estimate the directions-of-arrival (DOA), and finally (d) reduce the beam-width for transmission towards specific locations.

Although they have many applications in radar, sonar, radio transmission, seismology, and medical treatment now they can help arrayed microphones for a featured space-time acoustic signal processing. Typical new applications are hands-free acoustic human-machine interface enabling tele-presence, teleconferencing, high quality dictation, command-and-control, high-quality audio recording, computer games and dialogue systems. Figure 3.24 depicts a core adaptive filtering of the acoustic/audio beam former.

Using node replacement for removing the side lobe interference control can be shown in an example from (Ahmed and Vorobyov, 2009). This study demonstrates

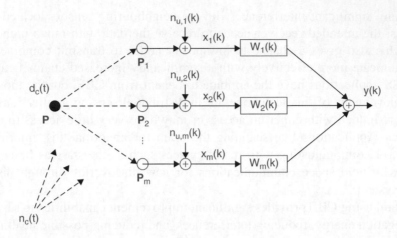

Figure 3.24   Core adaptive filtering of the acoustic/audio beam former.

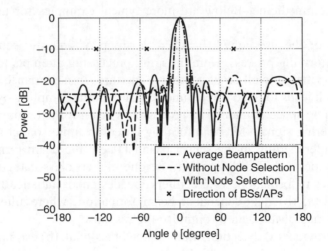

Figure 3.25   Graph in which a core beam former pattern can be seen. © 2009 IEEE. Reprinted with permission, from Ahmed and Vorobyov (2009). Node selection for side lobe control in collaborative beamforming for wireless sensor network.

how node selectivity and node replacement can be used in CBF to reduce the effect of a strong transmitting distributed source interfering with other nearby access points and stations. In this example we have ideal conditions of:

• wireless channels are ideal with no fading or interference effects;
• sensor nodes are randomly distributed over the area;
• each node is equipped with a uniform antenna;

- nodes are either ON (selected) with a unit power or OFF (not selected) with zero power;
- single source located at the centre of the coordinates;
- main destination located at zero degree angle only receives signals;
- unintended destinations (BSs/APs) use identification codes (ID) for their communication with the source procedure;
- the unintended destinations inform the source when receiving the message with power above an agreed threshold, Pthr;
- average power density at each angle is calculated independently.

Figure 3.25 shows the result of the CBF for a special case of the main intended receiver being located on the reference angle, at $\varphi_0 = 0°$, and three unintended neighbouring destinations, D1, D2 and D3, located at angles $\varphi_1 = -125°$, $\varphi_2 = -50°$, and $\varphi_3 = 80°$, respectively. The pattern of power density, in this figure, shows the improved pattern of the source with and without the node selection process, where the relative improvement of 20 dB, 10 dB and 30 dB for the neighbouring destinations shows the significance of the node selection process.

4

Monitoring Well Being

The term *quality of life* is used to evaluate the general well being of individuals and societies. It is used in a wide range of contexts, including the fields of international development, healthcare and politics. *Quality of life* should not be confused with the concept of standard of living. Standard indicators of the *quality of life* not only include wealth and employment, but also the built environment, physical and mental health, education, recreation and leisure time, and social belonging. The monitoring of any aspects of being well is critical for any government in order to get feedback about the society. This reason emphasised the necessity of providing efficient developments and systems which can provide tools for measuring such *quality of life*. This chapter is focused on monitoring being well from the point of view of health monitoring.

Thus, this chapter is structured as follows: Section 4.1 describes different sensing platforms for measuring health by means of vital constants. After that, Section 4.2 is focused on the monitoring of chronic diseases, since these kind of monitoring systems entail particular requirements in the design of the sensing systems. Sections 4.3, 4.4 and 4.5 describe particular use cases directly related to the well-being monitoring in which the reader is able to get real application scenarios in which distributed sensing platforms are being used for such a purpose. In particular, the use case described in Section 4.3 describes the usage of a new concept of smart shirts which contains a distributed sensing platform for retrieving real-time vital constants from patients. Section 4.4 provides a complete description of a distributed sensing platform used for geriatric care and, finally, Section 4.5 describes a system for carrying out outpatient monitoring. Note the special and particular requirements implicitly available in this kind of sensing platform in which the sensors are in the patient's home and they require a direct connection to the hospital in order to keep an update of information on a patient.

Distributed Sensor Systems: Practice and Applications, First Edition.
Habib F. Rashvand and Jose M. Alcaraz Calero.
© 2012 John Wiley & Sons, Ltd. Published 2012 by John Wiley & Sons, Ltd.

4.1 Measuring Health

The monitoring of health can be addressed from two different angles. The first angle is the design of a sensing platform focused on the measurement of environmental aspects, which can have direct implications over the health of people. Airborne monitoring, gas concentration monitoring and water quality monitoring are good examples of this kind of measurement of health. This kind of approach is fully addressed in Chapter 7 of this book.

The other angle for monitoring health is to determine the degree of well being in a patient based on the usage of a wide range of smart and advanced heterogeneous sensors, which are able to measure the different vital constants available in the patient. This is the approach tackled in this chapter. Depending on the aim pursued in the design of the sensing platform, different combinations of sensing devices can be joined together into a system to gather, process and correlate such data providing information about the patient's health. Let us enumerate the most common sensing devices used for measuring health and the vital constants associated with them.

First, *body temperature* is a very important often vital variable in order to identify infections and diseases directly produced by virus and bacteria. This information is achieved by means of a sensing thermometer, which is in direct contact with the patient enabling the sensing of the body temperature. Second, detection of *heart activity* is very critical information to get the real status of the patient. It is usually gathered using smart sensors for measuring the *electrocardiograph* (ECG) of the patient's hearing and also the *heart rate*. Third, *blood pressure* is another important physical parameter to be measured in patient monitoring. Fourth, *respiratory activity* is another important aspect of health monitoring. Many recent studies also use sensors for such a purpose. In general, they focus on measuring the respiratory rate and the pulsometry, that is, the level of blood oxygen. Muscular activity is also monitored by means of sensing devices to gather *electromyogram (EMG) data*. And finally, many sensing platforms used for monitoring health require some data for the *body posture* in order to infer more accurate information from it. This information can be used for sensing using a simple 3-axis accelerometer in which all movements can be registered.

Considering that various forms of body posture, positioning, placement and identification of a patient can help with many aspects of health measurements, clinical treatments, and so on, we look into some of the systems in here for three groups using three different source technologies of light, ultrasound and radio. For this we examine a survey from Yanying et al. (2009).

The infrared (IR) technology, with a room limited coverage range, is commonly used for absolute position estimation, but needs line-of-sight for signals without any serious weaknesses against other interfering light sources, but somehow weakened by direct light from fluorescent and sun. To name a few we can mention *active badge, firefly Optotrak* and *IRIS*.

Active badge system was designed by AT&T in the 1990s to identify an active badge location sensing covering the area inside a room using diffuse IR technology. In order to estimate location of the badges carried by people the sensor transmits an IR signal every 15 seconds. The position of an active badge can be determined by the information coming from the sensors, which are connected to a central controller to estimate the position of the person using them. This is an old project, which had been closed down sometime ago with no commercial products.

Firefly designed by the Cybernet System Corporation as an IR-based motion tracking system with a good accuracy through analysing the location of small tags emitting IR light and mounted on the object by animating complex virtual reality motions. As shown in Figure 4.1, the Firefly comes with a tag controller, several tags and one array of camera. For example in the setting three cameras are placed on a 1 m bar whilst tags, much smaller than coins, help with the 3-D estimation. With an accuracy of about 3 mm, measurement delay of 3 ms at a sampling rate of 30 scans per second, for example with 30 tags it can provide high-speed real-time tracking. Its limitations are limited in coverage area to within 7 m with a field of view $40° \times 40°$.

Designed by Northern Digital Inc. Optotrak uses three cameras in a linear array to track 3-D positions of various markers on an object. One example shown in Figure 8GY, the tracker comes with three cameras for covering a volume of $20\,m^3$ for a distance up to 6 m. The triangulation technique helps with calculating the position of IR light emitters.

This system has advantages of dynamic referencing, relative motions for example in Figure 4.2, three emitters A, B and C mounted on the surface of a car form a dynamic reference with static relative positions where the tracked emitter E on the

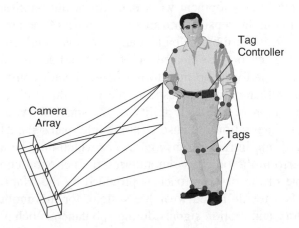

Figure 4.1 Firefly motion tracking system architecture for tracking human movement. © 2009 IEEE. Reprinted with permission, from Gu et al. A survery for indoor positioning systems for wireless personal networks.

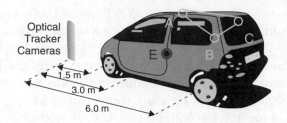

Optical
Tracker
Cameras

1.5 m
3.0 m
6.0 m

Figure 4.2 Optotrak tracking system integrated into cars and road-side devices. © 2009 IEEE. Reprinted with permission, from Gu et al. A survery for indoor positioning systems for wireless personal networks.

door can be measured with relative position changes with respect to the formed dynamic reference.

This system offers accuracy of 0.1 mm to 0.5 mm and 95% success probability. It works over an area of 20 m³.

For ultrasound technology, rather similar to the last hundreds of years in navigation, bats navigate in the night. One example is Active Bat designed by AT&T which provides a 3-D position and orientation information upon the tracking tags. The tag periodically broadcasts short pulses. The short ultrasound pulses received by the ceiling-mounted receivers estimate the distance between the tag and the receivers upon the time of arrival. The distance to the ceiling indicating the position of the patient is significant because when it is in the middle of a room or pre-defined positions it should raise an alarm. See Figure 1.3 in Chapter 1.

One of the most significant advantages of using ultrasound over radio and IR is its unique advantage for indoor coverage and limited interference with the neighbouring rooms, as shown in Figure 4.3.

Almost all the recent development work done for health monitoring is used with smart shirts to carry out the sensing information for most of the previously indicated parameters. This shirt is worn by the patient and contains a built-in sensing platform therein. Note the the sensors are located in the shirt and are in direct contact with the body in order to perform the monitoring of the health constants. There is an important area of research in this respect following three different approaches. The first one is to improve the wear-ability of the smart shirts doing research on miniaturisation and integration of sensors into textile materials. The second one is to innovate ways of non-intrusion sensing devices which may be available in daily life in order to perform the pervasive monitoring of such vital constants.

A good example of the first approach is provided by the MagIC (Rienzo et al., 2005) system. It is a textile-based wearable system, for the unobtrusive recording of cardio-respiratory and motion signals during spontaneous behaviour in daily life and in a clinical environment. In essence, this system is composed of textile sensors integrated in the shirt for ECG and breathing frequency detection, and a portable electronic board for motion assessment, signal pre-processing and wireless data

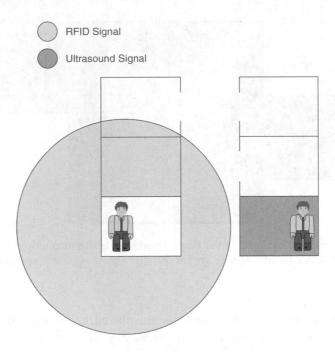

Figure 4.3 Combination of ultrasound and RF technologies in an example scenario.

transmission to a remote computer. Figure 4.4 shows the smart shirt developed in *MagIC*. Moreover, the reader can see the detail about how the sensors are built-in to the search and also how the electronic board is placed in one side of the shirt.

The MagIC vest is mainly made of cotton and lycra and is fully washable. Through a CAD/CAM process two woven electrodes made by conductive fibres are located at the thorax level so as to obtain an ECG lead. The contact between thorax and textile electrodes is guaranteed by the elastic properties of the garment, avoiding the necessity of gel or of any other medium. The vest also includes a textile-based transducer for the assessment of respiratory frequency.

ECG and respiratory signals feed a portable electronic module which is placed on the vest through a Velcro strip. The electronic board detects also the subject's movement through a 2-axis accelerometer and transmits all signals to a remote computer for data visualisation and storage on disk.

Another example is that provided by Loriga et al. (2005). They are investigating wearable systems which are able to monitor cardiopulmonary vital signs. The innovative technological core of the system is based on the use of a textile conformable sensing cloth, where conducting and piezo-resistive materials are integrated in the form of fibres and yarns, giving rise to fabric sensors, electrodes and connections. *Electrocardiogram* and *impedance pneumography* signals are acquired through the same textile electrodes, while to discriminate between abdominal and thoracic

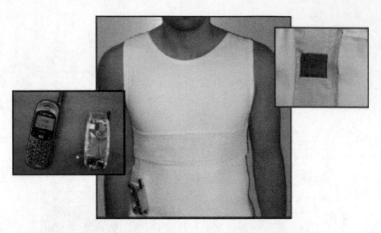

Figure 4.4 Smart-shirt developed in the magic system for health monitoring. © 2005 IEEE. Di Rienzo (2005) A testile-based wearable system for vital sign monitoring: applicability in cardiac patients.

activity, two piezo-resistive fabric sensors are placed below the lower end of the sternum and at the level of the navel for recording the thorax and the abdominal pattern of breathing. The use of impedance pneumography methodology reduces the artefacts due to the movement, phonation and rib cage expansions disjointed from respiratory mechanics. All the signals are acquired simultaneously allowing a comparative control of the cardiopulmonary activity and artefacts rejection. Figure 4.5 shows the positioning of the fabric electrodes used for ECG acquisition into the smart shirt.

Regarding the second approach, Lim et al. (2011b) study how to obtain vital constants using innovative non-intrusive methods. In particular, they propose a method for measuring heart rate and respiratory rate, that is, *ballisto-cardiograms (BCG)* using capacitive coupled electrodes without direct contact to exposed bare skin, and these electrodes are installed in chairs, beds, belts and toilet seats. Figure 4.6 shows the deployment of such sensors in a bed. Four load-cells sensors installed at the bottom of the bed legs detect force changes at each leg of the bed. With appropriate filtering, a measured composite signal can be separated into three different types of signals that represent a *BCG* and movement, weight changes and respiration.

Note that much high level and valuable information can be inferred from the information sensed using the previously described devices. For example, the angles of the body inclination can be easily determined using the 3-axis sensor previously introduced for determining the *body posture*. In fact, a more complex and valuable metric can be determined as the *activity index*. This index can be calculated using the aggregation of the different sensed angles in the 3-axis sensor over a period of time. Then, this information can be used to determine the degree of activity of the patient. Note that a continuous variance in the angle implies a constant movement of the

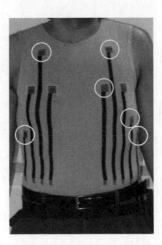

Figure 4.5 Pattern of shirt where placement of the fabric electrodes for the ECG acquisition are shown. © 2005 IEEE. Reprinted, with permission, from Loriga (2005) Textile sensing interfaces for cardiopulmonary signs monitoring.

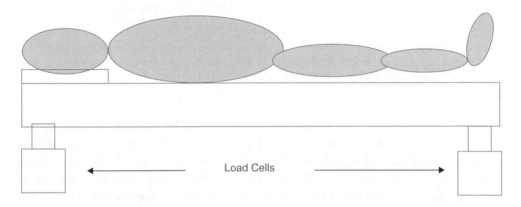

Figure 4.6 Deployment of 4 load-cell sensors used to measure ballisto-cardiograms.

patient, whereas a minimal change in such angles implies activities such as sleeping, relaxing, and so on. LOBIN (Lopez et al., 2010) is a good example of monitoring a well being system which infers such calculated metrics from the basic sensing information. This system is carefully addressed in a further case study of this section.

(Pandian et al., 2008) propose a wearable physiological monitoring system embedded in a washable shirt, which uses an array of sensors connected to a central processing unit with firmware for continuously monitoring physiological signals. The data collected can be correlated to produce an overall picture of the wearer's health. The wearable data acquisition system is designed using a microcontroller and interfaced with wireless communication and global positioning system (GPS)

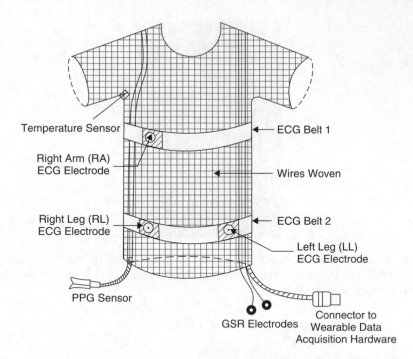

Figure 4.7 Diagram of the different elements integrated in typical health monitoring smart shirt (Pandian et al., 2008). Reprinted from: Pandian (2008) Smart Vest: Wearable multiparameter remote phisiological monitoring system, with permission from Elsevier.

modules. The physiological signals monitored are ECG, photo-plethysmogram (PPG), body temperature, blood pressure, galvanic skin response (GSR) and heart rate. The acquired signals are sampled at 250 samples/s, digitised at 12-bit resolution and transmitted by wireless to a remote physiological monitoring station along with the geo-location of the wearer. Figure 4.7 shows the deployment of the different sensors across the shirt. Note how the temperature sensor is placed under the axilla in order to get precise information. There are also two belts located at heart level and under the belly in order to achieve a direct contact with the skin at this location, which is necessary to monitor the different derivations available in ECG. These derivations enable doctors to monitor different aspects of the ECG.

Another important aspect of the monitoring well-being systems is the necessity of correlating data gathered from the different sensing devices in order to get them aligned and to perform data processing techniques over them. To cope with it, the sensing platform can rely on the usage of time synchronisation protocols. This problem is very close to other scenarios as in structure building health monitoring in which a wide set of sensors have to be time synchronised in order to correlate their sensing data together. In case the reader is interested in the protocols for synchronising time in distributed sensing platforms, Gautam and Sharma (2011) provide a very complete overview of the state-of-the-art of the topic. In summary, 12 time synchronisation protocols are analysed and compared. From all of them, we would like to define the intrinsic problems associated with monitoring well-being and thus using them to determine a suitable option for time synchronisation protocols. In this monitoring well-being scenario, there is not an important amount of deployed sensors composing the distributed sensing platform. Note that usually the sensing architecture is composed of 3 to 30 sensors. So, scalability is not an issue here. However, accuracy is the most important issue since the exact correlation of events is almost a vital necessity in monitoring well being since it can directly affect the quality of the sensed data and it could put the patient's life at risk. Thus, only those synchronisation protocols with a high level of accuracy can be considered in this scenario. Moreover, another very important issue is the fault tolerance since the whole system design has to deal with it in order to ensure that data is always accurate. Hence, fault tolerance is considered as another critical feature. Having all these features in mind, a good candidate protocol, which claims to fulfil all these requirements, is Time-Diffusion Synchronisation Protocol (TPD) (Su and Akyildiz, 2005). It allows the sensor network to reach an equilibrium time and maintains a small time deviation tolerance from the equilibrium time. To do so, initially, a set of master nodes is elected. For external synchronisation, these nodes must have access to a global time. This is not required for internal synchronisation, where masters are initially unsynchronised.

Master nodes then broadcast a request message containing their current time, and all receivers send back a reply message. Using these round-trip measurements, a master node calculates and broadcasts the average message delay and standard deviation.

Receiving nodes record these data for all leaders. Then, they turn themselves into so-called 'diffused leaders' and repeat the procedure. The average delays and standard deviations are summed up along the path from the masters. The diffusion procedure stops at a given number of hops from the masters. All nodes have now received from one or more masters the time at the initial leader, the accumulated message delay, and the accumulated standard deviation. Then, this information is used to estimate the current time. After all nodes have updated their clocks, new masters are elected and the procedure is repeated until all node clocks have converged to a common time.

Another important issue directly related to monitoring well-being is the incredible amount of data being collected from the patient due to the real-time nature of much of this system and also due to the number of sensors being used and the continuous monitoring involved in the process. Thus, effective data aggregation techniques are almost always necessary in order to increase significantly the life of the whole system due to the associated hard power requirements. A clear example of data aggregation is provided by Lopez et al. (2010). They aggregate all the sampling done in order to compose the ECG values. Then, they only send the information after

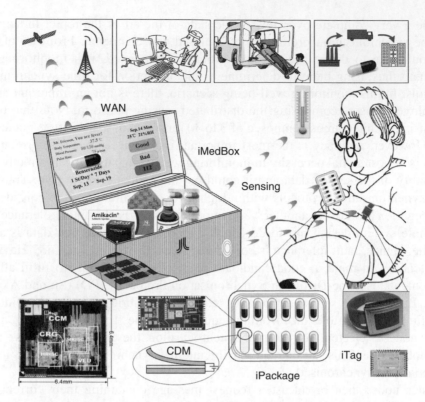

WAN

iMedBox

Sensing

CCM

CRG

CDM

iTag

iPackage

Figure 4.8 Diagram of the different hardware elements available in the smart medical box proposed by (Pang et al., 2009). © 2009 IEEE. Reprinted, with permission, from Zhibo (2009) A pervasive and preventive healthcare solution for medication noncompliance and daily monitoring.

25 sampled data. It enables a ratio 1/25 of packet sending in comparison with the sampling rate. Moreover, the sensed value can be significantly reduced in terms of size application in-network pre-processing of the data in order to reduce at maximum the size. It can be addressed in many different ways. For example, reducing the precision of the sensed data (reducing the number of bits used to represent such value) or applying seamless data compression techniques over the data.

An interesting system for monitoring one's well being and which acts as a preventive healthcare solution is the smart medical box proposed by Pang et al. (2009). This is a distributed system platform composed of a set of embedded sensing devices in health-based accessories. The overview of the whole architecture is shown in Figure 4.8.

The rapidly increasing demand for daily monitoring with onsite diagnosis and prognosis is driving homecare solutions to integrate more and more sensing and data processing capacities. A powerful system is needed not only to address the medication noncompliance, but also to be used as a pervasive healthcare station in the home. The authors propose a healthcare solution for medication noncompliance and daily monitoring using an intelligent package controlled by Radio Frequency Identification (RFID).

The smart medical box has many potential applications such as: the management of medicine inventory, registering, recording and querying all medicine utilities automatically being read bbye RFID tags; the provisioning of medication reminding by means of automatic parsing for the prescription according to patients' and medicines' RFID numbers and timely reminding can be presented to the patient according to the prescription, for example the recording of noncompliance related activities (e.g. destroying some medicines by mistake, taking or forgetting a dose, throwing some medicines away, and so on) can be detected and recorded automatically. Moreover, the prevention of noncompliance is really useful since patients cannot, by hand, open the dose package without opening a command from the medical box observing the prescription. Thus advanced and excessive medication is entirely prohibited. Another direct application is the remote monitoring of the patient's daily activities and health situation parameters, such as tri-axis accelerometer, body surface temperature (BST), ECG, blood oxygen saturation (SpO_2), respiration oxygen saturation, blood sugar concentration, blood pressure (BP), and so on. The onsite diagnosis and prognosis provided by the medical box performs a suit of basic diagnosis and prognosis examinations and gives the patient suggestions at the first time and place. According to the record of medication noncompliance, body situation parameters, subjective feelings feed back and on-site diagnosis and prognosis results, a doctor's suggestion could be feedback to the patients through the medical box. When an emergency situation happens, such as acute anaphylactic reaction and myocardial infarction, an alarm can be sent to caregivers automatically.

4.2 Managing Chronic Diseases

There is a wide number of use cases in which monitoring human vital constants can be extremely useful for diagnosis and treatment of patients. They could be divided into *short-term* and *long-term* monitoring. A clear example of short term health monitoring is a scenario in which patients have been recently operated on and the post-surgery requires a constant monitoring of the evolving health of the patient during a short period of time. Due to the short time associated with it, this kind of scenario could continue being managed directly at the hospital. This is the main reason why most of the research work done is focused on long-term health monitoring scenarios in which an in-home continuous monitoring of the patient needs to be done for a very long period of time. There is a wide range of scenarios which may require the usage of this kind of distributed sensing platform developed at a patient's home. Mainly, all the scenarios relate to older patients. They include, *incontinent elderly* directly related to the lack of control over sphincters, *chronic diseases* associated with the elderly which are not curable and need to be treated for long-living terms, *disabled elderly* in which either motion or mental processes are affected, *amnesiac elderly*, in which patients continuously have memory lakes, and *vegetative elderly* in which patients have a total loss of conscience. Each of them has associated a set of particular requirements when a new sensing platform needs to be designed for patient monitoring and this section is focused on those diseases directly related to the long living monitoring of the patient, that is, chronic diseases. However, they are not available only in older patients since a chronic disease can be caused by a wide set of external factors and can potentially affect any type of the population such as babies, youths and mid-age population. Any generic, inherited or acquired dysfunction in the heart can produce a cardiopathy which is a consequence of a wide range of chronic diseases, a chronic obstructive limb disease, or a chronic renal insufficiency are also good examples of chronic diseases in limbs and kidney, respectively. Moreover, virus and bacteria are also an important source of chronic diseases like, for example, *tuberculosis*. Note the wide range of the population affected by all these diseases and the need for the constant long-term monitoring of them in the population.

Note that traditional cable-based systems are usually fixed in a given location, generally the patient's home, and they hamper the patient mobility not only into different home rooms, but also outside of the home. Moreover, the continuous wearing of wired sensors usually producing a shortage of personnel is negatively impacting care quality, leading to elderly individuals suffering eczema and other skin diseases. Thus, the usage of disturbed wireless sensing platforms can increase significantly the quality of life in these patients.

A good example of distributed sensing architecture is described by Wu and Huang (2011). This architecture is depicted in Figure 4.9. In essence, it is divided into different conceptual layers. At the bottom layer, there is a *management layer* in which medical staff is in charge of interacting with the distributed sensing platform in order to configure and receive all the information monitored by the patient by means of a monitoring and controlling management platform. This layer is in charge of using a wireless sensor network layer, in this particular case ZigBee in order to interconnect the distributed sensing platform. This platform is composed of a set of sensing devices intended for monitoring the different information necessary for ensuring the patient's well being depending on the chronic disease being monitored. For example, humidity sensors, patient location, bed presence, physiologic parameters, fall detection and emergency report are some of this sensed information. The set of sensed data is directly determined by the chronic disease being monitored and according to the doctor's suggestion which decided the requirements of the deployed system.

Each sensing device has been carefully selected to provide useful information. Thus, the patient diaper with electronic humidity sensor provides suitable health care for incontinent and disabled elderly patients. When the electronic humidity sensor detects humidity exceeding the threshold, the system informs the care personnel to change the diaper. Under conditions of personnel shortage, the location monitoring system not only receives the active help requests from amnesiac elders

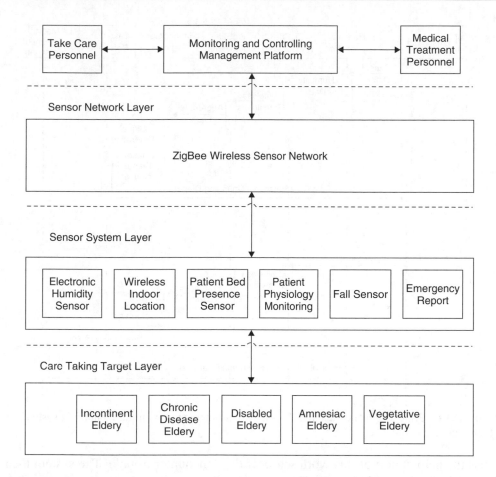

Figure 4.9 Generic architecture for elderly monitoring system (Wu and Huang, 2011).

but also detects the event and assigns the medical treatment personnel to provide support. The bed stress sensor sends a signal to the controller after the patients leave their bed. The safety monitoring system informs medical treatment personnel so they can verify whether the patient has fallen from the bed or left their bed voluntarily. Obviously, the physiology monitoring system monitors vital signs by wireless, sending body temperature, heart rate/pulse, blood pressure, and other data to the controller via wireless sensor networks. If the vital signs data exceed the defined threshold, the system reports to desk personnel and the data is recorded. The fall sensor device facilitates prompt provision of care to elderly patients suffering falls. When this sensor senses that the patient has been inclined beyond a certain point, the system uses indoor location technology to inform desk personnel to provide immediate support. Moreover, a clear feature of this kind of sensing platform is the necessity of an event report system from which the patients can

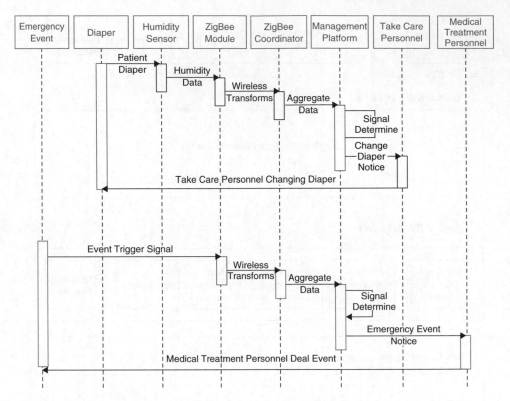

Figure 4.10 Typical sequence diagram for the patient interaction with the monitoring system for chronic diseases.

press the help button of this worn sensor if they encounter trouble. The system then integrates the indoor location technology to inform medical treatment personnel to deal with the problem.

In summary, the sequence diagram depicted in Figure 4.10 represents a common scenario for the monitoring of a chronic urinary incontinency patient's disease. In essence, the diaper can attach a wireless device in which a small humidity sensor determines if the patient has suffered incontinency or not. This sensor is continuously monitoring such humidity and when a humidity threshold is achieved, then an alert notification is submitted across the network to the care personnel indicating that the patient requires a change in the diaper. The same figure represents the case in which the patient presses the alarm button to notify the doctor directly about an emergency. This is received by means of the wireless network established between all the sensing devices involved.

Although the proposed architecture seems to be just another sensing platform, the design of a distributed sensing platform in this kind of scenario has associated a wide set of particular requirements which impose several challenges in the design of these platforms.

Firstly, supporting large-scale and easy to deploy networks is a challenge. Note that in many scenarios, including natural and man-made disasters; wireless sensing systems should be able to collect data from a large number of patients dispersed around the hospital. Therefore, the system should be able to scale to wide areas and be easy to expand even to areas without pre-existing wireless network infrastructure. This challenge can be easy tackled using totally pure ad-hoc wireless sensing platforms without infrastructure such as that provided in WSN. Secondly, there is an important aspect to take into account, the high data-rate networks required. Note that in addition to supporting a large number of users, medical sensing applications should be able to support high volumes of traffic generated from high data-rate sensors such as ECG and respiratory sensors. This requires efficient compression, aggregation, fusion and packaging techniques. Thirdly, patient mobility is very common in many clinical scenarios including patients in the ER and intra-hospital patient transport. This is especially important for chronic disease monitoring in which the mobility of the patient in inherently huge due to the time being monitored. For this reason, medical sensing applications should provide high reliability without restricting the patients' movements. Moreover, tracking patient locations is not only important during the transport of patients admitted to the hospital, but also simplifies the process of locating patients in the ER waiting room. Just keep in mind scenarios with amnesic chronic diseases in which the patient can be easily disorientated. Doing so would reduce staff workload and improve the efficiency of patient-care workflow. Fifthly, soft real-time data delivery with any-to-any routing medical applications requires low latency and high reliability to ensure that medical staff can make clinical decisions based on up-to-date patient information. Furthermore, the measurements must be reliably delivered to all the caregivers who are associated with a certain patient. Finally, security is almost critical and even mandatory in many governments. For example, according to US law, medical devices must meet the privacy requirements of the 1996 Health Insurance Portability and Accountability Act (HIPAA). To meet these requirements, the system must never broadcast identifiable patient data and guarantees the authenticity of the data it delivers.

(Ko et al., 2010) describe the *MEDiSN* sensing platform, a very good example of the real implementation of a healthcare monitoring system which has been intensively used in the Johns Hopkins Hospital, Baltimore, Maryland. This implementation identifies and copes with many of the challenges previously described. In essence, this sensing platform is composed of two different types of nodes: reference nodes which are deployed around the hospital in order to be reference points for the localisation of patients and patient nodes which are in charge of monitoring the vital signs of the patient. Figure 4.11 shows the different types of sensors previously described, (b) and (a) respectively and the base station sensor used as a touch point between the wireless sensing platform and the wired system which process the information and alert the nursery if necessary.

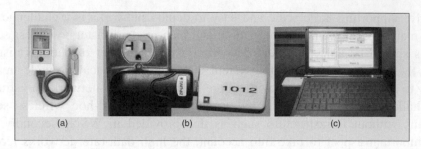

Figure 4.11 Different types of devices deployed by (Ko et al., 2010) at John Hopkins hospital for health monitoring. © 2010 IEEE. Reprinted, with permission, from Ko (2010) Wireless Sensing Systems in Clinical Environments: Improving the Efficiency of the patient monitoring process.

For this architecture, all the sensors are attached to only a main board in the patient node providing wireless connectivity to the network. They record pulse rate and blood oxygen levels and integrate actuators including a buzzer, a 1.8 in liquid-crystal display (LCD) and five light-emitting diodes (LEDs) (see Figure 4.11(a). It collects a patient's vital signs through a finger clip. Before transmitting these measurements using its on-board radio, the device encrypts and signs them using the security primitives that the radio API offers. The data is sent to the base station using the reference nodes as relay if necessary in a multi-hop networking. Data is collected and transmitted every 30 seconds to the base station in order to get a soft real-time status of the patient. The patient localisation is based on the RSS information received in the patient node from the beacon packaged periodically sent by the reference nodes. This information is calculated in the patient node and also transmitted to the base station in order to get an estimated position of the patient.

In fact, Hsiao et al. (2011) describes other sensing platforms, which also have been validated in a real scenario in the National Taiwan University Hospital in which they deploy a 45-node system in its elder care centre. They focused on the long-term mobility tracking of the elderly, using the potential of a WSN-based indoor-location system as a support for the already-overloaded nursing staff. Again, authors use a RSSI-based location algorithm in order to determine the accurate position of the patient. In this case, if the reader is interested in how a RSSI-based localisation algorithm works, this has been addressed further in this book. The university hospital's elder care centre used the deployed system for eight months, achieving very interesting results about the mobility track of elderly people: (1) each elder's daily mobility shows a reoccurring pattern. The pattern, however, differs from individual to individual. (2) The mobility level shows a significant variability, that is, not all elderly show reoccurring patterns in mobility levels. These suggest that mere quantity of how much the elderly move around the facility will not be a suitable target for behavioural modelling. However, the exact location of their presence becomes more relevant.

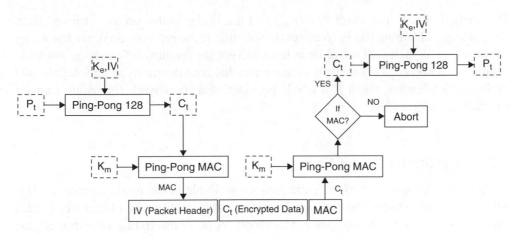

Figure 4.12   Main encryption and decryption steps in a typical security architecture (Sain et al., 2010).

Regarding security in healthcare environments Sain et al. (2010) propose a secure middleware to be used by ubiquitous healthcare systems. This middleware can be directly applied to the enforcement of the security in the information exchanged between nodes in monitoring and managing health diseases. In essence, authors base the whole middleware on the usage of PingPong-128 (Lee and Chen, 2007) which is a stream cipher with the addition of mutual clocked control structure. The algorithm is designed with both security and efficiency in mind to satisfy the need for lightweight algorithms. PingPong-128 is a highly secure algorithm, dedicated to hardware environment and easy to implement in software. It is a bit-based stream cipher. PingPong-128 accepts a key as 128-bits and 128-bits initialisation vector to feed the internal states. It generates an output block of 128 pseudo-random bits from a combination of the internal states for every iteration PingPong-128 has 257-bits of internal state. Then, authors proposed the work-flow depicted in Figure 4.12 in order to implement information security in healthcare environments. In essence, the sender inserts P_t (plaintext) into PingPong-128 cipher to produce the C_t (encrypted data) as output. To produce this output, the shared key (K_e) and the initialisation vector is provider (IV). This C_t is inserted again into PingPong-MAC with the key shared (K_m) and obtained an encrypted MAC as output. Now the encrypted packet contains the initialisation vector, the encrypted data (C_t) and the encrypted MAC. For the network authentication, when another node received this encrypted packet, then this receiver first confirms whether this data is sent by an authenticated node and if the package is it should be processed by the receiver. To do so, the PingPong-MAC uses the (K_m) shared secret key to decrypt the original MAC address and only for the case in which it matches with its MAC address, the process continues. Then, in decryption process, the C_t (encrypted data) is fed into

PingPong-128 with the share key (K_m) and the initialisation vector retrieved from the package, obtaining the P_t (plaintext). Note that if the receiver does not know any of these two keys, it will never be able to decrypt the medical information available within the message. This security scheme enables confidentiality, data integrity and entity authentication which is a whole provider of a framework for secure transfer of medical information.

4.3 Case Study: Smart Shirts

Despite the vast numbers of different sensors available in the market, there is a significant lack of sensors specially designed for either monitoring effectively human vital constants or for being practical enough to be comfortably wearable by the patient without disturbing his regular daily activities. Many research works have been done in order to improve the design of new sensing devices for healthcare. An example is from Park et al. (2006). They have created an active device especially designed for monitoring an electrocardiograph which exposes new features being more suitable for real and practical healthcare systems. They are ultra-wearable circled sensors sizing around 1 cm of diameter, which are using a wireless transduction built-in and with some processing capabilities. This device is also a very low-power device being able to be operative for a very long time period with a simple minimal buttons battery. Another example is from (Borges et al., 2010), they provide a primitive smart sensor platform composed of eight different sensor devices interconnected to monitor the foetal movements, which happens inside a pregnant woman. These sensors are distributed homogeneously in approximately 180° degrees in order to sensor the different angles of the foetal movements. To do it, special flex sensors are used to monitor the flexibility of the sensors which is changed when foetal movements succeed. They are especially designed to be attached to a belt attached to the patient's belly. Although these are really appreciated are valuable attempts for improving the wearability of new sensing devices a significant effort has to be done in the future in order to achieve really effective and practical devices minimising the disruption of the patient's daily activities.

Some early attempts are recently appearing in this sense. Concretely, an innovative way of placing sensors for health monitoring in order to significantly increase their wearability is into clothes and accessories. The idea is to create novel smart clothes and accessories in which sensors are directly in contact with the body and the environment where as they are cleverly placed into the clothes. In this sense, an imminent future revolution in clothes may come from the idea of the smart shirts. They are defined as shirts having a complete built-in distributed sensor network interconnected either wirelessly or using a bus plane which is also incorporated in the shirt. Generally, they usually provide an external wireless communication in order to provide the sensing capabilities directly to the coordination agent.

Their direct application may open new and novel ways of sensing covering a self-powered shirt to the innovative usage of smart sensor systems for helping in the performance of many daily activities. Smart shirts can open new trends and revolutions into medical applications such as health monitoring, military applications such as in-fight real-time reports of injuries, sports applications, especially for those elite players, fitness, and so on.

This case study is focused on the application of smart shirts for health monitoring. In this sense, several research works of smart shirts have been appearing recently in the health monitoring market. For example, Kemp et al. (2008) propose a smart sensor platform for monitoring the health status of police specialists in bomb deactivation. This special cover is very robust in order to protect the police against possible explosion of bombs, which is directly associated with the important lack of mobility due to the necessary excessive over width imposed by the cloth design. The policeman suffers very high temperature overheating inside of the cover, mainly due to the strong isolation necessary. Thus, authors insert a distributed sensing platform into the dress in order to perform a distributed sensing of the whole body temperature of the policeman in order to control possible stress situations, unexpected behaviour, etc. caused by high temperatures. The smart shirt is composed of several SSD temperature monitoring sensors spread over the whole body coverage and also of one actuator sensor, which is attached to the built-in cooling system available in the cover. The monitoring loop consists of gathering the temperature values for the whole body, which come from localised information. Then, the coordinator node analysed such collected information into its own network generating the thermal sensation model. This model, together with extra information such as contextual information that is, number of sensors, position of sensors and whether redundant sensors have been used, is sent to the police operation centre to give up-to-date information about the policeman. Figure 4.13 shows the proposed design of the architecture in which the reader can see what the sensors and actuator have deployed within the anti-bomb dress.

This case study exposes a real smart sensor system to provide solutions to practical applications in reality. Probably, in the future, the dress could be autonomous, controlled directly from the controller agent regulating the cooling system autonomously without any active intervention of the police operative centre.

Figure 4.14 shows the screenshot of the software, which renders the current status of the policeman in real-time. Note how different zones of the body are represented according to the temperature sensation perceived by the sensors deployed in such areas. It provides valuable information to control the fans available inside of the cover accordingly. Figure 4.15 also shows the implemented prototype for the dress.

Note that sensors in the anti-bomb dress have to be unnoticeable and this is a big challenge to tackle in the coming years. This is clearly a potential pitfall, which researchers may explore significantly. In case the reader is more interested in a

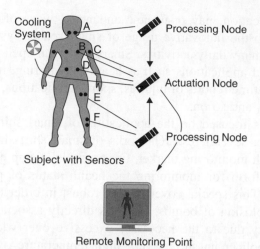

Figure 4.13 Design of an anti-bomb dress. © 2008 IEEE. Reprinted, with permission, from Kemp (2008) Using Body Sensor Network for increasing Safety in Bomb Disposal Missions.

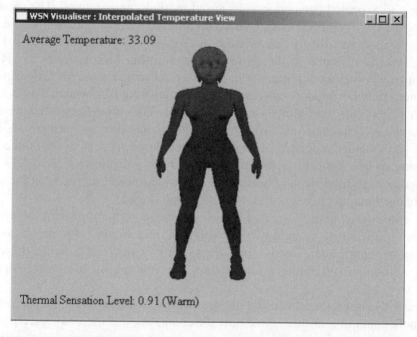

Figure 4.14 Screenshot of the software to control the status of the policeman in real-time. © 2008 IEEE. Reprinted, with permission, from Kemp (2008) Using Body Sensor Network for increasing Safety in Bomb Disposal Missions.

Figure 4.15   Implemented prototype anti-bomb dress. © 2008 IEEE. Reprinted, with permission, from Kemp (2008) Using Body Sensor Network for increasing Safety in Bomb Disposal Missions.

literature review in smart textile advantages, Pantelopoulos and Bourbakis (2010) provide a comprehensive viewof the topic.

Another scenario in which smart shirts are successfully used in a real scenario is that provided under LOBIN (Lopez et al., 2010) project. This project aims at a platform to provide remote location and healthcare-monitoring support for hospital environments based on the combination of e-textile and smart sensor technologies. To this end, the system is composed by a set of smart shirts worn by the different monitored patients. Each shirt has embedded a device to collect and process the physiological parameters and to transmit them wirelessly. Moreover, each patient also has another smart sensor device, usually worn in the pocket or in a bracelet used for localisation. Authors decide on simplicity, performance and reliability to keep both health monitoring and location systems, using independent hardware. Moreover, authors deployed a set of wireless sensor devices across the hospital in well-known positions. These stationary devices are used for two different purposes. Firstly, they periodically send beacon frames, which are used by the localisation of the patient using RSSI signal strength. Secondly, they are also used to collect the data gathered by the smart shirts and to transmit such data to the gateway node, which is in charge of transmitting this information using wired Ethernet to the final server in which the monitoring information is being tracked.

Regarding health monitoring systems, the smart sensor device embedded into the smart shirt is composed of the following sensors: i) An e-textile-based tape that measures the bioelectric potential of the human body. Its most relevant feature is the integration of two e-textile electrodes to perform the biomedical data acquisition. This e-textile-based tape is integrated into a wearable, non-invasive, comfortable, and washable smart shirt. ii) A three-axis accelerometer to detect patient movements and to determine if the patient is lying down or moving about in order to aid appropriate diagnosis of received signals. iii) A thermometer that measures the body temperature with direct contact with the skin of the patient. All the physiological parameters are sampled every 4 ms. However, they are processed locally in the smart sensor device aggregating them and as a result only one message is transmitted after collecting 65 electrocardiograph samples. This amount of aggregation is calculated according to the size of the frame resulting from the healthcare-monitoring frame together with the additional networking information in order to minimise fragmentation.

Regarding a localisation system, authors use an algorithm based on triangulation of the received RSSI signals periodically sent by the stationary sensors located across the hospital. In case the reader is interested in how this kind of algorithm works, they are described later in several sections of this book.

4.4   Case Study: Geriatric Care

There is an important percentage of the worldwide population requiring geriatric care due to physical and/or mental impairments. These people are condemned to live depending on others, who meet their long-term care needs. Many of these needs are related to the constant monitoring of the vital constants of the patient, which traditionally entails the necessity of being attached to the instrumentation available in the hospital bed. The usage of wireless sensing technologies can improve significantly the quality of life for such persons, improving the mobility of the persons in a controlled geographic area either in-hospital or at home. These smart sensors enable the continuous monitoring and localisation of the patient using them whereas it also allows a significant freedom for the mobility of the patient, in many cases more than enough covering the whole apartment or hospital boundaries. This scenario reveals some challenges from the perspective of the design of an efficient and practical smart sensor platform.

As was previously revealed, there is a necessity to design more practical and wearable sensor devices. This necessity also applies to the geriatric care scenario in which the patient has to be long-term monitored, probably for the rest of his life, and improvements on this area may be directly related to improvements in the daily life of the patient.

The in-body location of the sensors also entails new challenges for the inter-sensor communication since the body may act as a source of interferences for the wireless communications. Thus, new models of radio propagation are required in order to really know how the human body propagates the different radio frequencies and how to deal with efficient propagation models, which do not cause additional dangers to the patients due to the very long and constant exposure to an attached-source of radiation. In this sense Chen et al. (2011) cover the different aspects which have to be addressed for the effective design of new body sensing devices. They identify a complete set of features such as the development of more effective antennas with lower specific absorption and better coupling to the dominant propagation modes, evaluation of the performance of the PHY layer proposals, prediction of link level performance in alternative sensor deployment configurations, and so on. This is work to be done in further years in order to make the smart implanted sensors a reality.

Another challenge associated with the design of efficient and practical smart sensor platforms for health monitoring is the creation of real smart, distributed sensor devices which are able to perform in-sensor data analysis of the data directly received from the sensors and to act accordingly. Shnayder et al. (2005) say that '...[u]nlike many sensor network applications, medical monitoring cannot make use of traditional in-network aggregation since it is not generally meaningful to combine data from multiple patients...'. While this affirmation is possibly true, it is imperative that the usage of in-network processing capability may open the doors to new and novel applications of further sensor platforms. A clear future example over the same scenario can be the usage of the data provided by a smart sensing system composed of different sensing capabilities such as electrocardiographs, electrical conductivity, motion and localisation of the patient, and so on. in order to apply an agent-based distributed inference process executed in-network to determine the disease from which the patient is suffering in real-time directly from the sensed information, and thus, acts in consequence, for example, by alerting the hospital centre immediately and pervasively.

Another challenge directly associated with healthcare is the management of sensible information and the associated level of security required in the sensed information since it is directly related to the patient health status with its identity. Thus, information privacy is critical in this scenario. In this sense, some innovative attempts of an in-network sensor authentication and authorisation proposals have been recently provided. In case the reader is more interested in this aspect of healthcare, Huang et al. (2009) provide a comprehensible architecture for healthcare monitoring in which security is fully covered. However, novel light-weight fully decentralised privacy-enabling protocols are demanded for a new generation of smart sensor platforms in order to cope with the necessity of power-aware algorithms which minimised the usage of the sensor resources, including processing capabilities.

The information managed in a healthcare environment has to be received potentially by a number of different receivers. Note that it is an unusual scenario to find a hospital in which there is more than one centralised place to monitor patients, usually governed by nurseries, and then this information may be notified to more than one central place. Even when this issue has been fully addressed in the wired world by means of publication/subscription protocols, novel implementations and optimisations of such protocols are required in order to provide a fully distributed approach of publication/subscription systems which are imperatively power-aware and scalable according to the number of subscriptions in a shared medium such as the air.

Moreover, although mobility nodes are really intensively covered in the literature such as our patient in the case study, smart sensor platforms may provide not only mere static implementations of a given software, but also adaptable algorithms to cope with changes in the context information. For example, what happens if the patient is going outside the hospital? Traditionally, the sensors may not be able to change their algorithm adapting to different alternatives for localising the patient, however, a novel distributed sensing platform may cope with this and other adaptive features by means of the use of advanced in-network adaptive algorithms.

Another challenge associated with healthcare is the imperative reliability of the information exchanged between the sensing platform and the controller agent. In this scenario, the use of light-width traffic control protocols and acknowledge-based protocols across all the communication layers involved, possibly doing a cross-layer optimisation is necessary. These aspects have been intensively covered in the literature since researchers have focused more on network layers rather than on the application layers. Many traffic control protocols at the transport layer are provided to the WSN world, moreover by several hop-by-hop and end-to-end packet recovery protocols, which have been provided at link and network layers. Some of these techniques are explained in further chapters.

This case study is focused in geriatric care, which is directly applicable in several scenarios. Among others, the real-time patient monitoring performed in the hospital using a distributed sensor platform can open new possibilities by the usage of smart sensors, which can infer that a patient is located at a particular place in the hospital and he is suffering an imminent heart attack which requires prioritised assistance. This information can be sent directly to the closed geo-localised nurse in order to receive the most rapid assistance. This smart sensing system may change the medical care structure, as we already know it is enabling most effective and reliable patient information and has direct consequences for the medical implications.

An early attempt and interested architecture for performing patient monitoring and localisation is implemented in the Code Blue (Shnayder et al., 2005) project. This project provided a distributed sensing platform for device sensor discovery; publish/subscribe multihop routing; and a simple query interface allowing caregivers to request data from patients. The data gathered is not only vital for the patient, but also RF-based patient localisation is an ideal system for geriatric care. Authors assumed the following requirements were directly associated with an in-hospital monitoring case study: reliable communications, patient mobility and wearable sensors, one-to-many communications and security. Regarding vital signals monitoring, the sensors employed in the distributed sensing platform are: i) The currently smallest and lowest-power OEM module available for measuring pulse oximeter, sizing around

40×20 mm and provided by BCI Medical Micro-Power Pulse Oximeter. This SSD sensor is connected to a *MicaZ* mote and acts as one of the agents, which compose the architecture proposed. ii) A SSD composed of two or three wired electrodes used to provide continuous electrocardiographs telemetry suitable for evaluating a patient's cardiac activity for an extended period. This sensor is also connected to a *MicaZ* mote providing wireless connectivity. Regarding patient localisation, there is another sensor used in the DD, which is in essence a *wearable tag* that can store patient information and track location using RF signals. This sensor is also analogously connected to a *MicaDot* mote.

The proposed architecture relies on TinyOS for implementing *a publish/subscribe routing framework*, allowing multiple sensor devices to relay data to all receivers that have registered an interest in that data. Moreover, *a discovery protocol* is implemented to allow end-user devices to be adaptable to the usage of new information provided by other sensors. Additionally, *a query interface* allows a receiving device to request data from specific sensors. A query request message can include two predicate expressions at maximum; this limitation allows the predicate to fit in a single query message. For example, the query predicate (HR < 50) OR (HR > 200) would trigger event-based data transmission only when the patient's heart rate falls below 50 (beats per minute) bpm or exceeds 200 bpm. Note the distributed smart selection done over the information provided by the different sensors. Then, when the coordinator node receives triggered information from a given sensor, it relays the monitored information to the central nursery location. Regarding localisation, the architecture contains a SSD sensor, which provides a SSD: a robust decentralised built-in RF-based localisation system, called *MoteTrack* (Luinge et al., 1999). It is designed to operate using only the low-power radios already incorporated into the sensors available in the distributed sensing platform and thus, in real scenarios, it achieves an 80th percentile location error of about 1 m, enough for localising a patient in a particular room. In MoteTrack, the mobile node listens to beacon signals sent by stationary sensors spread around the building and acquires a signature that consists of the average received signal strength indication (RSSI) for each beacon node, frequency and power level. The signature is compared to a database of pre-acquired signatures (each labelled with a known location) and a 3D location is determined. Note that this localisation method requires an important collaborative coordination between sensors and also many in-network heuristic techniques need to be performed in order to estimate the patient localisation. A more comprehensible explanation about localisation techniques is explained in Chapter 7. Figure 4.16 shows a snapshot of a possible Code Blue graphical interface in which it can be appreciated how the different patients are localised within the area of interest. Moreover, run-time information about the patients is being shown in the upper part of the screen. Concretely, Figure 4.16 shows a heart rate story about a given patient. Note as well how different sensed data can be retrieved from the patient using the left part of the interface in which a list of all the available information is shown.

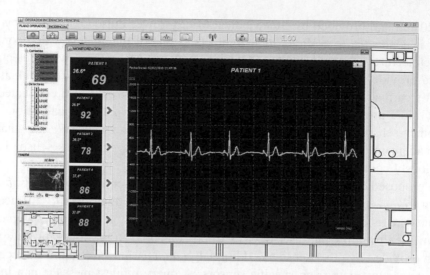

Figure 4.16   Code blue graphical interface screenshot. © 2010 IEEE. Reprinted, with permission, from López (2010) LOBIN: E-textile and Wireless-Sensor-Network-Based Platform for Healthcare monitoring in future Hospital Environments.

While the analysed case study is a useful and prominent attempt of distributed sensing shape, future distributed sensing systems might increase the level of in-network processing capabilities. For example, note that once the coordinator received the events from other sensors, it acts merely to forward such events to the final centralised nursery location. However, more intelligence could be placed into the distributed sensing platform performing a first aggregation and correlation of the different sensor's information and then performing a classification and inference algorithm to determine the causes for which the heart beats are out of control, then, and only then, notifying the nursery centre if and only if it is a real emergency. For example, the excessive number of heartbeats can be due to a heart attack or due to running across the corridor looking for the cookers for the nightly food. Thus, a simple in-network correlation of the data with the difference between pre- and post-localisation points could be enough to determine the cause of the heart beat acceleration. Thus, smart sensing can provide a more accurate sensor with a small percentage of false positive and false negative alarms.

Another interesting system to carry by geriatric patients is proposed by (Fraile et al., 2010). In this case, authors propose a multi-agent-based system (MAS) that uses smart wearable sensors for the care of patients in a geriatric home care facility. The system is based on a distributed sensing platform based on ZigBee, which provides location and identification microchips installed in patient clothing and caregiver uniforms. The idea is that every smart sensor is based on RFID technology. These devices, which identify and contain information for the residents, are hidden within their clothes and uniforms, so as to not inconvenience the residents in any way. RFID readers and mobile devices with NFC technology are used to read the RFID tags. The use of radio-frequency identification and near-field communication technologies allows remote monitoring of patients, and makes it possible for them to receive treatment according to preventive medical protocol.

Figure 4.17 shows the proposed system in which different agents are involved:

- *Catcher agents*: These agents manage the portable RFID readers in order to obtain information from the RFID tags. There is one catcher agent for each type of information (location, identification, security, access control, and so on.) captured in the system.
- *Control agents*: Supervise activities of each of the catcher agents with the help of control directives and validates the information that is provided. They also classify the information received from the catcher agents, organise it and send it to the data agent and organiser agent.
- *Data agent*: Stores the information that it receives from the control agents in the information system.
- *Organiser agent*: Automatically generates supervision plans and simple patient diagnoses. These plans are sent to the head agent.
- *Head agent*: It controls the rest of the system agents, including the activation and deactivation of the control agents. Additionally, it is responsible for notifying about incidents, according to their priority level.
- *Applicator agent*: It manages the communication with the medical personnel.

A differentiating point of this architecture from others is the usage of an expert system in order to provide patient diagnosis from the sensed data. This smart data processing is done by means of the *Organiser Agent*. Figure 4.17 shows the internal architecture of this agent. In essence, all messages received by the organiser agent are the first steps toward the internal reactions, and the deliberative and reasoning mechanisms processed by the organiser agent. The most significant feature of the organiser agent's design is the integration of a case-based reasoning (CBR) engine and a reactive system that gathers data from the sensors and control systems. Then, based on the results from the CBR reasoning engine, the organiser agent sends plans to be executed immediately as events, generated and stored to be executed at a future time. The executions for these plans are for much functionality like disease detection, smart notifications, localisation prediction, and so on. It enables an effective smart and ubiquitous tracking and care of the patients available in the geriatric centre.

4.5 Case Study: Outpatient Care

It is unquestionable that long-term and chronic diseases directly affect the quality of the patients' daily activities. In particular, many diseases require constant monitoring

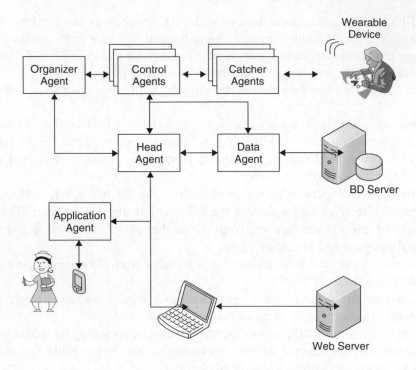

Figure 4.17 Agent-based wearable service architecture (Fraile et al., 2010).

and caring of patients to control physiological constants of the patient. Initially, this fact may require a continuous hospitalisation of the patients for long periods and even, in the worst case, for the rest of their lives. Outpatient care tries to externalise the monitoring of such patients in order to perform it at home while it also minimises the number of times that patients have to be hospitalised, normally, when a really significant event happens in their body due to the disease they suffer. This scenario entails several requirements, which make it very peculiar. Among others, it requires a reliable and secure constant transmission of the patient monitored data to a central repository, which is controlled by a doctor. Note the presence of the wireless area network (WAN) interfaces in the outpatient monitoring system to send the data. Moreover, the localisation of the patient at home is also important to get critical context information from the patient when a significant event is being detected in the central repository. Distributed sensor platforms can contribute to providing solutions to such technologies. Moreover, the wireless nature of such platforms enhances the quality of the patient since he is being monitored using a wireless device, thus, increasing his mobility. Much research work has been done in this field trying to provide distributed sensing architecture for enabling an efficient and secure architecture for remote health monitoring. (Lopez et al., 2010) provide an overview and comparison of several systems for such a purpose.

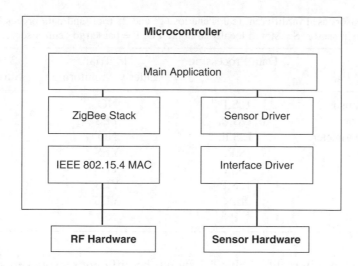

Figure 4.18 Sensor architecture showing the protocols involved (Junnila et al., 2010).

This case study is focused on a general-purpose in-home remote monitoring system, which aims at adaptability of the system in order to be suitable for monitoring several diseases. This distributed sensing platform has been implemented and validated in real scenarios by (Junnila et al., 2010). In summary, the authors propose a distributed sensing platform wirelessly interconnected using ZigBee technology where low-power consumption, relatively simple structure, network topology management and large network size, are some features of the method. The authors implemented a complete network layer stack, as well as the underlying MAC-stack, which are interrupt-driven. The main reason is that there was not any open source stack available at the moment in which they did the deployment. The sensor nodes' main application code communicates with the ZigBee network layer directly. They also use a sensor interface, which is used to abstract the sensor device from the communication interfaces. Thus, each sensor comes with its own sensor driver, which is controlled by the sensor node application. Figure 4.18 shows the architecture composition of each sensor. Note how this architecture fosters the adaptability of the system to new scenarios since it potentially makes the insertion of new types of sensors in the sensing platform easy.

The network topology used in the different real deployments is a star topology. The sensor nodes are directly connected to the network coordinator. The range of this network can be extended by increasing the number of routers. However, in the small test apartment, the sensor nodes are able to connect directly to the coordinator from anywhere within the apartment, therefore, no routers are required. To reduce the amount of transmitted data, some sensors need to perform data processing locally before sending. A wide set of sensing devices are used in the two trial

Table 4.1 Sensors used in different real scenarios (1° and 2° trial) and data processing done in the device. L = in-sensor, S = stored locally in home PC, P = passed to central server

Sensor	Data Processing	1° Trial: Elderly Monitoring	2° Trial: Hip Surgery Patient
Heart Rate Sensor	L,S,P	NO	NO
Infrared Sensor	L,S,P,	YES	NO
Intelligent Pedometer	L,S,P,	NO	NO
Blood Pressure	S,P,	YES	YES
Weight Scale	S,P,	YES	YES
Bed Sensor	L,S,P	YES	YES
Floor Sensor	L,S	YES	NO
Video Call	P	NO	YES

scenario developed in reality. Table 4.1 shows the different sensors supported in the architecture, the sensors used in the different real scenarios deployed and the kind of data processing done for each kind of sensor. Note that those sensors which need information almost in-real time like heart rate and a pedometer require a high data sampling rate and thus, the data is collected and aggregated in the node before being sent to the network coordination in order to increase the network efficiency whereas it provides latency times acceptable to detect any emergency in less than 10 seconds.

The network master node is (network coordinator) connected to (and powered by) the home PC. This PC is running a Java application and decode the incoming data into a readable format, according to the device type. Based on the device type, sensors may also require some additional signal processing before the data is saved into the hard drive for a backup and sent forward. Depending on the application, the required post processing of the sensor data is done at the home PC or at the central server located at the hospital. Note that if the system is used for dementia or fall patient monitoring, the desired acute alarms can be generated, or if the system is used, for example, in the monitoring of independent rehabilitation, there may be more need to derive indexes related to the long-term progress of the rehabilitation process. The home PC provides two levels of limits for generating alarms: recommendation limits and alarm limits. If a sensor value exceeds the notification limit, then a notification is sent to, for example, the user interface. If such value is crossing the alarm limit, then an alarm message is triggered to the central server being received instantaneously by the medical personnel. Moreover, the home PC operates as a gateway to the Internet using a wired DSL connection. Sometimes a redundant connection is configured using a mobile broadband network per redundancy in case the first one fails. All the communication with the server is encrypted with the secure sockets layer (SSL) protocol.

The first real deployment was for elderly monitoring whereas the second deployment was for hip surgery patients. In both cases the following sensors were deployed: 1) The bed sensor measures basic physiology, such as HR, movements, and respiration rate during the bed time. The bed sensor is installed below the user's mattress. 2) Infrared sensors are developed to detect the presence of a person or use of heat-producing appliances, such as a stove or coffee maker. The sensor array consists of seven sensors made of polyvinylidene fluoride (PVDF) and can be directed toward the area of interest. Note that this sensor is used to perform the patient localisation in the apartment. 3) The floor sensor consists of a thin, laminated mat installed under the floor mat and sensor electronics developed by UPM-Kymmene Corporation. The measurement is based on the change of stray capacitance between two electrode tiles caused by movements of a conductive material near the laminate mat. Since the human body is fairly conductive, the technology measures the presence of the person itself. This method is that one does not need to carry any kind of tag to be positioned. On the other hand, this makes the identification of a person and the tracking of two or more closely. 4) Finally, the blood pressure sensor is a wearable sensor, which is in charge of measuring this physiological parameter. Localisation sensors, that is, 2) and 3) previously introduced were not used in the hip surgery patient monitoring since localisation is not relevant in this scenario, however, in this case, a video call system was provided in order to get a face-to-face doctor chat in case of emergency.

As a result, an efficient outpatient monitoring system is provided and validated in real scenarios. In fact, the usage of a smart sensor platform this time has significantly increased the quality of life in patients who only have to be disturbed when they are suffering real episodes which require doctor intervention.

Data acquisition is one of the most relevant aspects in tele-monitoring systems. Information fusion helps these systems to better unify data collected from different sources. In this sense (Tapia et al., 2010) presents a case study that consists of a tele-monitoring system aimed at enhancing remote healthcare for dependent people at their homes. The system deploys a service-oriented architecture over a heterogeneous distributed sensing platform to create smart environments. Such architecture allows both services and service directories to be embedded in nodes with limited computational resources regardless of the radio technology they use. Furthermore, the system allows the interconnection of several networks from different wireless technologies, such as ZigBee or Bluetooth. This approach provides the system better flexibility to change its functionalities and components after deployment.

The idea of the architecture proposed is to have a base station with several heterogeneous interfaces. For example, authors provide for example a dual interface ZigBee and Bluetooth. This base station also provided the touch point with the Internet services and enables the interconnection between the two different underlying technologies acting as package gateways. Then, the heterogeneous sensors provide a wide set of functionalities like fire alarm notifications, fall sensors, respiratory monitoring, ECG monitoring, smoke sensors, light sensors, lamp relay controllers and alarm buttons. An example of the system operation can be shown in Figure 4.19.

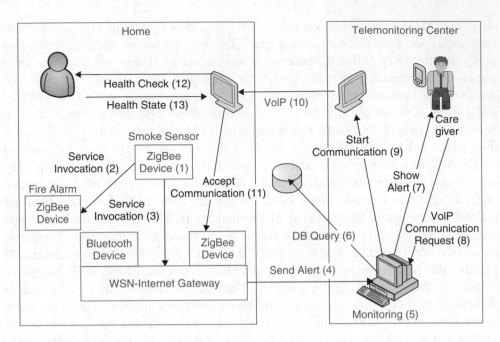

Figure 4.19 Example of the remote monitoring system provided by (Tapia et al., 2010) in a scenario where a sensor can detect the smoke at home.

In the case shown in Figure 4.19(a) the smoke sensor detects a higher smoke level than a specified threshold (1). Then it invokes actively a service offered by the node which handles the fire alarm, making it ring (2). At the same time, it also invokes a service offered by the computer that acts as both ZigBee master node and Internet gateway (3). Such a gateway sends an alert through the Internet towards the remote healthcare tele-monitoring centre (4). At the remote centre, the alert is received by a monitoring server (5), which subsequently queries a database in order to obtain the information relative to the patient (6) (that is, home address and clinical history). Then, the monitoring server shows the generated alert and the patient's information to the caregivers (7), which can establish a communication over VoIP (Voice over Internet Protocol) or by means of a webcam in the patient's home in order to check the incidence. As the gateway in the patient's home accepts the call automatically (8), now the caregiver can see the patient by means of the webcams deployed through the patient's home to assure the chance of establishing the communication with the patient. If the patient is conscious, he can also talk with caregivers and explain the situation (13). If necessary, caregivers will call the fire department, send an emergency ambulance to the patient's home and give the patient instruction about how he should act.

The same authors provide a more complete description of the architecture internals in (Corchado et al., 2010). In general, the protocol stack per wireless sensor available

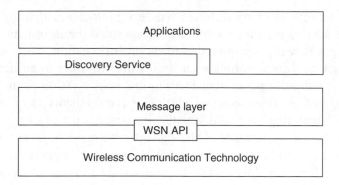

Figure 4.20   Protocol stack available in each smart sensor device (Corchado et al., 2010).

in the distributed sensing platform is composed of four different layers. At the bottom side, there is a Wireless Sensor API to interact with the concrete wireless technology available in the smart sensor. In order to make this access to the wireless technology smooth, a message layer is placed on top of it offering the upper layers the possibility of sending asynchronous messages between two wireless sensing devices. The messages are being sent using a customised internetworking protocol called service protocol. It works in a similar way to Internet Protocol (IP) and allows sending packets of data from one node to another node regardless of the wireless technology to which each one belongs. The messages specify the origin and target nodes, and the service invocation. Then, applications can directly communicate between devices, using the message layer. Moreover, the proposed system also has a directory sublayer that uses in turn the mentioned message layer. The directory sublayer offers functionalities related to the discovering of the services offered by the network nodes. A node that stores and maintains services tables is called directory node. This architecture can be seen in Figure 4.20.

Nasser and Chen (2010) provide an interesting comparison between several elderly monitoring systems in which several common aspects to this kind of system for remote monitoring being specially designed for monitoring for elderly people is shown. All the analysis solutions are not implemented using a middleware platform in order to enable a more flexible development of the infrastructure. Moreover, almost all the solutions only cover the outpatient monitoring inside of his home leaving aside scenarios in which the patient is in outdoor scenarios. The comparison emphasised the real-time nature of this type of application, which requires a constant monitoring of the patient and also shows the lack of support for quality of service assurance in this kind of system.

Security and privacy in long-term monitoring and remote monitoring systems are very critical due to the several nodes and gateways associated to such a kind of system and the diverse types of medium and networks involved. Note that packages can be sent in many different wireless protocols and using not just the local area, but also wide area network technologies and the information carried out in such packages is really sensitive due to its intrinsic nature. Then, the design of this kind of system requires strong security features in order to ensure the privacy of the critical health information managed. In this respect, (Compagna et al., 2010) have identified and analysed a set of particular requirements associated with the remote health monitoring scenarios. Firstly, the smart shirt in charge of monitoring all the health vital constants of the patient requires a high rate of reliability since any failure due to both malicious (that is, someone intentionally changes the configuration, buffer over flow on the diagnosis software, and so on) or accidental (that is, software bug, Smart shirt not correctly installed, and so on) would compromise one's safety. So, it is necessary to provide monitoring services that ensure that the software and hardware infrastructure are functioning correctly. Secondly, the identification of the patient's localisation is critical for the patient's safety, thus a secure and reliable service for identifying the patient's location (and a close doctor) is a clear challenge for this kind of system. The main threats to reliably identify patient's location concern: 1) failure of the in-home server in charge of receiving the updates about the patient's localisation and submitting them to the medical centre, 2) corrupted information stored on a doctor's repository, 3) incorrect information provided by a localisation system like the GPS, and 4) failure on the doctor's front-end device. Any of these threats would make the safe determination of the doctor's search impossible. Some fault tolerance techniques should be deployed to ensure the information collection, storage and communication decreasing the risk level related to the identified threats by means of the increase of the resiliency capabilities of the system. Thirdly, non-repudiation of commitment by the doctor and proof of fulfilment of the emergency call are almost necessary in order to provide efficient attestation services in which the life of people is at risk. It could be easily addressed using a public key infrastructure authentication signing confirmation message sent by the selected doctor. However, it only provides a non-repudiation of commitment, but it is not enough to ensure that the doctor actually has gone to visit the patient. So, another proximity-based authentication between the patient's Smart t-shirt (or his health terminal) and the doctor satisfies the proof on fulfilment of the rescue. Fourthly, integrity, confidentiality and mutual authentication have to be addressed during the communications between patient and doctor and also between doctor and the medical system. An encryption-based solution could be used. Moreover, an authorisation solution managing patient's medical data needs to be deployed in order to control the access to such personal and private information.

5

Clinical Applications

Considering medical budgets, national or private, allocated for handling new disease and common illnesses is making a distinct mark in this century, where finding new solutions for the old problems through adoption of a more effective approach for upgrading our medical equipment for significant enhancement through potentially available integration of wireless and intelligent devices under ubiquitous technologies needs serious attention. To this effect, we suggest deployment of DSS technology built upon interconnected smart sensors and other intelligent sensing devices in the workshops and theatres of the hospitals and clinics in crowded cities; also, if possible in remote places, as the most urgent program for advanced civilised societies. Clinical treatment and surgical operations, running at horrendous expense and often with poor results have been hindering deployment of sensor-centric smart and intelligent spaces required for the most sensitive places that our lives depend on.

In order to achieve its objective, this chapter provides selective examples to demonstrate a need for new, more efficient technology solutions for integrating the information, with communication by smart sensors and miniaturised actuators, where adoption of new cost-effective shrinking equipment and sophisticated professional tools can help the experts obtain higher success rates, more accurate decisions and rich data-centric medication and to perform more reliable operations.

To this end in some cases we may not be able to determine the exact location of the devices being used, but it should be easy to imagine how, by wireless or with ubiquitously interconnected smart and intelligent devices, we can supply or even interact with the anatomical media for abnormality and intervene, if required, as easily as we deal with other aspects in life. To extend this to DSS solutions we include: (a) research materials to highlight typical scenarios and areas requiring application and (b) discuss the feasible possibility of using smart sensor solutions.

Distributed Sensor Systems: Practice and Applications, First Edition.
Habib F. Rashvand and Jose M. Alcaraz Calero.
© 2012 John Wiley & Sons, Ltd. Published 2012 by John Wiley & Sons, Ltd.

5.1   Surgical Applications

Due to its historical development and sensitivity to one's life, clinical surgery dominates the medical profession and is heavily technology intensive, where smart sensor technology should be regarded as an indispensible component of the practice. Now, smart sensors are finding their sensitive place at the heart of medical engineering interfacing three other distinct fields of: (a) the well-established world of the medical profession; (b) exciting forthcoming new precision engineering and; (c) an exciting world of information-communication technologies. Smart sensors have shown their uses in surgical application being in the hands-on part of the practice or provision of smart space where we expect clinics, hospitals and operational theatres.

However, apart from a few exceptional cases, clinics generally consider minimum use of technology in their surgical and operational theatres. This low use of technology is due to two practical complications that is, (a) lack of mutual cooperation between medical and non-medical (mainly technical and engineering) experts and (b) extreme shortage of technology-rich cost-effective products. We therefore see these theatres virtually empty of any latest proven, superior technology, as most of their existing equipment and tools are old, bulky and impractical for any agile and practical ad hoc surgical operations.

Use of medicine as a practice of preventing, diagnosing and treating disease can be broadly categorised into two main groups of surgical and nonsurgical areas. Sensors used in the management of nonsurgical methods for example, provide medical treatment using drugs, diet, psychology and similar methods integrated into the management and administrating systems are much more popular. However, due to many critical and vital issues attached to clinical and operative aspects of surgery the indication is that one of the serious problems is lack of communication between various disciplines, medical and engineering practitioners. There is some evidence that integrated solutions for adopting technology for medical applications may be achieved in the academic and research levels with none being potentially implemented for real. For example, use of autonomous and intelligent sensors can help tense hands-on surgical operations providing previously listed conditional information from a bank of data storage archives, as required.

Surgical operations vary extensively in complexity, and precision and risks from a simple dental clean up or low risk amputation to a life-sensitive complicated operation on a patient with a serious heart condition. In all these cases the risk elements associated with possibilities of contamination, human error, accessibility and control over the operating tools and so on play a significant role in the success of the operation. Whereas, under a reliable sensor enabled integrated system equipped with automated or semi-automated devices, we are able to turn the clock around and change a poorly conducted, unprepared, risky surgical operation into a successful one.

Surgery as the most delicate part of medical practice is marked by factors of 'life', where experts can be helped with the supply of a regular flow of sensing data, raw or pre-processed, through sophisticated equipment and precise tools under clean, safe and reliable smart spaces criteria.

Recent developments in DSS can be easily adopted into surgical operations and hi-tech theatres for many reasons including:

- better control over the process;
- better prepared operation by taking away boring lists and check points;
- better information in the working environment;
- real-time information on sensitive organs and parts of the patient's body;
- monitoring progress confirming information and signals;
- useful and timely critical warnings;
- coordinated operations assisted by pre-selected lists of operation;
- higher confidence and trust for risky operations.

In order to succeed in our mission for a better lifestyle we all need to improve our clinics and surgeries for up-to-date technology and best practice. The main difficulty we face these days is complexity in an emerging global world and its conflicts with national and domestic measures, which may affect one's efforts. However, there are signs of progress around the world and many motivated professionals lead on encouraging others for better use of the technology for a better life. For example, these days every mobile telephone is now equipped with a quality camera and movie recorder so there is no excuse for our clinics not to make use of or abandon technologies up to their full capability. The usefulness of a camera in clinics was discussed by Palomo et al. (2004) about a decade ago. For any further discussions on this area the viewer should look at the Mayo Clinic Programs.

Wireless Access, Implant

One of the bottlenecks for most surgical operations is the problem of gaining access to the organ or RoI hidden under the skin and often beyond to other parts of the body where invasive intervention always causes a serious problem for the practitioner as well as a risk of bodily harm to the patient. To this effect minimally invasive therapeutic intervention is increasingly used in surgical tools with embedded miniature sensors providing real-time information to improve surgical results whilst use of nanotechnology devices requires permission. However, non-implantable sensors can be advised without permission. Wireless sensors, however, in the individual form or clusters can provide easy access to various organs or areas of the body without any harm to the patient's body. They can be integrated on to the patients' body and scan the body for useful information. In surgery, tools and professional gadgets can be regarded as the vision power for an automatic intelligent sensing agent or have follow-up orders which can be an easy integral part of a larger operation, able to detect or remove a cancer infected area or just a monitoring device collecting data for statistical characterisation of the organs' behaviour.

Currently available fourth generation implantable cardioverter defibrillators (ICD) use a single ventricular electrogram (EGM) for sensing and detection of tachyarrhythmias. Detection is accomplished by classifying each interval between successive sensed ventricular events as fast or slow according to programmable rate criteria. Some devices offer more than two rate zones. Rate zone counters keep track of progress toward detection. Detection is complete when a zone counter exceeds a programmed limit. It is reported that between 27% and 41% of ventricular tachyarrhythmia detections are inappropriate even with added programmable features such as stability and onset.

Cancer

Cancer is a complex form of malignant disease where an uncontrollable growth of cells behave like an invasion, disturbing normal behaviour of an organ or other parts of the body as a malfunction which in most cases ends up in the termination of life. Smart and DSS-style sensor systems, however, can be used in very different ways to help with an operation. In practice, however, identification of the location is based on some disorderly behaviour of the organ with the suspicious location, the identification of the source cells are estimated resulting in either loss of a large proportion of the patient's organ or many repetitive actions attempted on the same source.

Lung cancer is the leading cause of cancer death worldwide. Despite many discoveries an introduction of new agents and schedules, chemotherapy still obtains unsatisfactory response rates, rarely completed and remissions come with relatively short duration responses.

This is from a review report pointing out the progress being made in characterising the inhibitor of the apoptosis (IAP) family, with a focus on the available data relevant to the treatment of lung cancer. The World Health Organisation (WHO) estimates that there are 11 million new cases of cancer per year which will rise to 16 million by 2020. Lung cancer is the leading cause of cancer death worldwide (17% overall, 23% in males and 11% in females), with the major histological types being small cell lung cancer (SCLC), adenocarcinoma, squamous cell carcinoma and large cell carcinoma.

Unfortunately, the presentation in the majority of patients is with advanced, metastatic disease for which there is currently no cure. Chemotherapy, most commonly a platinum agent in combination with another cytotoxic, for example gemcitabine, vinorelbine or a taxane for NSCLC or etoposide for SCLC can prolong survival and palliate the symptoms. However, even with modern chemotherapy most patients survive less than two years following the diagnosis. Therefore, new therapeutic approaches for lung cancer are urgently required. Tumour formation is a multistep process involving the progressive transformation of normal human cells into highly malignant derivatives. Six alterations in cell physiology form the hallmarks of malignant growth: self-sufficiency in growth signals, insensitivity to growth-inhibitory signals, evasion of programmed cell death, limitless replicative potential, sustained angiogenesis, tissue invasion and metastasis. Chemo- and radiotherapy have traditionally been central to cancer therapy and work by causing irreparable genomic damage.

One of the problems frequently encountered during cancer surgery or other regular clinical operations is the difficulty of gaining access to a place in the body without retracting neighbouring tissues. Further complications may occur due to tissues obscuring the blood supply resulting in ischemic damage. For the oxygenising process (Fischer et al., 2006) integrate the oxygenation sensors on the surface of surgical tools which augment critically important information on the saturation of oxygen on the surface of local tissues to the surgeon during the operation in the RoI. The tools are commonly called retractors and graspers, where the embedded sensors provide a means for applying minimally invasive laparoscopic activities and higher precision, whilst the health of the retracted surrounding organs is measured for oxygen saturation level (SO_2).

As an example of the oxygen problem one may consider a damaged liver in Figure 5.1. Having the oxygen should be delivered to the tissues by haemoglobin via proteins in red blood cells. For a healthy patient the arterial blood oxygen saturation (SaO_2) is around 90% or above when blood vessels to an area are constricted the decrease in blood flow leads to reduced SO_2 and creates an unhealthy organ. The human liver, for example, has only 15 to 20 minutes ischemia tolerance and if it goes beyond 90 minutes the prolonged ischemia may cause irreversible damage to the affected tissue.

In order to check patients for ischemia, or lack of oxygen, in an inaccessible location one should first monitor and help prevent any further ischemic damage by

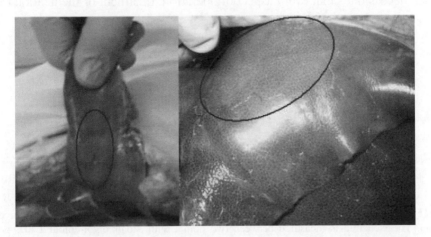

Figure 5.1 Ischemic damage induced in porcine liver during retraction. © 2006 IEEE. Reprinted, with permission, from Fischer (2006). Ischemia and Force Sensing Surgical Instruments for Augmenting Available Surgeon Information.

using an instrument measuring the local oxygen saturation directly on the working surface. The next step is to interact with the forces using the tools to enable the surgeon to measure the forces and torque, including the load cells and force with the sensors and strain gages.

For designing their new tool authors investigated making use of all 6 degrees of freedom, which come with force/torque sensors (ATI, Inc.). However, due to the cost, size and ease of use they considered using reduced foil strain gages configured in Wheatstone bridge configurations located at the end for sensing interaction. Whereas the use of strain gages makes the device inexpensive and slim, whilst providing the required high quality force measurement.

Another similar three-fingered tool is shown in Figure 5.2. Here, a standard five-fingered 10 mm laparoscopic fan retractor is modified to integrate the force and sensors, in a full-bridge configuration and bonded to the central finger. The laparoscopic tool capable of sensing forces and tissue oxygenation can help to prevent any tissue damage. It is comprised of three components: (a) a gripper to accommodate all sensing components, (b) a metal sheath to cover the instrument shaft and (c) an endoscopy instrument interfacing the surgeon with the gripper.

Figure 5.3 shows the result of using a laparoscopic grasper for oxygenation sensing. In this experiment, a section of a bowel is isolated and the supply of blood to that section periodically changed, on and off.

Heart

The heart, as the centre of the human body's anatomical activities, counts as an important organ, and needs to work well even if some other organs fail to respond. This fact necessitates that every aspect of operation, treatment or medication to keep

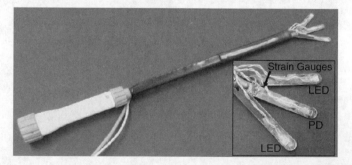

Figure 5.2 Modified standard fan retractor for measuring laparoscopic retraction forces and ischemia. The ischemia is measured between two points with a pair of bi-colour LEDs and one photodiode (PD). The force is measured by the strain gages. © 2006 IEEE. Reprinted, with permission, from Fischer (2006). Ischemia and Force Sensing Surgical Instruments for Augmenting Available Surgeon Information.

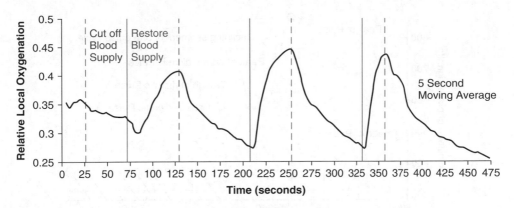

Figure 5.3   Demonstrating local oxygen sensing results and the ability of a laparoscopic grasper in a controlled experiment restoration of the bowel segment's ischemia. © 2006 IEEE. Reprinted, with permission, from Fischer (2006). Ischemia and Force Sensing Surgical Instruments for Augmenting Available Surgeon Information.

an eye on monitoring the heart, should ensure us of a continuously healthy state of a working heart under any circumstances. For example, a patient going under an emergency scalp surgery needs to be accurately monitored for his heart's behaviour. Obviously, if the surgery is for the heart disorder then this measurement becomes a critical part of the process. Using smart sensor systems such as Smart shirt then, apart from very occasional circumstances, measuring the heart's behavioural analysis in the control panel should assure experts of any reaction from the body and any action being made in the course of operation should reveal a better understanding of the heart. To this effect we look at (Gong et al., 2011) for their proposed system for a better understanding of the heart's behaviour and its interactions with other parts or organisms of the body. Commonly, the human heart generates pulses with a main peak and two minor peaks, called dicrotic peaks, one dicrotic trough and one trough for the pulse as shown in Figure 5.4. The process of course does not end with the pulse measurement, as it is just the beginning. For good use of the device a powerful process to accommodate a thorough statistical analysis of the pulse for the time, periods and the level of each signal point in comparison with the same person's average and larger populations becomes an important part of the process. Some may use these data for peace of mind, but it is mostly used for those under intensive care or under surgical operations.

A stand-alone health diagnostic for general use includes various components including a powerful remote computer, wireless connectivity and data log for a working system as shown in Figure 5.5.

This work makes use of traditional Chinese medicine to develop a new analytical method based on a pulse diagnosis theory used in the clinic for thousands of years with proven values. The potential for this new method is significant in

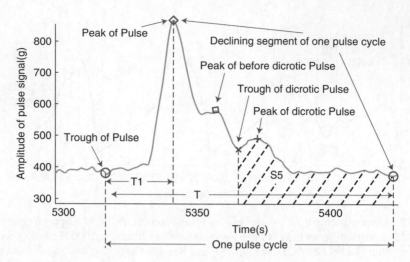

Figure 5.4 Five important physiological points in human heart pulse (Gong, 2011). © 2011 IEEE. Reprinted, with permission, from Gong (2011). PDhms: Pulse Diagnosis via Wearable heatlhcare Sensor Network.

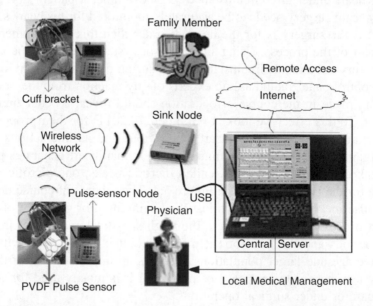

Figure 5.5 Schema of a health diagnosis system for both local and remote monitoring (Gong, 2011). © 2011 IEEE. Reprinted, with permission, from Gong (2011). PDhms: Pulse Diagnosis via Wearable heatlhcare Sensor Network.

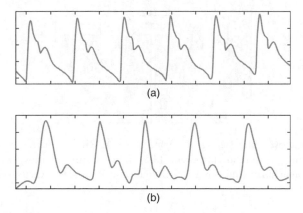

(a)

(b)

Figure 5.6   An illustration of different pulse patterns (Gong, 2011). © 2011 IEEE. Reprinted, with permission, from Gong (2011). PDhms: Pulse Diagnosis via Wearable heatlhcare Sensor Network.

many non-invasive health-monitoring methods and is becoming essential. The pulse contains an abundant source of physiological and pathological parameters and information about the human body. Here each pulse comes with a different pattern, within which one can identify different cases for different groups of people with a specific symptom or disease. For example, in Figure 5.6, the Normal Pulse pattern of a normal woman is significantly different to the Slippery Pulse of an expectant mother.

Experts believe that a light-weight pulse requires measuring algorithms, a robust collection of pulse data, real-time pulse analysis mechanism and most of all an intelligent pulse diagnosis with further challenges ahead for the following: (a) difficulties associated with a stable and appropriate external pressure on the three radial artery positions and (b) resource shortages at pulse-sensor nodes of the DSS.

One application for using smart sensors in surgery is for the treatment of abdominal aortic aneurysms (AAAs). AAA is a weakening of the aortic wall near the junction of the aorta and iliac arteries in the abdomen with a number of cases causing death, and the third leading cause of death among men over 60. The prevalence is about 1.5 million cases with 200,000 new cases being diagnosed annually. Repair typically takes place in a surgical fashion, in which the weakened section of the aorta is replaced with some surgical graft at the expense of traumatic pains with high morbidity and mortality rates. Using the traumatic procedure the surgery often yields lengthy hospitalisation for a full recovery with a combined morbidity rate of 15%.

The endovascular therapy, a catheter-based approach used for repairing blood vessels from within, is a more familiar treatment for repairing coronary vessels using endovascularly-delivered stents rather than open-heart surgery. In the case of AAA, a stent-graft is used instead of the stent which reduces the treatment by half, see Figure 5.7, left. Unfortunately, there are other complications leading the research to try to find a better monitoring method by placing a permanent wireless pressure sensor implant in the aneurysm sac during the implantation of the stent

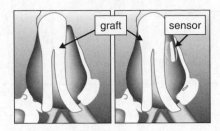

Figure 5.7 Left: Endovascular slent-graft repair of AAA. Right: Smart sensor incorporated into AAA sac to monitor the pressure in sac (Allen, 2005). © 2005 IEEE. Redrawn with permission, from Allen (2005). Micromachined Endovasculary-Implantable Wiressless Aneurysm Pressure Sensors: From Concept to Clinic.

graft. As shown in Figure 5.7, right, the smart micro sensor monitors the intra-sac pressure and communicates with an external RF receiver for further interactive functions (Allen, 2005).

Mayo Clinic

The Mayo Clinic is a symbol of a success story of using the technology for medical care, which stems from the early days of development and then regenerated with the advancement of an extensive image database for medical progress. For these the best reference work is Robb's guest editorial paper (Robb, 2001).

This historical brief provides a brief description emphasising the progress and milestones over the past three decades of development upon a successful Biomedical Imaging Resource (BIR).

In the late 1960s, 'Wood's Group' had begun looking at the ways to quantitatively analyse video fluoroscopic images, primarily of the heart, lungs and circulation. In the early part of the 1970s the remarkable advent of medical computer-assisted tomography (CT) started. With CT imaging scanning of the head becomes both challenging and interesting that leads to a new cross-sectional imaging technique for three-dimensional (3-D) images of the heart.

The video digitising system produces another milestone resulting in the first digital subtraction angiogram (DSA). Then, the electrocardiogram images of the heart could be digitised and subtracted, highlighting the contrast media filled coronary vessels. These unique digital video and computing capabilities enabled several new applications such as video densitometric analysis of bone porosity, nerve morphometry and 3-D digital reconstruction.

Robb considers the following points as the Mayo's BIR key success points:

- Vision: The leader(s) of imaging science groups.
- Convert Ideas to Tools: Both good ideas and practical working tools are part of a successful enterprise.

- People: A unit like the BIR has benefited from a stable, skilled staff of committed and dedicated engineers and programmers and scientists who have worked together.
- Environment: A supportive institution and administrative resources.
- Physical Resources: Modern facilities and equipment.
- Commitment to 'Open-Ended' Development.
- Commitment to Support.
- Funding: Multiple sources of funding are usually required to make an enterprise like the BIR successful.
- Collaboration: This includes both internal and external collaboration.
- Luck: This would probably be on anyone's list of keys to success.

The Mayo Clinic has over 500,000 patient visitors every year at three sites of Rochester in Minnesota, Scottsdale in Arizona and Jacksonville in Florida.

The Mayo Clinic is a not-for-profit medical practice and medical research group. It is a progressive national motivator for public health. Its long history of service as a symbol of quality for the national and international community has been a model to the nation. Its continuous update of important aspects of life and health information has been a truly inspiring cornerstone for the globe as well as the USA. One of the missions of the Mayo Clinic is quality and professionalism in their practice. As a university it educates the best graduates and leads with new disciplines and new innovations for practical ends in the profession of medical practice. To this effect the Mayo Clinic has said: 'In 2009, the U.S. News and World Report magazine named Mayo Clinic a best hospital for 20th straight year, and ranked number 2 in the Honour Roll'. In this chapter, the basic reason for the success of Mayo Clinic is studied by use of the concept of systems ethics. The initial development of systems ethics in Taiwan is presented first. Then, a brief review of the history of the Mayo Clinic is made. Finally, my own actual experiences as a patient in the Mayo Clinic are used to support the study.

Due to the vast number of clinical cases, for example Mayo Clinic's list of diseases is so far 1153 (MayoClinic, 2011) and growing. For our limited volume we resist mentioning, let alone discussing all possible or complicated specialised operative cases of medical applications of DSS for which we limit our discussion to selective cases in four distinct application areas of post operation, heart surgery, dental surgery and scalp surgery. Where the significance of the DSS solution upon featured sensors, actuators in integration with other common and specialised equipment and devices can be appreciated. In order to develop a systematic style unified solution for integrating sensors, actuators used for medical surgery and other clinical surgical applications we need to examine, individually and collectively, their effectiveness and their requirement to possess certain features considering the following aspects of the solution.

Broadly speaking, the whole system needs to be reliable. Trust in a system to help with a clinical operation or post-operative observation of a critically important health process is vital. It should possess the security requirements of data against being tampered with or abused, resilience against any errors caused by the media, fault tolerant and free from human-errors. It is prepared and able to reject or tolerate any possible or unforeseeable error, or interference from the media, or other devices or associated systems.

Designed to fit the application well, any extra, redundant or non-fully utilised device could become a bottleneck for a tedious operation whilst it consumes/ occupies the limited space (and time) available to the experts and their team for the main operation. Also, any little mistakes or emission of an irrelevant action or information in their use of a function or device could have a major impact on the operation for which ultimate tests for their usability, complexity and compatibility of Roger's criteria (OpenUniversity, 1996) should be considered.

Spacing of the operation theatre is almost always an important issue problem for an operation room. Expanding the size of the operating theatre does not help with approaching the limited space available at the vicinity of the patient's organ or operative point. Optimisation of the space plays a significant role throughout the operation, though the usage does very much depend on the surgeon's capability and preferences. Some are able to carry out their best performance within a crowded condition and some need their full concentration as they could be easily irritated by any unwanted information or unnecessary interrupting interference.

Having all bulky equipment in the operation room consumes the space considerably, but with the recent significant development of the low power wireless connectivity virtually all equipment could be moved away from the operation room and maintained under direct supervision of technical experts whilst medical experts and surgeons can focus on their own tense activities into full use of the minimum space occupancy for important images and other critically important information for the operation. Some required equipment is still bulky and cannot be reduced in size for decades. Under a new integrated solution it can be scheduled to move in and out per requirement in either a fully or semi automated style by technical experts sitting behind the glass.

In most operations access to patient's medical details as well as some specific detailed medical information, normally available in the local or global databases, is needed for the operation. Recent developments of distributed intelligent MAS under new integrated DSS solutions (see Section 1.4), automated or assisted, can help, guide or inform the experts to make the right decisions by providing effective detailed information usually unavailable to the operation team or help them achieve a smooth and thoroughly reliable step-by-step operation at their own pace depending on the circumstances and patient organs' reactions, activities leading to an operation with upmost accuracy and reliability. This is particularly important in heart or cancer surgery operations where accuracy of the operation can be optimised through an interactive dialogue for timely actions against the ongoing reaction of the patient's body.

Though mostly still in their infancy period and in short supply, use of miniature devices such as MEMS and NEMS under a centralised MAS technology control unit can bring a much more reliable automation into complex processes for unreachable organs and count as a great breakthrough in medical surgery providing superior solutions to many tedious and dangerous operations.

Finally, through the ultimate power of new DSS unified approaches, hospitals, consultants, professional bodies, medical clinics and surgery experts can enjoy an integrated infrastructure to increase their success rate by reducing their critical everyday problems and helping them with emergencies, their pressing and urgent issues of medical education, management hands-on clinical treatments and professional surgery operations.

5.2 Dental Applications

Due to the nature of dental practice, often called orthodontics, dental surgery may be regarded as the most used surgical practice. Being desired for looks as well as health issues this surgery can easily adopt plenty of technology in its practice. There are, however, some who do not take advantage of the features though some local restrictions can limit their freedom for using Smart sensors integrated into their equipment and working environment.

The enclosed nature of this practice dictates few basic, but very important forms for adoption of new technologies into their system, including:

- Smart environment;
- wireless enabled networking;
- agile database;
- data rich imaging facilities;
- agile management system.

Out of these five requirements we examine some as indicators of the potential of technological applications of DSS. Furthermore, other common technological adaptations such as precision integration of MEMS and miniaturised silicon technologies into the surgical tools and equipment can benefit dental surgery with advancing smart, intelligent and autonomous sensing and actuating technology. A smart environment can be easily adopted using commonly available smart sensors to maintain a healthy working environment for all surgeons, medical staff and the patient. Lighting and air condition are critical for a fault-free surgical operation. In support of this claim we have an example from Helmis et al. (2007) who report an in depth analysis of air quality in an indoor dentistry clinic. Chemical pollutants identifying the sources of dental activities such as CO_2, VOCs, PM10, PM2.5, SO_2 and NOx during the operation and non-working hours are recorded. Due to the nature of dental clinics, the materials and the ventilation schemes lead to high concentrations and far above the limits set by international organisations for human exposure.

As far as the importance of using an agile database is concerned we should look into the practice itself. Dental clinics are normally small and dynamic. Due to the moderately low cost of surgical operations their real-time activities are limited to a few items in the database, where a simple and light user-friendly program can easily make all basic data and images available in the form of an interactive style also accessible throughout the operational network. However, a relatively more complex data system can integrate all functions of dental surgery into a single, rich but still relatively light and agile program. For this we look at a management system developed by Niziałek et al. (2008) using an enterprise Java programming. This system can be applied to all types of dental surgery using a multi-tier program.

As well as being able to access distributed data storage located in different places it is equipped with a modular structured set of management tools for the graphics via a user-friendly interface.

A doctor or a member of staff can use the menu, buttons and colour palette to draw patient's various details and diagrams integrating other details including the visit report and store any other information. Using such a system makes the visit smooth, saves time, reduces errors and provides a user-friendly working environment. Using the Java Enterprise Edition development package the system structure is very simple to pass on to and simplify the work of the users. The access is a standalone client, but comes with sophisticated graphical capabilities. In the middle layer, concentrating the whole business logic Enterprise Java Beans tier is located providing clients with access to the database. The block diagram of the application is shown in Figure 5.8.

This application is built on the foundation of Java language and comprises various technologies that support development of large enterprise systems upon a multitier infrastructure, shown in Figure 5.9.

Other examples suitable for dental surgery are production of orthodontic radiographic images, where a suitable networked set of computers can provide good results. With a suitable network security and wireless technology one can boost the

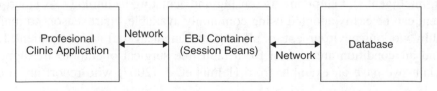

Figure 5.8 Basic structure of the dental management package.

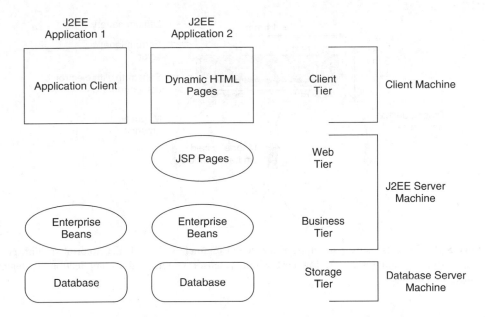

Figure 5.9   Multitier infrastructure that enables an embedded flexibility.

efficiency of an orthodontic clinic. Although digital radiography is widely accepted today, radiation has been always an issue. Using new low distance smart sensing devices, however, digital radiography has cut radiation doses and contributed to the overall efficiency of orthodontic clinics because of the ease of acquisition, storage, access, duplication and transfer.

Our final example is related to an interesting use of RFID technology for dental retainer solution. Nowadays, in order to control the growth of the jaws and keep teeth in the jaws in a proper growing condition we use removable or tight retainers as a normal solution. For younger children, for example between age of 6 to 10 years, the position is corrected using some bracelets or removable retainers. In order to succeed, the removable retainers should be kept in position regularly for many hours. But, it is hard to ensure the children do wear the retainers over a necessary daily routine as prescribed. The evidence shows this is not usually the case and irregular use affects the outcome of the practice and in most cases we end up with poor results, that is, many improper jaws and waste of time and resources.

To this effect, (Brandl et al., 2009) have designed a RFID-equipped dental retainer to monitor and supervise the use of the retainer for various factual information about the use of them, measuring the regularity of the patient's essential sensing information of the condition of the jaws, mouth and associated parameters for better understanding of the process and ensure more effective treatment. See Figure 5.10 and Figure 5.11 for the device schematic and the prototype in position. It can be used in one of two options:

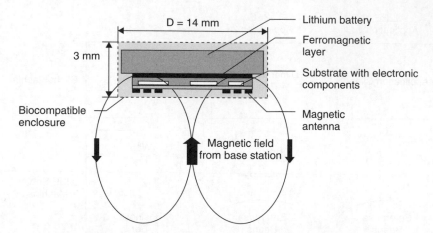

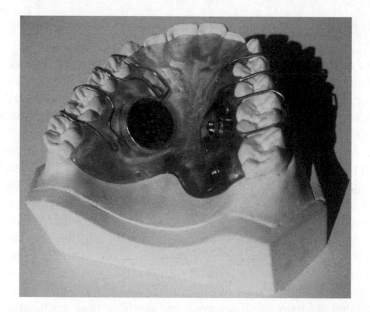

Figure 5.10 Wireless sensor structure of the dental retainer. © 2009 IEEE. Reprinted, with permission, from Brandl (2009). A low-cost wireless sensor system and its application in dental retainers.

Figure 5.11 RFID-equipped dental retainer in place. © 2009 IEEE. Reprinted, with permission, from Brandl (2009). A low-cost wireless sensor system and its application in dental retainers.

1. Active operation with inclusion of a battery, but with sensor works without an RF field from an interrogation unit.
2. Passive operation, the RF field of an interrogation unit powering the sensor device.

For a self-powered wireless data transmission from the sensor to the interrogation unit a RFID transponder, operating in the 13.56 MHz ISM band, is used, and equipped with a CR 1216 lithium coin battery.

5.3 Scalp Applications

Human scalp and whatever is inside it is as mysterious as ever because it contains the most complicated and most valuable part of our biological life and it is best not to tamper with it. However, in many cases responsible professionals may study or in some cases may examine it using non-invasive methods to discover new insight for understanding its biological structure or equally responsible practitioners may try surgical operations or treatment to save the life of a seriously ill patient under a visible operative environment, where availability of maximum data and sophisticated sensing and visionary tools and equipment ensure a minimum risk to the patient.

Cases for interacting with the head or scalp, passively or actively, happen, unfortunately, quite often. They can be classified into three groups of (a) those injuries which occur in accidents, (b) those being developed through genetic or biological disorder and (c) those caused by unpredictable viruses or common diseases.

Head injury, most common in the first group, can be defined as an injury to the scalp, including swelling, abrasion, or contusion as well as laceration or simply a blow to the head. Head injuries, mostly in children, most commonly result in death following trauma where their lives normally remain in the hands of the general and orthopaedic surgeons. Many head injuries are as a result of road traffic accidents or falls. Other head injuries are domestic, sport-related or of an isolated nature. Most head injuries are associated with skeletal, facial and spinal injuries usually distributed equally regarding the seriousness of mild, moderate and fatal.

The scalp treatment in any of the groups normally requires a logical sequence of (a) understanding the nature of deficiency or injury and finding its location and then (b) advice for a solution, which usually includes professional care and follow up treatments.

In their broad meaning sensors and actuators cover all professional tools and gadgets used for collecting information from and interacting with the media, PoI. Therefore, smart sensors and actuators in the form of distributed or autonomous are regarded here as the main technology interacting with the surrounding objects. For example, an epilepsy seizure can be detected from some spikes in the brain using an electro-encephalography (EEG) device recording electrical activity along the scalp. EEG can be counted as a sensor, which when equipped with powerful signal processing to analyse the brain waves and recognise them becomes a smart sensor.

Epilepsy affects some 60 million people in the world and is known as the second most common neurological disorder after stroke. This serious disorder is associated with recurrent, unprovoked epileptic seizures, which result from a sudden disturbance in the brain characterised by abnormal firing of cortical neurons engaging neighbouring cells into a critical mass surge of energy. EEG is an important tool for diagnosing epilepsy normally characterised as epileptiform, spike or sharp-wave. Over the past 30 years of development automatic spike detection includes expert system, mimetic method, context-based, template method, artificial neural network, Kalman filter, morphological filter, wavelet analysis, independent component analysis, support vector machine, fuzzy C-means, deterministic finite automata and multi-stage system.

The epileptic spike can be characterised by 'a distinct transient brain surge of activity with pointed peak lasting from 20 ms to 70 ms'. It is usually hard to separate the detection of spikes amongst other EEG signals, electromyogram (EMG) signals, blinks and abnormal EEG activities. The EEG is generally described in terms of activity in different frequency bands, reflecting different properties of the symptoms:

- δ-band for (0–4 Hz);
- θ-band for (4–7 Hz);
- α-band for (8–12 Hz);
- β-band for (12–30 Hz);
- γ-band for (30–100 Hz).

For example, an increase in ratio of θ/γ could indicate patient's memory loss.

Another example for use of EEG is for more intensive use of a computer for EEG signals for diagnosing epileptic behaviour. Here, EEG features are divided into two groups of (a) classical spectral features using the power spectrum analysis and (b) dynamic feature extraction using time series analysis.

An interesting area of development is in therapeutic and applications of RF and microwave using medical implants transferring information at non-ionising frequencies. The resonance of implanted antennas at a frequency band of 402–405 MHz outside the body favours implants working at short-range radiation. This enables continuous monitoring of intracranial pressure, which helps with brain disordered and trauma patients. Due to the place of *intracranial contents* within a rigid vault (skull), any direct pressure measurement requires neurosurgical intervention.

Considering invasive monitoring methods can result in fatal infections and brain injuries from existing invasive methods such as fluid-filled manometer and external strain gauge, which should not be continued. Also, all the current non-wireless methods using implants for intracranial pressure monitoring have limited mobility, cause patients discomfort and impose infection risks to the implants. There are also some battery-less passive means using MEMS near-field sensor technology which must be placed within a few centimetres of the body and often do not work very well in place of use. A better solution proposed by (Kawoos et al., 2008) uses industrial–scientific–medical (ISM) band of 2.4 GHz.

Originally assembled in a small 12-mm cylindrical case the device used a piezoresistive (PZT) pressure sensor, but due to high power consumption and sensitivity

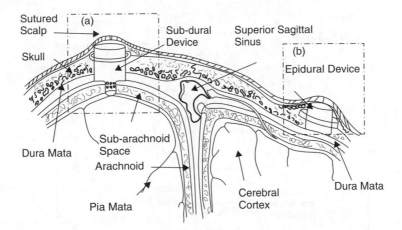

Figure 5.12 Device placement of MEMS in skull: (a) Sub-dural device implantation and (b) Epidural device in contiguous sections of the meninges. The sensor is exposed to the cerebral spinal fluid in sub-arachnoids space for sub-dural pressure detection. In epidural detection, the sensor maintains a contact with the dura mata and relies on dural deflection. (The sub-arachnoids space is the space between the innermost and the middle protective covering of the brain). © 2008 IEEE. Redrawn with permission, from Kawoos (2008). In-vitro and in-vivo trans-scalp evaluation of a intra-craneal Pressure Implant at 2.4 Ghz.

to the temperature it had to be replaced by a MEMS capacitive pressure sensor and after development of the planar inverted-F antennas, as in Figure 5.12, it has been successfully used for intracranial pressure monitoring experiments.

In biology we have two experimental studies being conducted under a condition where components of an organism being studied in isolation from the whole

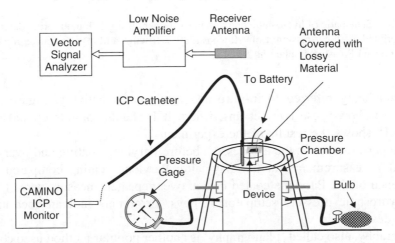

Figure 5.13 Schematic of the setup based on sphygmomanometer technique for monitoring air pressure. © 2008 IEEE. Reprinted, with permission, from Kawoos (2008). In-vitro and in-vivo trans-scalp evaluation of a intra-craneal Pressure Implant at 2.4 Ghz.

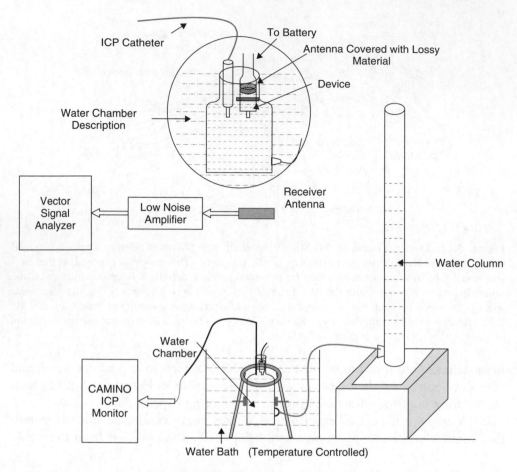

Figure 5.14 Schematic of hydrostatic pressure measurement using ICP monitoring device. © 2008 IEEE. Reprinted, with permission, from Kawoos (2008). In-vitro and in-vivo trans-scalp evaluation of a intra-craneal Pressure Implant at 2.4 Ghz.

and within being referred to in-vitro and in-vivo, respectively. Figure 5.13 and Figure 5.14 show these two set ups, where ICP stands for intracranial pressure. Figure 5.15 shows the results of the experiment.

Device performance in the air pressure before Parylene coating and over repeated (five trials) measurements for five days after Parylene coating compared with the performance of the Parylene coated device over repeated measurements (12 trials) in the hydrostatic pressure setup for 12 days. Error bars indicate measurements' standard deviation.

Tomography, also called 'planography' is another popular method used commonly for scanning the brain through the scalp that can be counted as a powerful sensor using the X-Ray, gamma, electron, ion or a similar radiation technology with the

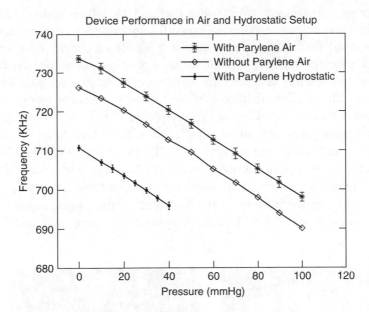

Figure 5.15   Results of the average experimental pressure showing air pressure in terms of pulse frequency displayed by the measuring device.

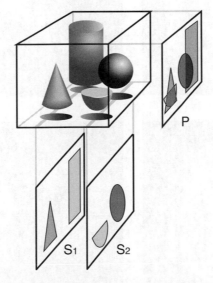

Figure 5.16   Basic concept of tomography image processing [wiki]. This photo is distributed under Creative Common License. http://en.wikipedia.org.wiki/file:TomographyPrinciple-Illustration.png.

basic image process concept indicated in Figure 5.16, where various reflective and refractive images are combined using powerful computers to reconstruct the scalp.

It can be used for building an image of a whole or specific part of the brain or in many cases of medical treatment, as a powerful actuator, where a Monte Carlo proton tomography suggested by Jarlskog et al. (2006) scalp, is interesting to examine. The Monte Carlo method is a well-known statistical method for solving problems upon their probability specifications and is used to build a probabilistic model of the biological content of a patient's scalp whilst being treated. In this work the combined computing tomography (CT) image processing and Monte Carlo method enables us to generate a very powerful radio surgery tool for brain treatment and other clinical operations. This idea uses the principle of a steep fall of the irradiation fields, called isodoses, off the centre of the target without interfering with other vascular and neural structures around it. Figure 5.17 and Figure 5.18

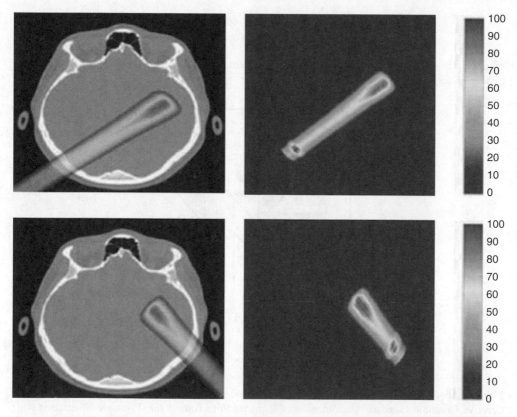

Figure 5.17 Isodose curves for two fields (out of 4) for a radio surgery case. Left: pencil beam dose, right: Monte Carlo energy deposition. The isodose intensity areas are for into six prescription doses of >90%, 90%–80%, 80%–60%, 60%–50%, 50%–30% and 30%–10%. © 2006 IEEE. Reprinted, with permission, from Jarlskog (2006). Proton Monte Carlo in the Clinic.

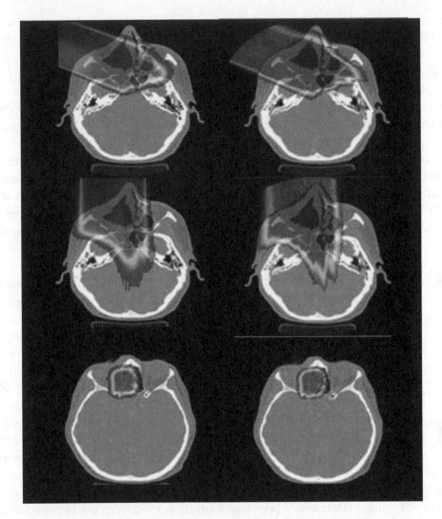

Figure 5.18 Isodose curves for three fields for a paranasal sinus cancer. Left: pencil beam, right: Monte Carlo. The isodose intensity areas are for into six prescription doses of >90%, 90%–80%, 80%–60%, 60%–50%, 50%–30% and 30%–10%. © 2006 IEEE. Reprinted, with permission, from Jarlskog (2006). Proton Monte Carlo in the Clinic.

show the isodose curves for two and three field radio surgery respectively with the dose coloured red.

5.4 Post-Operative Applications

Patients of a surgical operation or complex treatment go through a very distressful experience of being traumatised for a period usually much longer than the length of the operation. In order to maximise the result of a surgical operation it is important for the patient to be awake and rested for a pacifying period under serious intense care. Often, due to internal distress and psychological side effects of the operation combined with other sources further complications can result.

In practice, however, a successful medical operation depends on two other phases of preparation and post-operation. Preparation period, if not an emergency, usually has the nature of intensive administrative or management. This period is normally regarded as important and sometimes a lengthy process due to the time required for analysing patient's records, audit data for non-technical aspects, or it could be of a technical nature for checking and calibrating the equipment or critical devices for their technical aspects, which often for non-intelligent and dated systems is normally tedious and time consuming. These preparatory activities, however, are usually routines and are carried out by administrators, nursing and technical staff with the limited role of technology in saving time and reduced risk operations.

The post-operation period, however, is of a different nature and depends mainly on the level of risk involved in the operation. This can be divided into three groups of (a) routine operations with no imminent risks or surprises, (b) operations with limited risks with a need for intensive care and (c) highly volatile operations with unstable health conditions and risks of further complications.

In all cases, normally the body of a patient is quite tense mainly due to being subject to various drugs, injections, cuts and scratches, where bodily bruises, common in most cases, require naturally healing and often regular treatment. Patients are normally distressed and worried so they may need or demand the continuous attendance of nurses and specialists. However, highly stable and sophisticated systems showing detailed activities and status of their organs, particularly for those closely related to the operation, should enhance their confidence, ease up their nerves to help with a speedy recovery based on use of the body's inherent defence energy.

Post-operational recovery could be analysed upon specific cases applicable to the above groups. For the first group though the operation is not life critical but the patient requires attention or poor post-operation procedure could provoke many unpredictable issues. Individually they may not be important, but as they count for the majority of cases and as these cases are due to increase rapidly in the future we should consider their effect seriously.

In principle, poor care or negligence could easily reduce the speed of recovery. This in turn will increase the overall cost of the operation, which could be significant in some budget-limited cases. More important is, however, that lengthening the recovery period could raise further issues such as psychological complications and unpredictable infections jeopardising the overall result of the operation for a proper recovery process or returning to an ultimate status of normal health.

For the other cases of more life critical operations one can easily visualise the importance of post-operative care. Although it is quite hard to put a finger on any accurate figures it is easy to realise the importance of post-operative care by comparing relatively far higher success rates occurring in private clinics in comparison to the state's clinics running with limited budgets and limited facilities.

For example, under a properly built-in post-operational service or adequate monitoring system any signs indicating a build up for a dangerous symptom could be easily picked up by an integrated warning system built upon a set of inter- linked Smart sensing devices distributed around or located in various places of a Smart space, such as Smart shirts can trigger an alarm or generate a special signal for attendance or an alert as an indicator for a fatal or very risky condition that may result in follow up surgery or immediate special care or medication. That is, without a proper post-operation intensive monitoring system a successful operation could easily end up in a patient's death, lifetime scars or poor health due to damage or loss of important organs simply because of short delays in intensive care.

Considering that much of the existing healthcare equipment and systems are very basic, bulky and usually make very little use of today's widely researched and developed technologies such as smart and intelligent autonomous technologies very few patients actually are comfortable and many cannot properly communicate with busy and overwhelmed nurse and medical staff. Therefore, reliable automation and use of technology rich upon extensive use of DSS technology is urgently needed in both private clinics and national hospitals.

Now, before discussing some examples or recent developments we would like to emphasise the technological importance of intensive care for post-operative conditions, often located in a common space or block of an intensive care unit (ICU). An ICU whether public or private is one of the critical places that should not be kept away from new technologies. We need sensor-dominated data-rich smart space technology for these places, which can make the upmost use of wireless technology to provide Smart space services as basic. For this end there follows a few general features for using DSS in WSN style:

Recording – with Smart sensor rich and connected to a large database are all critical as well as many other partially related features of patient's health behaviour and should be recorded regularly throughout the recovery and stored in secure databases.

Care – under availability of continuous recording data systems should enable both real-time examination and regular visit of the readings for further analysis using a feature extraction program.

Monitoring – with a central processing capability the system continuously provides user-specific information made available for all essential parties involved: (a) selective monitoring information for the patient, (b) care related detailed information for operational staff, (c) health significant information for professional and managerial staffs and (d) alerts and emergency signals for various groups and warnings.

Irregularities – professional decision-making programs should provide reliable information on patient's health variations such that real-time variations are compared with the expected and long-term improvement and provide regular and interactive information as required.

Evidence – all health parameters for various significant symptoms and dangerous cases are documented for further examination, if required.

Intelligence – about a vast amount of information being made available in the system a professional program should be able to suggest optimised medication and follow-up treatments.

Follow-ups – about a professional program the system should be able to run an optimised automated schedule for check-ups, injections and other time significant activities.

Normally the overall cost of an operation counts as the sum of the cost of three main stages that is, of pre-operation, operation and post-operation. Most clinics do not consider the cost of failure seriously, which is an invisible cost that they should allow for and invest in its reduction using automated and data-rich wireless systems widely available at little cost in comparison to the cost of the risks some take.

There is a study from (Rienzo et al., 2006) for using a wearable medical shirt for heart disease rehabilitation, where a Smart sensing system, MagIC, is composed of a vest of textile sensors made of cotton and lycra with a portable electronic board. The conductive fibres in the garment carry ECG electrodes and a piezo-resistive transducer measures the breathing frequency. Fastened to the vest a 3-axis accelerometer which communicates with a remote controlling device using Bluetooth wireless technology. One of the functions of this system is its clinical use for patients with heart problems. For calibration of the ECG readings it needs to be set up for the test in comparison with a traditional ECG device under various patient's conditions, see Figure 5.19.

In a routine test of 31 patients, 10 of them lay on the bed. Then the remaining 21 patients go through various physical activities of resting (4 min lying, 1 min standing), 10 min mild physical exercise, 15 min pedalling on a cycloergometer and 6 min walking (Ade et al., 2009).

This test, in comparison with the traditional ECG recording provides the following properties for the system: (a) the quality of signal, (b) heart beat rate and (c) identifies abnormalities including atrial fibrillation, flutter, supraventricular and ventricular ectopic beats, left and right bundle branch blocks, atrio-ventricular blocks. See Figure 5.20 for a Sample MagIC ventricular ectopic beating test signals.

One good example is from a survey from Pantelopoulos and Bourbakis on the use of a wearable health-monitoring system (WHMS). Following an operation such as an abdominal surgery in the clinic the patient can choose to adopt being monitored at home for various important reasons such as (a) to continue with his normal life rather than staying in the hospital or special care unit and (b) excessive cost of living

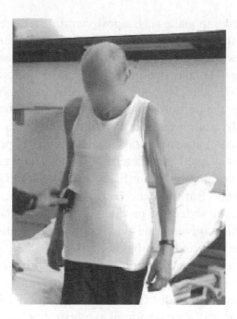

Figure 5.19 Clinical application of MagIC heart conditioning wearable system. © 2006 IEEE. Reprinted, with permission, from Di Rienzo (2006). Applications of a Textile-Based Wearable System in clinics, exercise and under gravitational stress.

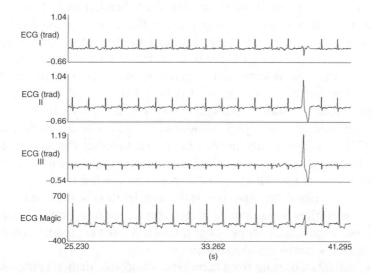

Figure 5.20 Sample MagIC ventricular ectopic beating test signals. © 2005 IEEE. Reprinted, with permission, from Di Rienzo (2005). A textile-based wearable system for vital sign monitoring: applicability in cardiac patients.

away from home. In order to make such a postoperative recovery system possible the patient needs to wear the WHMS which comes with an intelligent algorithm design system that recognises the type and the intensity of the user's body movements and associated activity. This has two purposes of (a) monitoring patients' activities continuously through the wireless connectivity network connecting a set of sensor devices spread all over the body for analysis and decision-making and (b) ensuring that the patient actually follows surgery's instruction for a routine exercise. The measurement devices can vary from one case to another depending on the required body functions and associated parameters positioned at various locations of the body. Usual devices for most cases are ECG, EMG and EEG for heart, nerves and brain pulses, respectively. But there are many other sensors, as listed Table 5.1.

This survey provides a variety of products available in the market with specification on the controller hardware and wireless communication modules as shown in Table 5.2 (Pantelopoulos and Bourbakis, 2010).

Another interesting example which proved very helpful for many cases of post-operation patient care is when a patient needs to remain in a certain position for a certain period of time until full effect of the operation takes place. Work from (Dlouhy et al., 2007) provides a case for an eye operation patient who needs to hold their head in the right position for a length of time. This is a simple device but it features two big steps in the right direction. Firstly, it is up-to-date technology so that its usability life is sufficient for a competitive global market and it addresses real problem solving issues in the most wanted application area of medical surgery. The device uses MEMS accelerometers under the control of an RISC microcontroller to inform the patient whenever his head moves out of the correct position. The specific care for the design is for post-operation care of vitreo-retinal surgery, where a specialised eye microsurgery operation in the area of the retina and vitreous in the rear part of the eye saves a patient from partial or full blindness and enables him to gain his normal eyesight (Dlouhy et al., 2007). In this operation the vitreous is removed and the eye is temporarily filled with an expansive gas such as SF6 or C3F8. As gas gently presses the retina together the damaged part starts to recover. For the process to take place a reference angle against the gravitational acceleration vector is estimated. Using a position sensing method based on a tilt sensing micro machined for dual-axis static acceleration a static sensor creates an open-loop acceleration measurement architecture detecting both positive and negative positions. Figure 5.21 shows how the MEMS sensor works and Figure 5.22 shows the device's hardware block diagram.

Similarly, patients suffering from hemolytic glaucoma, diabetic retinopathy, vitreous hemorrhage, central-vein occlusion and other retinal detachment disorders may develop eye disease, where the jelly-like substance that fills the inner eye becomes a routine repair surgical procedure. The healing process, however, is both painful and inconvenient for the patient because he must contend with the monotony, stress and discomfort of a 23 hours facedown position.

Table 5.1 Sensor and associated sensing normally used for remote monitoring. (Pantelopoulos and Bourbakis, 2010)

Type of Bio-signal	Type of Sensor	Description of Measured Data
Electrocardiogram (ECG)	Skin/Chest Electrodes	Electrical Activity of the heart (continuous waveform showing the contraction and relaxation phases of the cardiac cycles)
Blood Pressure (systolic and diastolic)	Arm cuff-based Monitor	Refers to force exerted by circulating blood on the walls of blood vessels, especially the arteries
Respiration Rate	Piezoelectric/ Piezoresistive Sensor	A measure of the body's ability to generate and get rid of heat
Oxygen Saturation	Pulse Oximeter	Indicates the oxygenation or the amount of oxygen that is being "carried" in a patient's blood
Heart Rate	Pulse Oximeter/Skin Electrodes	Frequency of the cardiac cycle
Perspiration (sweating) or skin conductivity	Galvanic Skin Response	Electrical conductance of the skin associated with the activity of the sweat glands
Heart sounds	Phonocardiograph	A record of heart sounds, produced by a property placed on the chest micronophe (stethoscope)
Blood Glucose	Strip-base glucose meters	Measure of the amount of glucose (main type/ source of sugar/ energy) in blood
Electromyogram (EMG)	Skin electrodes	Electrical activity of the skeletal muscles (characterizes the neuromuscular system)
Electroencephalogram (EEG)	Scalp-placed electrodes	Measurement of electrical spontaneous brain activity and other brain potentials
Body movements	Accelerometer	Measurement of acceleration forces in the 3D space

© 2010 IEEE. Reprinted, with permission, from Pantelopoulous (2010). A survey on wearable sensor-based systems for health monitoring and prognosis.

Table 5.2 Wearable systems product specifications

Hardware Description	Communication Modules	Measured Signals	Medical Applications
PDA, Microcontroller board	Wires, 2,4 Ghz, radio, GPRD	ECC, BP, R, T, Sa0$_2$, EMB, GSR	Parkinson symptom and epilepsy seizures detection, behavior modeling.
Wrist-worn device	GSM Link	ECG, BP, T, SaO$_2$, A	High-risk cardiac respiratory patients
Custom microcontroller-based device and commercial bio-sensors	Serial cables, Bluetooth	ECG, BP, R, T, SpO$_2$, A	Medical monitoring in extreme environments (space and terrestrial)
PDA, Textile and electronic sensors on clothes + heart belt	Conductive yarns, Bluetooth, GSM	ECG, R, other vital signs, A	Prevention and early diagnosis of CVD
Textile and electronic sensors on jacket	Conductive yarns, Bluetooth, GPRS	ECG, R, T, EMC, A	Monitoring of rehabilitation and elderly patients, chronic diseases
Vest with textile sensors, custom electronic board, PDA	Bluetooth	ECG,R,T	Recording of cardio respiratory and motion signals during spontaneous behavior in daily life and in a clinical environment
Garment with knitted dry electrodes, PDS	Conductive yarns, RF link	ECG, R, T, A	General health monitoring
Vest with woven sensors, microcontroller	Woven wires, 2.4 Ghz ISM RF	ECG, BP, T, PPG, GSR	General remote health monitoring
Sensot motes with custom processing boards	ZigBee	ECG, SpO$_2$, A	Real-time physiological status monitoring with wearable sensors

Table 5.2 (*continued*)

Hardware Description	Communication Modules	Measured Signals	Medical Applications
Zigbee-based motes and Zigbee-based custom base device	ZigBee, Wi-Fi, GPRS	ECG, BP, R	Detection and prediction of human physiological state (wakefulness, fatigue, stress) during daily activities
Custom tiny motes, cell phone and commercial sensors	ZigBee, CDMA	ECG, BP, SpO_2, A	Health monitoring and remote identification of suspicious health patterns for further evaluation by physicians
Miniature Low-power BAN nodes, energy scavenging	ZigBee	ECG,EEG,EMC	Enable autonomous wearable sensor networks for general health monitoring
Custom sensing board, communication sensors and cell-phone	Bluetooth	HR, SpO_2	Monitoring users during their sleep to detect sleep apnoea events
Cell phone and communications available in BT bio-sensors	Bluetooth, GPRS	ECG,A	Individualized remote CVD detection
Cell phone and communications available in BT bio-sensors	Bluetooth, GPRS	ECG,BP,A	Hear-attack self-test for CVD patients
Microcontroller board, PDS	Wires, ZigBee, GPRS	ECG	Remote detection of cardiac arrhythmias
Mask, glove, chest sensor	Wires, Bluetooth, Wi-Fi	ECG,R,A	All-day remote health monitoring
Chest Belt	Bluetooth or ISM RF	ECG,R,T,A,P	Remote monitoring of human performance and condition in the real-world

© 2010 IEEE. Reprinted, with permission, from Pantelopoulous (2010). A survey on wearable sensor-based systems for health monitoring and prognosis.

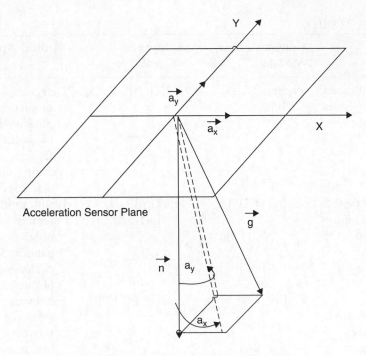

Figure 5.21   MEM device using gravity to show the position.

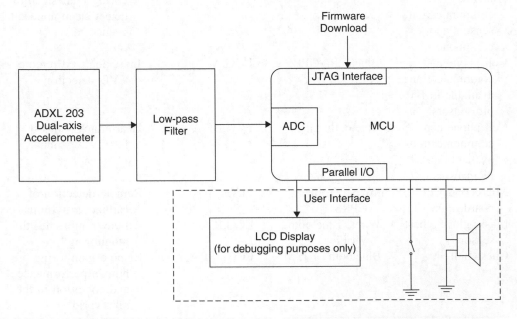

Figure 5.22   Block diagram showing min components of the accelerator.

5.5 Emergency Case Studies

The incidental health conditioning and emergency health issues can be related to many cases from road accidents to the outbreak of disease and natural disasters. In most of these cases shortages or inaccessibility of experts and professionals count for highest priority to save lives and reduce the damage caused by an unpredictable event.

Usually preparations on many occasions in both natural and manmade frequently occurring events (FOE) using advanced professional algorithms can help with (a) lesser losses due to lesser surprising events by being prepared for something of the kind to happen, (b) setting up common sets of alternative strategies would increase efficiency of experts and medical resources and (c) higher efficiency due to availability of trained experts for professional attendance familiar with a few basic and essential sets of agile tools and gadgets.

Although all the above-mentioned factors can help us with general preparedness for higher statistical effectiveness, however, they can be counted as basic infrastructure for emergency operations rather than real-time cases of the events. For example, similarity can be found between infectious outbreaks and road accidents. The use of technology and smart sensors in particular, however, can help to bring these cases closer together. They can help with the more accurate prediction of an earthquake and at the risk of the discomfort of possible false alarms far more lives can be saved and much less damage achieved through more reliable statistical information gathered through advanced sensing systems.

In order to shed a light on some of these cases we scan a few recent research works for:

- rich information data fusion and management prospective and their potentials in the future;
- urban associated health incidents including road traffic accidents;
- health and life issues associated with emergency cases and disasters commonly categorised under telemedicine.

The infrastructure for dealing with the health aspects of emergencies consists of three areas of (a) human resources, (b) technology rich information and sensor technology and (c) specifically designed user-friendly packages by professional developers.

For professional human resources front-line teams interact with humans in very poor conditions often traumatised by the event. These staff should follow their professional ethics, trained and possibly experienced in the field. With today's rich software simulation packages available at little cost virtual experience and use of augmented reality devices would help. For the ethic of the front-end professionals we expect all responsible governments should have ethical guidelines for the ambulance services of the practice. For the technology for example, (Vincen et al., 2009)

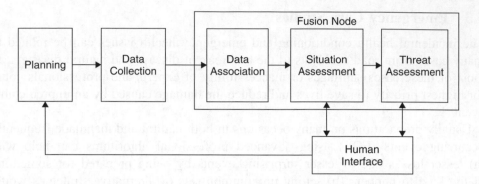

Figure 5.23   Data fusion model for situation and threat assessment (Vincen et al., 2009).

provide a Centralised Intelligence Fusion Framework for integrating Emergency Services.

With the fusion of sensing data in view the raw reading signals are organised for objective sets of data which had been searched, analysed and explained to become knowledge. Intelligence is a special selection of knowledge required to convert into tasks. The term intelligence fusion obtained from fused heterogeneous sensors information provides the Common Operational Picture (COP) allowing situation awareness to aid the decision-making process. This generic role of Intelligence Fusion node is shown in Figure 5.23, where a generic data fusion model supports the centralised intelligence fusion framework.

In order to take advantage of IoT technology many nations have taken sensor-based EMS on board. Wireless sensor network and intelligent video can be used to detect abnormal situations and emergency conditions for various circumstances such as chemical leak, pollution, security of mines, subway safety, square and other regional fire emergency and earthquake disaster. Then, precise on-site information could trigger the emergency event of a multi-dimensional information acquisition using an emergency database model out of an event situation-map for auxiliary decision-making, effectively coordinated directional rescue including associated information such as fire spread direction, spread speed and risky areas. For these systems Zhang et al.'s device uses RFID based IoT technology in various parts of the system. Using the RFID electronic tag can carry on the automatic diagnosis and track to the dangerous chemical and the transport vehicle, combining communication technology to realise the real-time monitoring and the network management, strengthens the dangerous chemical transportation safety supervision dynamics, to ensure transportation security, and reduces traffic accidents. Using RFID tags on goods and containers, when coordinated with a video one can mark and activate automatic identification and global tracking, to record in real-time the switch which seals time and location. For medical emergency cases it can track and manage the blood identification, blood transfusion, blood donation, where the RFID tag carried

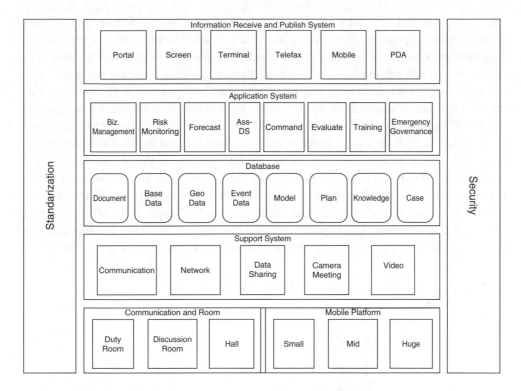

Figure 5.24   Technical structure of EMS in China (Zhang and Anwen, 2010).

with the blood keeps the record and associated tracking services up to a patient's home or the hospital (Zhang and Anwen, 2010). Figure 5.24 shows the technical structure of EMS in China.

Another emergency case is associated with clinical aspects of busy towns and cities. One interesting work is from (Shnayder et al., 2010) which uses sensor enabled cloud style ubiquitous access for implementing emergency medical systems (EMS) for such places, which is due to a considerable increasing interest in the adoption of personal health records (PHR). PHRs carry patient information, which can be used for building an infrastructure to integrate both healthcare services as well as the tax. The key difference between PHR and electronic healthcare record (HER) is that in PHR the patient keeps control over the access.

In the USA alone automobile crash is the source of over 40,000 deaths per annum. However, a timely response to these crashes can reduce this figure to much less and reduce consequent disability. The data shows that most of the casualty rate could be improved by availability of health information technology (HIT), the emergency medical response times and the degree of care provided to crash victims. However, due to various factors the figure is steady but still quite high and varying significantly (Benjamin et al., 2009).

It is shown that one key point of improving the figure is due to a clinical decision-making process where many basic factors such as the speed of provision of complete information about the patient when he is not able to give it. The time is the sum of time from calling 911 followed by the paramedics to being accepted in the trauma centre for rehabilitation and treatment. The best practice is provision of inte-grated information available through a multi-organisational information sharing sys-tem based upon a case study analysis in the Minnesota Mayo Clinic trauma system, from both local Mayo Clinic practitioner perspectives and State level trauma sys-tem perspectives, a high-level descriptive architecture of an integrated crash trauma information system. The study conducted in the Mayo clinic for measuring the total travelling time for the emergency medical service (EMS), also called end-to-end information process accounts for (a) incident reported through a call, (b) incident information acquisition, (c) dispatch call and routing, (d) response coordination and eventually definitive care. The results are shown in Figure 5.25 and Figure 5.26.

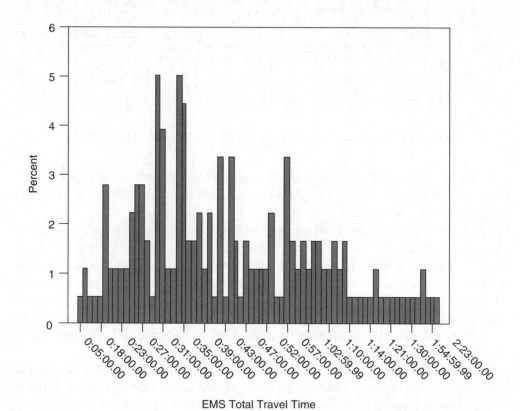

EMS Total Travel Time

Figure 5.25 The percentage of total travelling time (Benjamin et al., 2009). © 2009 IEEE. Reprinted, with permission, from Schooley (2009). Integrated Patient Health Information System to improve Traffic Crash Emergercy Response and Treatment.

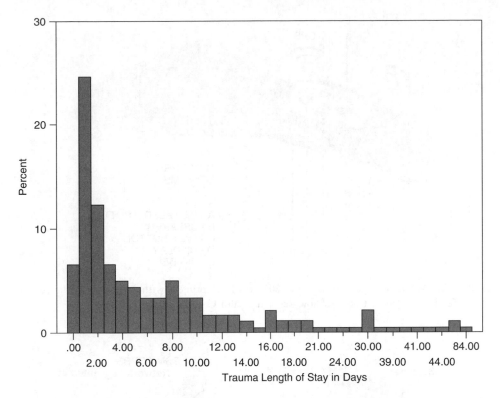

Figure 5.26 The percentage of the trauma period in number of days (Benjamin et al., 2009). © 2009 IEEE. Reprinted, with permission, from Schooley (2009). Integrated Patient Health Information System to improve Traffic Crash Emergercy Response and Treatment.

Finally, the last case of emergency is for those who may have been taken by total surprise. This could be simply travelling through an unknown dangerous area or being caught in unexpected weather, resident in an infected area, and so on. Such places and situations normally happen in remote places where neither experts nor any proper clinic or drugs are expected. Main support could come from privileged units called Telecasters being upgraded to health or e-health centres. The process in these cases is like being lost and therefore found and rescued and being taken to a place where some supply of medication mostly applies. One source for this case is from Ade et al. (2009) who indicate heart failure in such a situation as the main danger or propose the use of a wireless style device for automatic connection through the satellite or other access covering capability being connected to a central controller for periodical transmission of information, for example the heart beat signals, as shown in Figure 5.27. Figure 5.28 shows the connection scenario by which the person in need can be located by GPS.

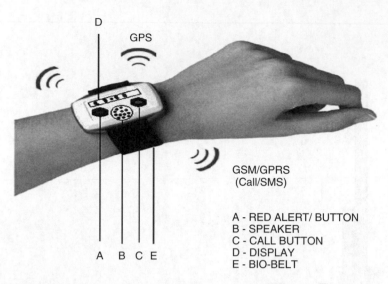

Figure 5.27 Wearable Tele-Bio watch. © 2011 IEEE. Reprinted, with permission, from Ade et al (2011). Telehealth: HealthCare technologies and Health Care Emergency System.

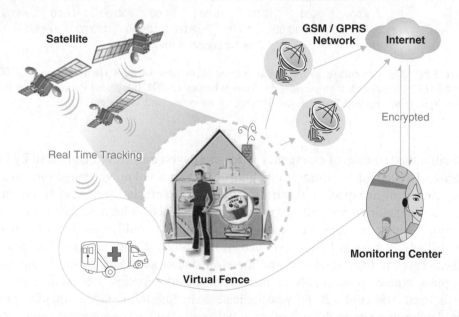

Figure 5.28 Tele-Health emergency, system architecture. © 2011 IEEE. Reprinted, with permission, from Ade et al (2011). Telehealth: HealthCare technologies and Health Care Emergency System.

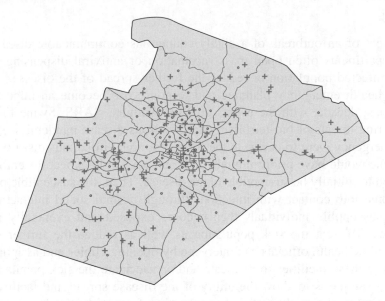

Figure 5.29 Map of Wake County, NC, highlighting census tracts, their centroids, and public elementary school locations. © 2010 IEEE. Reprinted, with permission, from Carr and Roberts (2010). Planning for infectious deseases outbreaks: a geographic disease spread, clinic location, and resource allocation simulation.

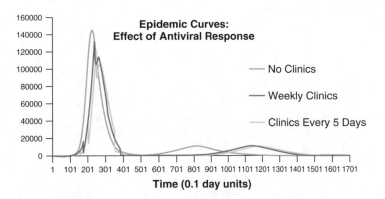

Figure 5.30 Comparison of infection spreading dynamics over time. © 2010 IEEE. Reprinted, with permission, from Carr and Roberts (2010) Planning for infectious deseases outbreaks: a geographic disease spread, clinic location, and resource allocation simulation.

Outbreak

In the event of an outbreak of a highly contagious communicable disease, public health departments often open mass-vaccination or antiviral dispensing clinics to treat the infected population or reduce the further spread of the disease.

Public health emergency planning and response has become an important topic over the last decade, with the events of 9/11, Anthrax, SARS, Swine Flu (H1N1) and recurring threats of bio-terrorism and infectious disease pandemics. If a public health emergency were to occur, in the form of a novel infectious disease outbreak, then no one would have prior immunity to it and we would expect the entire general population to initially be susceptible to the disease. Then, the susceptible population would come into contact with infections through normal social interaction. Some of these susceptible individuals then become exposed and eventually infectious themselves. To treat the sick population as well as reduce the further spread of disease, public health officials can intervene by opening clinics and assigning health resources to these facilities to medicate and/or vaccinate the sick population.

A case example is to show the utility of the disease spread and facility location simulation to public health officials in planning for a potential large-scale public health emergency such as an infectious disease outbreak. This case looks at the region of Wake County, North Carolina, as highlighted in Figure 5.29. This county contains the capital city of Raleigh. The population of Wake County exceeded 750,000 people in 2008. Moreover, Figure 5.30 shows a comparison of infection spreading dynamics over time.

6

Smart Home, Smart Office

The smart intelligent devices are finding their way into our homes and into all other spaces to help us to interact with such spaces for a healthier, less wasteful and more comfortable lifestyle. It is exciting to see so much effort being put into this area through smart environment projects coming under various terms such as smart house, smart office or ambient intelligence. There are incredible efforts in various societies providing more intelligent houses able to handle a large number of new smart services enabling better utilisation of basic resources like water and electric energy and also helping to increase the quality of life for people of all ages. The main debate, however, is to design new houses or to provide old houses with new technology so that the owner of the house can automate services to make better use of power consumption by means of a collaborative smart grid of electric devices with coordination between them in order to reduce the amount of energy invested. Houses in which an important part of the power comes directly from the house by means of solar or wind energy fostering environmental-aware resource reutilisation would play a more significant role. In this context, the usage of distributed sensing platforms for in-home or in-office applications become of great interest mostly due to their intrinsic nature of sensing data associated with a wider range of application scenarios.

With the recent move towards smart space as a generic move to making more effective use of smart devices such as sensors, actuators, systems and other application centric adaptations of distributed intelligence we consider the concepts of 'smart home' and 'smart office' as the first steps for the deployment of smart space as a techno-economical enabler. We, however, value the potential of smart home to be greater than smart office due to the following facts:

- In the office, use of smart devices varies extensively from an ordinary building site which can make do with minimal embedded intelligence to high-tech and flashy sales offices of new and expensive medical or domestic goods where

Distributed Sensor Systems: Practice and Applications, First Edition.
Habib F. Rashvand and Jose M. Alcaraz Calero.
© 2012 John Wiley & Sons, Ltd. Published 2012 by John Wiley & Sons, Ltd.

either (a) sensors provide routine functions, already adopted in different styles or (b) specialised and professional sensors to match the profession and the service the office may provide. This would offer a limited use of any potential to satisfy the market specifications.

- In homes under new global average age growth the fabric of societies is changing rapidly towards low cost, high quality info-rich media.
- With the population of wealthier people over 60 increasing, the existing boundaries of the housing market are being pushed towards a care intensive living environment.
- With social networking and growing use of the Internet, homes are becoming the main hob of living which in turn demands energy efficient homes.
- New initiatives of smart space, smart home and smart office should trigger new investments into technology intensive smart home projects.
- Integrated health and well-being initiatives require smart space systems and deployment of smart and intelligent sensors for automation.

With the above requirements in mind we can consider a three level deployment process of providing some smart yet flexibly structured networkable services such as the foundation of building new homes and new offices:

- Infrastructure: also called backbone, enables experts to ensure availability of basic resource requirement for a successful project implementation.
- Networking: Provision of agile interactive protocols interwoven upon ad hoc style networking architecture to support a potential predefined set of featured objectives (Ambient intelligence).
- Services: an overlaying style service interworking for a selectable set of basic and popular services (such as automation, security, medication, and so on).

We therefore, in the rest of this section, provide a number of recent developments to visualise the feasibility of a deployment using the above-mentioned three levels. But before doing this it is worth mentioning that this chapter first introduces the basic application requirements associated with this kind of scenario. After that, it provides the state of the art in energy and resource optimisation because it is one of the primary goals of this kind of sensing platform. Finally, the chapter ends with further details for a set of case studies related to smart home and smart office environments and systems.

6.1 Application Requirements

Following our earlier structure for smart homes and offices we can attend to some details. For example, Figure 6.1 shows a typical architecture for providing a distribution-sensing platform in a smart home. In essence this is the architecture

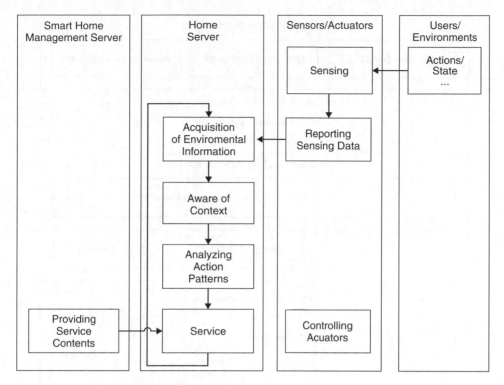

Figure 6.1   Common architecture for implementing Smart Homes, proposed by (Kim et al., 2008). © 2008 IEEE. Reprinted, with permission, from Kim et al. (2008). The Study of applying sensor networks to a smart home.

provided by Kim et al. (2008). In the bottom layer, the user and the home environment are interacting with the sensing layer. This sensor/actuator layer is composed of a wide set of smart wirelessly interconnected devices which are sensing the environment needed to provide such data to the home server. This home server is then acquiring all the environment information available in the home and is really aware of the context. This context information is analysed in order to infer and achieve new knowledge not specifically sensed, like analysis data and action patterns, multiple even correlations, data processing techniques, and so on. Then, the prepared information is provided to a module, which is really in charge of implementing the business logic associated with the service being provided at home. This service shown as a black box receives as input the information sensed and produces as output the required set of actions to be executed in the sensor/actuator layer to enforce the service into the house. The scheduling plan of actions is then enforced providing the automatic response of the house to the context sensed. Note that from the user perspective, the house is intelligent enough to act according to the environmental information sensed.

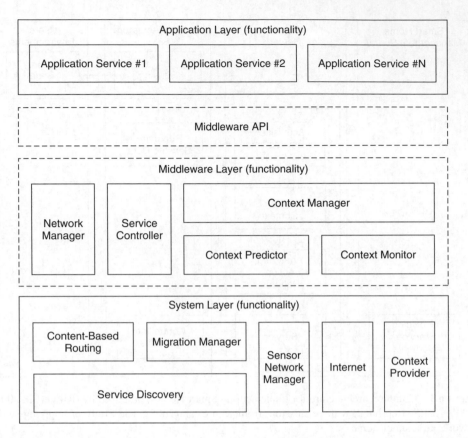

Figure 6.2 Common architecture for the home server, proposed by (Kim et al., 2008). © 2008 IEEE. Reprinted, with permission, from Kim et al. (2008). The Study of applying sensor networks to a smart home.

Figure 6.2 shows the internal structure functionality based on the sensor networks in the home server. According to module types, we used layer models and divided the layers into four by functionalities: Application, Middleware APIs, Middleware and System Layer. All of these modules should run collaboratively together.

The *System Layer* exposes several modules specifically designed to cope with the requirements associated with the Smart Home and Office. Firstly, the Smart Home requires an efficient service for programming, reprogramming, customising, configuring and in general, managing all the sensor nodes available in the architecture. These functionalities are usually achieved by a *Sensor Network Management System*. The Smart Home may need a wide set of services providing in-home information not only sensed directly in the house but also providing any source available on the *Internet*. This module can also be used to implement various context providers with the information required by the home server. In general, these context provider modules are in charge of providing the required information from the network and also serve as a gateway from upper layers up to the application layer. Another important aspect, which is an important requirement is the routing layer used in order to interconnect the different sensing devices available in the system. In this case, the content-based routing protocol is proposed as underline routing protocol. This kind of protocol is based on the examination of the message content and hence routing the message onto a different node based on data contained in the message. It enables a smart way of routing among all the heterogeneous sensing nodes involved in a smart house.

At the Middleware layer, we have *Network Manager, Service Controller, Context Manager* and *Context Predictor* and *Context Monitoring* modules. The *Network Manager* facilitates communications with a *Smart Home Management System* plus taking *Application Layer* data transfers, adaptive applications and so forth. Other duties, lying ahead of the middleware are management support sensor networks for relaying various commands of Service Controller to actuators. The Service Controller, on the other hand, generates controlling commands for actuators by processing querying messages, which are part of XML.

Helal et al. (2005) see the application of pervasive computing in smart places by interconnecting sensors, actuators, computers and other devices in the environment. They recognise that many first-generation pervasive systems lack the ability to emerge into application domains easily. This is mainly due to a lack of flexibility for integrating heterogeneous elements under an ad hoc process. The environments are also closed, limiting development or extension to the original implementers. Their approach in developing programmable pervasive spaces in which a smart space exists as both a runtime environment with a software library features service discovery and gateway protocols integrating the components using generic middleware and maintains the required service definition for each sensor and actuator in the space. For this they provide a descriptive model, for which a smart home application is shown in Figure 6.3.

At the heart of the operation are sensors and actuators, which have to be integrated. The integration can become complex due to the various types of sensors, software, and hardware interfaces involved. Consider, for example, climate control in a house. Normally, you would have to hard-wire the sensors to each room, connect these sensors to a computer, and program which port on the computer correlates to which sensor. To control climate in a home, for example, one would install a wireless sensor platform node in each room, connect both a humidity sensor and temperature sensor to each node, and program the firmware for each node. In addition to the firmware, the sensor nodes would contain the sensor driver that decodes temperature and humidity data. The individual node architecture, shown in Figure 6.4, provides for an alternative and a flexible configuration.

Another work is from (Hasswa and Hassanein, 2010), proposing a context aware architecture to capture multiple properties from an environment of different sources.

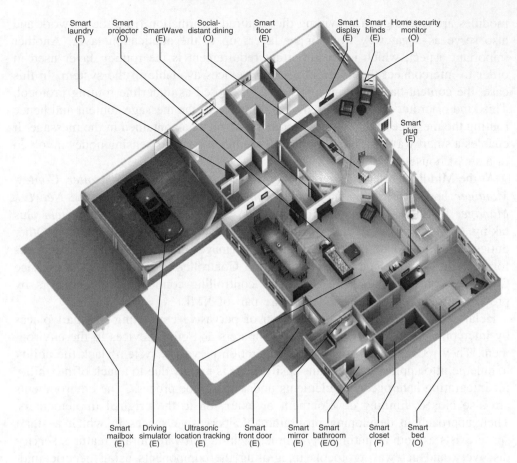

Smart laundry (F)   Smart projector (O)   SmartWave (E)   Social-distant dining (O)   Smart floor (E)   Smart display (E)   Smart blinds (E)   Home security monitor (O)

Smart plug (E)

Smart mailbox (E)   Driving simulator (E)   Ultrasonic location tracking (E)   Smart front door (E)   Smart mirror (E)   Smart bathroom (E/O)   Smart closet (F)   Smart bed (O)

Figure 6.3 Gator Tech Smart House. The project features numerous existing (E), ongoing (O), or future (F) 'hot spots' located throughout the premises (Helal et al., 2005). © 2005 IEEE. Reprinted, with permission, from Helal et al. (2005). THE Gator Smart House: a programmable pervasive space.

The captured contexts are heterogeneous in nature and range from social contexts acquired from social networks to physical contexts acquired from wireless sensors. Table 6.1 presents various contexts captured via a pervasive computing framework.

Sleman and Moeller (2011) suggest a Service-Oriented Architecture (SOA) Distributed Operating System for Managing Embedded Devices in Home and Building Automation. Their proposed Distributed Operating System (DOS) manages both data and events by using SOA-messages based on either XML or JSON (JavaScript Object Notation) formats that can be exchanged between the embedded devices in a home network or between them and the Internet or Smart Grids. The architecture uses a four-layered structure, shown in Figure 6.5 as including the hardware layer, bridging layer, SOA-DOS-layer and application layer.

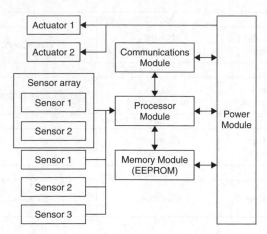

Figure 6.4 Sensor platform architecture. The modular design provides a flexible configuration (Helal et al., 2005). © 2005 IEEE. Reprinted, with permission, from Helal et al. (2005). THE Gator Smart House: a programmable pervasive space.

Table 6.1 Some heterogeneous contexts used in the pervasive computing framework (Hasswa and Hassanein, 2010)

Context	Definition	Source
Social Context	Information about people within the environment	Social Networking Services
Location Context	Information about a device's current and future physical location	GPS satellite tracking, Cellular tower, Wi-Fi or Wi-Max networks
Networking Context	Information about the networking capabilitites of entities within the environment. This info can usually be used to infer more information about an entity.	Heterogeneous networking interfaces such as Bluetooth, Wi-Fi, Wi-Max, 3G, Cullular Networks, etc.
Device's Physical Context	Physical information about the orientation of devices within the environment	Accelerometers, Gyroscopes, etc.
Environment's Physical Context	Physical information about the environment and entities within the environment, such as pressure, temperature, humidity, light intensity, etc.	Digital Thermometers, Pressure pads, etc.

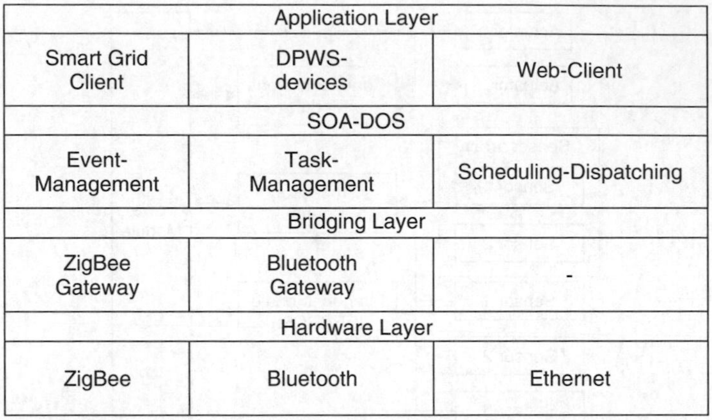

Application Layer		
Smart Grid Client	DPWS-devices	Web-Client
SOA-DOS		
Event-Management	Task-Management	Scheduling-Dispatching
Bridging Layer		
ZigBee Gateway	Bluetooth Gateway	-
Hardware Layer		
ZigBee	Bluetooth	Ethernet

Figure 6.5 Software architecture of the SOA distributed operating system (Sleman and Moeller, 2011).

The main components of the proposed system are shown in Figure 6.6. Here the system provides two types of service discovery to discover the services of the embedded devices: dynamic and static service discovery. Using dynamic service discovery requires enabling the dynamic service description in the μWS- Discovery module on the configured embedded device. In this case the μWS-Discovery will send a service description file in JSON format including the description of all device services to the services management module in SOA-DOS. This file can be sent each time the device is being turned on or upon receiving a discovery request from the SOA-DOS.

One of the most desired applications of smart home or office is associated with the finding of a hidden object. For this we look at a work from Kawashima et al. (2008). It is quite common to forget or not be able to find some items needed urgently, such as keys or other small item. This is more serious for small children and older people since it happens more frequently and sometimes could create a critical health issue or a dangerous situation. The u-object finder (UOF) is an application for a smart space in which robots move around to search for a lost object. Figure 6.7 shows the concept of the work.

The U-Object Finder (UOF) system is composed of the UOF-Manager functioning as a central controller including robots and a set of UOF-Robots in charge of searching the u-objects in the room. The UOF-Robots are in charge of searching a u-object in the room space to carry out two functions. The first is that the Robots move around the space and the second is that, the robot carries a RFID reader to detect a RFID tag within a certain distance, for example 10 cm. The UOF-Robot consists of an R-Controller, an R-Machine and an R-Reader, as shown in Figure 6.8. The Robots coordination process is shown in Figure 6.9.

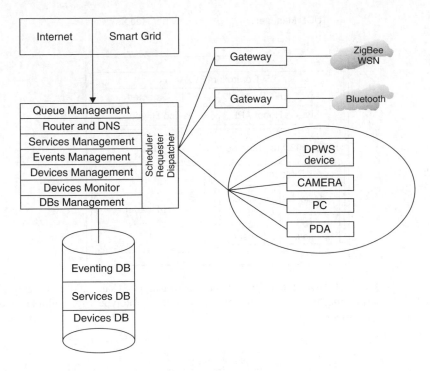

Figure 6.6   Main components of the SOA-DOS system (Sleman and Moeller, 2011).

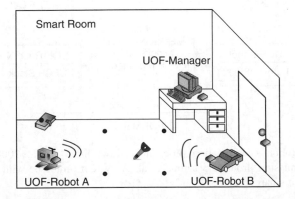

Figure 6.7   Diagram of the U-Object finder environment (Kawashima et al., 2008). © 2008 IEEE. Reprinted, with permission, from Kawashima et al. A system prototype with multiple robots for finding u-objects in a smart space.

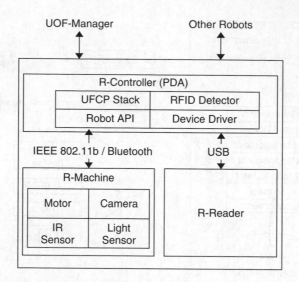

Figure 6.8 Diagram of UOF-Robot components (Kawashima et al., 2008). © 2008 IEEE. Reprinted, with permission, from Kawashima et al. A system prototype with multiple robots for finding u-objects in a smart space.

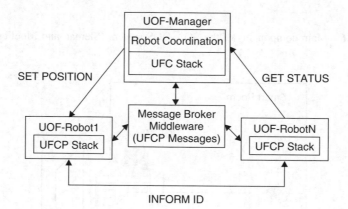

Figure 6.9 The Robots' coordination procedure (Kawashima et al., 2008). © 2008 IEEE. Reprinted, with permission, from Kawashima et al. A system prototype with multiple robots for finding u-objects in a smart space.

6.2 Energy and Resource Optimisation

One of the most common problems dominating the homes and small offices is the interference with communications and media based signalling activities. Being a classic IEEE 802.11 wireless LAN or more powerful sources such as Mobile or WiMAX the constant interferences with sensing systems always count as a bottleneck.

A study from (Yao and Yang, 2010) is quite interesting. For the IEEE 802.15.4 working with other wireless technologies on the same 2.4 GHz frequency band for a satisfactory bit error rate in a IEEE 802.15.4 system can be achieved with normally a loss equivalent ratio of signal to noise (SNR) of 3 to 4 dB. However, they suggest the noise level generated in practice is normally unstable and changes randomly all the time. For this reason the IEEE 802.15.4 wireless sensor networking integrated into a smart house application, needs to take on an adjustable complementary so that (a) it can avoid heavy interference situations and (b) take on a strategy to make maximum use of other radio activities for low data rate performance by dynamically adjusting the transmission frequencies suffering from heavy interference upon better understanding of the interfering networks. In this study the case of analysing sensors working with IEEE 802.15.4 in an environment dominated by IEEE 802.11b is proved to be effective. The setup scenario for the measurements is based on Figure 6.10 and demonstrated in Figure 6.11.

There are two existing approaches to monitoring the energy and similar resources: 1) non-intrusive load monitoring (NILM) where aggregate energy usage is measured by a single meter (e.g. a 'smart meter') as power enters the home and 2) complex instrumentation systems where each device's energy consumption is individually metered. Individually metering devices can provide more accurate data, but it is not practical to monitor every device in every home or office.

One interesting approach is similar to the self-adjusting method proposed by (Byun and Park, 2011) called Self-adapting Intelligent Sensor (SIS). The Intelligent Sensor for Building Energy Saving and Smart Service provision under a dynamic reconfiguration procedure uses both the middleware and network topology

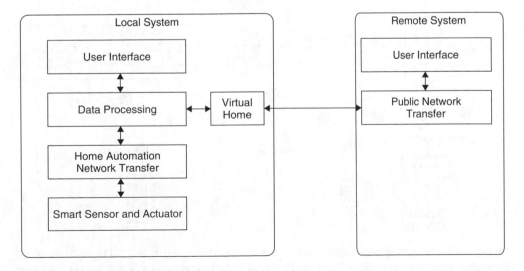

Figure 6.10   Example of home automation system structure.

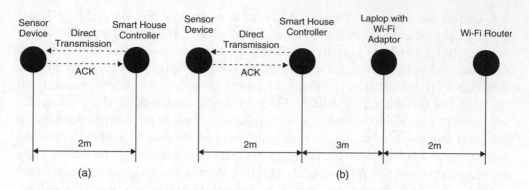

Figure 6.11   IEEE 802.15.4 Deployment (a) Without interference and (b) With interference.

as the environmental conditions change. The system works under the architecture with the main part SIS as shown in Figure 6.12. The system is generally structured to save energy. The SIS establishes an Energy-efficient Self-clustering Sensor Network (ESSN), where a building management server (BMS) initially decides whether it employs clustering topology or not based on a location characteristic. If the clustering scheme is employed, the BMS selects a cluster-head from the sensor considering the type of power supplied for the node.

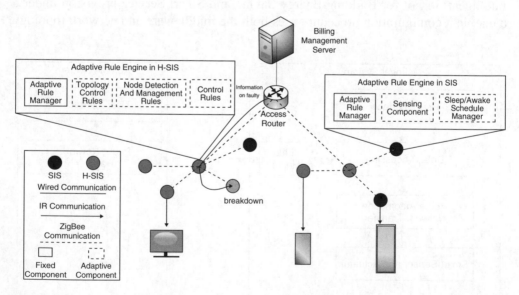

Figure 6.12   Overall system architecture of the system (Byun and Park, 2011). © 2011 IEEE. Reprinted, with permission, from Byun and Park (2011). Development of a Self-Adapting Intelligent Sensor for building energy saving and smart services.

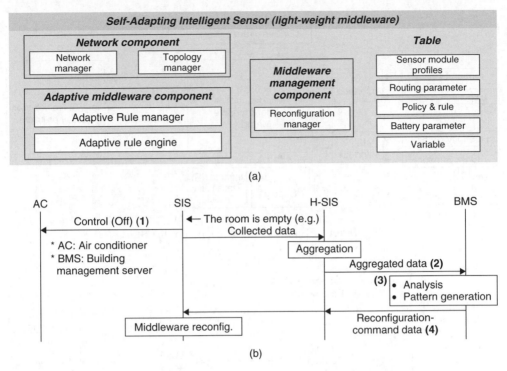

Figure 6.13 Middleware architecture. (a) Adaptive light-weight middleware; (b) Middleware reconfiguration flow. © 2011 IEEE. Reprinted, with permission, from Byun and Park (2011). Development of a Self-Adapting Intelligent Sensor for building energy saving and smart services.

As demonstrated in Figure 6.13, the node with a direct power source is selected as a cluster-head at first. In case of the node with a battery, our system utilises the simplified LEACH-based protocol for the cluster-head selection.

To demonstrate low power wireless this prototype's hardware block diagram, shown in Figure 6.14, where a wireless sensor uses the IEEE 802.15.4 (ZigBee) standard as a low-power low-cost solution. The SIS comes with various sensing modules such as humidity, motion detection, light intensity and temperature measuring functions as basic and able to be equipped with some optional modules such as fire detection, gas and air pollution detection. The SIS operates using a direct power source or battery.

For the last example we consider an interesting work from (Wang et al., 2010). In this work under a Multi-agent Control System with Intelligent Optimisation for Smart and Energy-efficient Buildings they provide a very interesting approach for an overall control of the resources. As shown in Figure 6.15, a layered control system with many interconnected agents is classified into two levels. The first level interacts with the power grid and the micro source called the central coordinator-agent. At the second level the customer activities are controlled and these are called local

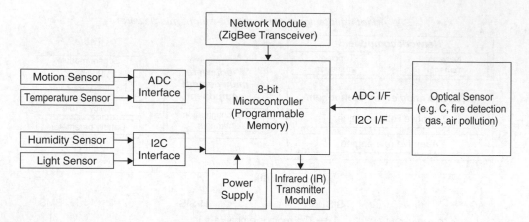

Figure 6.14 Hardware block diagram of the SIS.

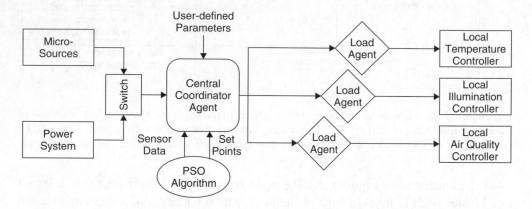

Figure 6.15 Structure of the proposed multi-agent system (Wang et al., 2010).

controller-agents. The users' comfort in the smart and energy-efficient buildings is decided by three main factors:

- the environmental temperature;
- the indoor air quality for CO_2 concentration;
- the lighting control under illumination.

The overall control target is to achieve the maximum comfort with the minimum power consumption. The central coordinator-agent works with the intelligent optimiser to accomplish its control goal. The load agents are used to shed loads to keep the customers' comfort at a high level when the power is not sufficient in emergency mode. There are distributed energy resources in the control system to

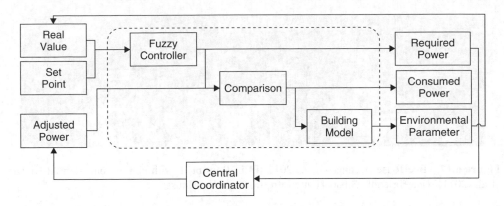

Figure 6.16 Structure of the subsystems for local control of the proposed multi-agent system (Wang et al., 2010).

support the building operations, which are usually renewable energy sources such as solar panels and wind turbine generators. Figure 6.15 and Figure 6.16 show the structure of the proposed multi-agent intelligent control system and the subsystem for local control.

6.3 Smart Home Case Studies

The smart home means computing sensing data for home automation and services to users by running the ubicomp applications. In this context, several case studies can be analysed among the wide set of automation and services available at home.

(Mohammed et al., 2011) propose a smart home infrastructure that offers the base platform for modular wireless sensing nodes interconnected using ZigBee technology. They describe a distributed sensing platform called *BeeHouse* for collecting data, sending information and controlling almost any aspect of the house given that a proper interface is established, as well as the ability to access those nodes and their information through a cross-platform graphical user interface. Figure 6.17 shows the different sensing nodes involved in the architecture.

The architecture is prototypically implemented in a three nodes sensor platform in which each node is equipped with temperature sensor, light sensor, motion sensor and a RF remote control for controlling appliances and light. The authors did not provide any real application of the software, but they demonstrate the feasibility of gathering all the information in a ZigBee platform in a scenario especially perturbed by interferences from other wireless devices.

The architecture of the proposed system and the graphical interface associated with it is shown in Figure 6.18. This figure described the hard architecture of the three sensors, which can be seen as a peer-to-peer communication among the sensors in the right side of the figure. Note that they are sensing heterogeneous

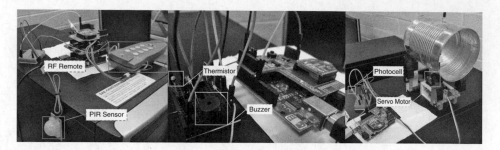

Figure 6.17   BeeHouse Architecture. © 2011 IEEE. Reprinted, with permission, from Al-Kuwari et al. (2011). User Friendly Smart Home Infrastructure: BeeHouse.

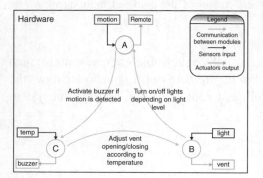 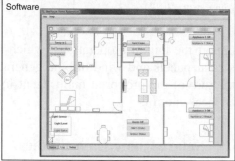

Figure 6.18   Proposed system and the graphical interface associated for BeeHouse. © 2011 IEEE. Reprinted, with permission, from Al-Kuwari et al. (2011). User Friendly Smart Home Infrastructure: BeeHouse.

environmental features and they share information in order to perform the automatic control of the house services. In the right side of Figure 6.18, it can be seen in the associated software in which the different sensing values and the actuator are shown to the user by means of an easy-to-use graphical interface which acts as a human-machine interaction to the whole distributed sensing device.

In smart homes and office environments, distributed sensing platforms are generally used to increase the inhabitant's comfort. As the current energy grid is evolving into a smart grid, where consumers can directly reach and control their consumption, distributed sensing platforms can take part in domestic energy management systems as well. Thus, Erol-Kantarci and Mouftah (2010) propose the Appliance Coordination (ACORD) scheme that uses the in-home distributed sensing platform in order to reduce the cost of energy consumption. The cost of energy increases at peak hours, hence reducing the peak demand is a major concern for utility companies. The way in which they address this balance is by means of shifting consumer demands to off-peak hours. Appliances use the readily available in-home distributed sensing

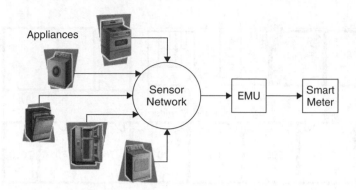

Figure 6.19 Domestic appliances collaboration in a Smart Home, architecture proposed by (Erol-Kantarci and Mouftah, 2010).

platform to deliver consumer requests to the Energy Management Unit (EMU). EMU schedules consumer requests with the goal of reducing the energy bill. The proposed architecture is shown in Figure 6.19.

Appliance coordination schemes require consumer negotiation and is challenging because of changing user behaviour is hard and also user comfort plays a key role in accepting new technologies. Reduced costs and awareness of environmental benefits can be effective in motivating consumers. Nevertheless, in appliance coordination schemes, the consumer demand should have high priority so that energy management systems do not disturb the consumer.

Another challenging topic in energy management is dynamically updating the price information. Currently, peak hour slots and prices are fixed. Dynamic rates depending on the demand may enable better billing. This is challenging because dynamic rates with load shifting capability of the appliance coordination may result in load oscillation.

The EMU needs to take these challenging issues into account during the scheduling of the requests received by the different appliances available in the home. Authors show that ACORD decreases the cost of electricity usage of home appliances significantly.

Another case study for a Smart Home is the air conditioning monitoring and managing. Tao et al. (2010) describe a real deployment used at the *National Engineering Laboratory* building in the *Beijing Jiaotong University*. In essence, they install some nodes with temperature sensors, humidity sensors and light intensity sensors along the corridor of size $50\,\text{m} \times 1.87\,\text{m} \times 2.6\,\text{m}$ for monitoring real-time building environmental conditions. Such sensors are chosen to collect parameters that affect indoor human comfort. A gateway and data server is located in the testing room near by the corridor. The environmental parameters of the building are monitored and transmitted to such data servers. The distributed sensing platform is deployed on the seventh floor of the laboratory building with a concrete floor and

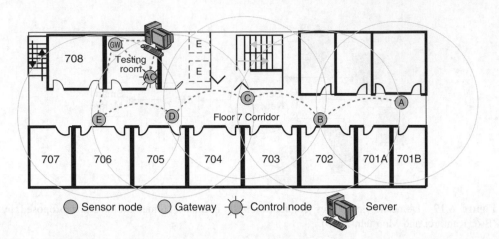

Figure 6.20 Deployment done at seventh floor of National Engineering Laboratory building in Beijing Jiaotong University. © 2010 IEEE. Reprinted, with permission, from Tao et al. (2010). Environmental Monitoring and Air Conditioning Automatic Control with Intelligent Building Wireless Sensor Network.

metal doors. Four different types of devices are deployed to collect data from the monitored area and control electronic devices. They are 1) sensor nodes, 2) control nodes, 3) gateways and 4) the servers.

The sensor nodes consist of a CC2420 transceiver chip, an Atmel AVR-Atmega128L microcontroller and the connector modules. The transceiver uses the IEEE 802.15.4 communication protocol, and Atmega128L is a low power 8-bit micro-controller based on the AVR enhanced RISC architecture.

They are deployed across the floor as depicted in Figure 6.20. This deployment has been carefully studied using an intensive study about signal interference between the IEEE 802.15.4 and the current Wi-Fi network available in building 802.11. Thus, this deployment allows the nodes to receive the signal from neighbouring nodes. The five sensor nodes are labelled from A to E, the gateway as GW, and the control node as AC.

The five sensor nodes with temperature, humidity and light intensity sensors measure in-building parameters and communicate the measurements directly with the neighbouring nodes or the gateway in the testing room. The gateway plays a coordinator role transmitting the measured data from the sensor nodes to the server and feeding the information to the control node close to the air conditioning units.

The underlying multihop routing protocol used is called the micro sensor routing protocol (MSR). The MSR sensor nodes enable the transmission of compressed Internet Protocol version 6 (IPv6) packets over IEEE 802.15.4 networks. The transmission time interval is set 20 seconds under the normal condition. When an emergency event occurs, the data is sent immediately not less than 30 ms to ensure a lower packet loss rate. In order to avoid adjacent channel interferences from the WLAN, the wireless sensor channel is set to 20 in a frequency band from 2.449 GHz to 2.451 GHz with 0 dBm transmission power whereas the Wi-Fi access control is fixed at channel 11 ranging from 2.451 GHz to 2.473 GHz.

The user interface consists of two parts: the data server and the front-end interface. The server manages the network events and enables users to monitor the network status and feeds back to control command. The front-end performs the following functions: displaying real-time data information including temperature, humidity and light intensity, showing the network topology based on the sensor nodes status updated in real-time, managing the network that contains the setting of the transmission time interval, data type and sleeping time, and controlling air-conditioning units based on the average temperature data.

The air-conditioning automatic control operation is composed of two modules: remote control and local temperature threshold control. The remote control consists of the direct submission to the control node in simple network management protocol (SNMP) commands from the server. After receiving control commands, the control node sends an infrared signal to manage an air-conditioning operation. The authors decide the usage of control instructions using the SNMP. Another air-conditioning control method is based on a pre-set temperature threshold. A sensor module with temperature and humidity sensors is responsible for collecting real-time temperature and humidity information and comparing them with the preset temperature and humidity thresholds. The sensor nodes monitor temperature once per second. When the temperature reaches the threshold, the sensor node begins to code an infrared signal and control the air-conditioning status. As a result, a complete automation of the air-controller is achieved using a distributed sensing platform in order to get very accurate information about the temperature available in different zones of the building and acts differently over each of such zones.

The use of distributed sensing platforms for enabling tools in human comfort solution is really attractive in society. Wireless sensors are used as cooperative smart objects that closely monitor the living space's environmental comfort (thermal comfort, visual comfort, indoor air comfort, acoustical comfort and spatial comfort) towards improving human comfort without sacrificing energy needs of a passive house. (Rawi and Al-Anbuky, 2009) proposed a distributed sensing platform for achieving the thermal comfort of a living space where mobile people and static nodes work together in calculating thermal comfort Predicted Mean Vote (PMV) value. The architecture is composed of a set of nodes spread across the house in order to sense parameters such as air temperature, mean radiant temperature, relative humidity, air velocity, clothing, metabolic rate, luminance level, shading level, CO_2 concentration and sound level. The proposed architecture works react following the next steps: the system senses the environment, ranks the environment, obtains the occupant's need, gets the energy usage and puts into action the corresponding adjustment needed to achieve the desired goal.

A human thermal sensation is mainly related to the thermal balance of his or her body as a whole. This balance is influenced by the environmental parameters as well as physiological parameters. Environmental factors include *Dry Bulb Temperature* (DBT), *Mean Radiant Temperature* (MRT), *Relative Humidity* (RH) and *Air Movement* (Vel). On the other hand, physiological factors consist of two factors, namely the person's *Metabolic Rate* (Met) and *Clothing Level* (Clo). When these factors have been estimated or measured, the thermal sensation for the body as a whole can be predicted by calculating the *PMV* by means of a linear relationship between the previous parameters. After that, this ranking is collected from all the sensors available in the network aggregating the information. As a result, a global rank determines the new values to be set in the thermal controller in order to improve the home comfort. Moreover, in this architecture each person wears a mobile sensor to measure many physiological parameters and also to balance the measured data of the rest of the sensors according to the distance to the given target (person). Its balance enables a better aggregation of the sensed values given as more important to those sensed directly from devices which are closer to the target.

Another innovative case study in smart homes is the integration of 3D graphics contents as the representative of guidance information in a real-time indoor environment. Lee et al. (2009) describe a system which enables the tracking of nearby targets by utilising exactly the same system within the wireless sensing range. The deployment of the proposed architecture is depicted in Figure 6.21.

Three different types of nodes are identified. Firstly, the different people available in the building wear target nodes as a blind node to transmit the packets data to the mobile handheld device to update their location in real-time in a certain period. The exact type of sensor nodes are also attached in each room's ceiling set as the reference nodes send their own coordinate and RSSI data to the blind nodes which appear within their sensing range. In order to receive the packets for a mobile device, a sensor node is configured to serve the purpose. TinyOS is the operating system that specifically designs for the sensor node to furnish its low processing capability and memory spaces.

The proposed system is composed of five parts:

1. 3D virtual viewer manages the 3D graphic models residing in the mobile device 3D databases to visualise the scene in 3D contents according to the location.
2. The indoor tracking system plays the role of a performing refinement algorithm to estimate the targets' position in an acceptable range of errors.
3. A stand-alone server is developed to monitor all the packets transmission flow for debugging purposes and store the latest version of 3D graphics models to be downloaded to the mobile device.
4. A digital magnetic compass (DMC) dynamically updates the user's orientation.
5. A mobile device supports the interface environment to synchronise workflow for better system management.

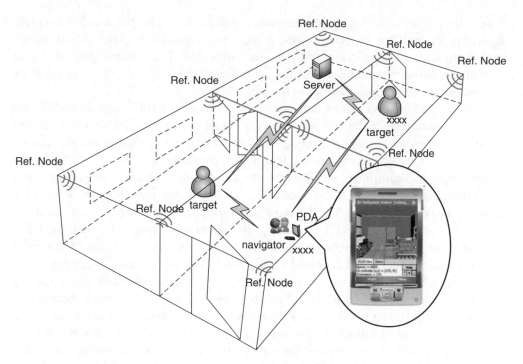

Figure 6.21 Architecture proposed by (Lee et al., 2009) for in-home localisation service in smart house environments. © 2009 IEEE. Reprinted, with permission, from Lee et al. (2009). WSN 3D Mobile Indoor Multiple User Tracking.

During the tour, union of 2D and 3D graphics provide even clearer path navigation while supporting textual information service to the targets, synchronising their location with real physical indoor scenes. The virtual world is modelled by 3D graphics software known as the Virtual Reality Modelling Language (VRML). VRML is a standard file format representing 3D interactive vector graphics, designed particularly with this in mind and has been accepted as the international standard by the International Organisation for Standardisation (ISO) in 1997. Similar to other modelling packages, VRML can be modelled through a dedicated text editor.

Due to the nature of RSSI the signal can be easily affected by the surrounding environment such as multipath fading and a shadowing effect. The obstacles in an inner building also make a big impact on providing an accurate position (x, y, and z). Instead of investing new tracking techniques, another alternative is observed to overcome the limitation. One of the solutions is to focus on designing high mobility RSSI-based distance estimation techniques to cater for such errors. Accordingly, an accuracy refinement algorithm is proposed to filter out the noise resident in the RSSI signals during the transmission. Nevertheless, due to the low-capability of the mobile device's processing, a complicated algorithm is not suitable for operation especially if involving lots of mathematical calculation. The proposed refinement algorithm is built up of four stages, which are: calibration process, smoothing algorithm, distance prediction and position estimation. Each stage involves direct computations that practically affect the following stage output's precision given its current output as the input to the next stage.

At the calibration process, an analysis of the behaviour of RSSI signals in different indoor environments is carried out. Generally, this pre-processing step is to gather the divergence average signal propagation constants' value. This is usually caused by human activities and obstacles appear in the surrounding area.

Maximum and minimum signal strengths are recorded at each measuring point to analyse the pattern of RSSI signals as a blind node may appear in either a dynamic or static situation. The average signal propagation constant is successfully obtained to reduce the signal noise during the estimation stage. After that, the RSSI smoothing algorithm is practically used to cater for the disadvantages of the radio signals dynamic fluctuation received by the blind node from closet reference nodes. It subdivides into two sub-stages, that is, estimation stage and prediction stage. As a result, the output (filtered RSSI value) obtained is applied for distance estimation.

Distances between a blind node and each reference node can be calculated. Next, by referring to the computed distances, the position can be estimated with the trilateration model. It requires at least three reference nodes to detect the signal strength transmitted by the blind node. It estimates the object's location based on simultaneous range measurements with a minimum of three known location reference nodes.

As a result, the hand held mobile shows a 3D representation of the house which matches with the current perception of the human in the building. Thus, an over-lapped set of different smart services can be offered in the home, acting as a smart 3D human-house interface. Figure 6.22 shows some screenshots of the implemented system.

Another interesting case study is the architecture proposed by (Goh et al., 2011), they developed a smart personal wardrobe. Either male or female of all age groups are spending more time choosing the right apparel to appear more presentable to impress their colleagues, superiors, customers, family and friends. However, to deal with choosing the right garment for the right occasion could be frustrating and time consuming. The proposed system is able to better manage their collection of garments by considering the users' choice of colours, styles, events and emotion. Target users are busy entrepreneurs and colour blind people. Radio frequency identification (RFID) technology was integrated into the garment and a prototype system was developed to trace the movement of the garments in the wardrobe. The RFID tag contains data regarding the clothes which will be captured and stored in the system. With the use of RFID technology, the system is able to assist users in better decision-making and assists the users. The prototype implemented can be seen in Figure 6.23. Note how there is a RFID reader and all the clothes have attached

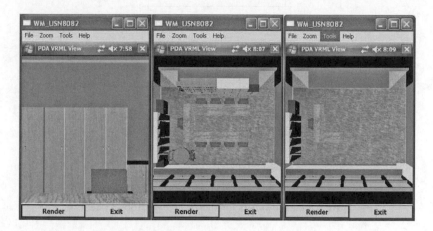

Figure 6.22 Overview of the Interface based in 3D models proposed by (Lee et al., 2009) for rendering of in-home services. © 2009 IEEE. Reprinted, with permission, from Lee et al. (2009). WSN 3D Mobile Indoor Multiple User Tracking.

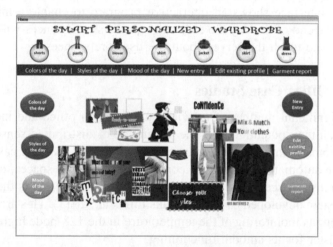

Figure 6.23 Smart wardrobe prototype. © 2011 IEEE. Reprinted, with permission, from Goh et al. (2011). Developing a Smart Wardrobe System.

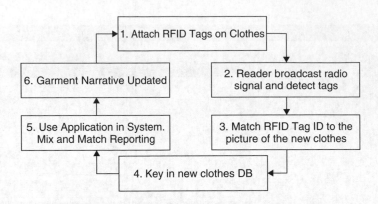

Figure 6.24   Flowchart for the smart wardrobe system.

an RDIF tag which acts as a server to the reader in order to discover the different clothes available in the wardrobe.

Figure 6.24 shows the proposed flowchart of the system. Essentially, at the first step all the clothes are tagged with a RFID tag. After that, the reader periodically sends a broadcast package in order to discover all the clothes available in the wardrobe. When, the reader receives such a clothes collection, it is sent to the data server in which the picture associated to the RFID tags is retrieved. After that, the graphical user interface shows an interactive program to enable a mix and match reporting. At the end of the day the user can report its experiences with the selected clothes in order to teach the server about the user preferences.

6.4 Smart Office Case Studies

There are a significant number of activities available in office and industrial environments which may potentially get the advantage of distributed sensing platforms in order to automate, improve and enhance such activities. A clear case study to be analysed is the automatic temperature control over working spaces, computer data centres and industrial scenarios. In this respect, a good example is the temperature monitoring system developed by (Baghyalakshmi et al., 2011). This architecture performs a continuous monitoring of the temperature in the 128-node high-performance computing cluster for its smooth functioning.

The temperature in the 128-node high performance computing cluster has to be maintained at 16 °C for its smooth functioning. If the cooling is not correct and the temperature rises beyond 20 °C, the cluster will auto shutdown due to overheating. Hence continuous monitoring of the temperature is essential. The data collected at the base station is made available to the administrators of the cluster for necessary action. The distributed sensing platform is composed of two sensing nodes equipped with temperature sensors, two extra sensing nodes used as routing nodes and the base

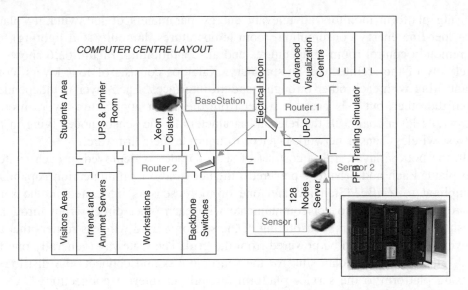

Figure 6.25   Smart office deployment controlling data centre temperature (Baghyalakshmi et al., 2011). © 2011 IEEE. Reprinted, with permission, from Baghyalakshmi et al. (2011). WSN based temperature monitoring for High Performance Computing cluster.

station node to connect the network. Figure 6.25 shows the deployment implemented in the computer centre using the floor map in order to indicate where the sensors are located and how they interact in order to deliver packages to the base station.

Initially, the sensor nodes provide sensing with a default sampling interval (2 sec). Whenever the timer gets fired the nodes transmit the data to its parent node for forwarding. Then, the data collected at the router is transmitted to the gateway or BS for further processing. This continuous monitoring is really a starving process which required an important amount of energy. For this reason, authors experiment with a different sampling interval from 15 sec to 240 sec. The battery value is also transmitted to BS along with temperature values and it has been logged in the database. The performance of the network is explained using the battery life versus time. It has been experimentally observed that the nodes battery life gets exhausted soon when the data rate is set high, managing ranges between 40 and 100 hours, respectively. This leads authors to get power directly from the HPC cluster in order to solve the problem of the energy. However, this is a very good example about real numbers of battery performance using a continuous sampling approach with a simple 3V battery for powering a sensing device.

Genova et al. (2009) propose another good case study for smart offices. They propose a distributed sensing platform called *Kaleidos* used in Telecom Italia (TI), being one of the top Italian electricity consumers, deployed for real-time energy management based on wireless sensor network technology WSN. The distributed sensing platform monitors all relevant energy parameters of its switching plants (e.g. per-line energy consumption, room temperatures, humidity and lighting) and to remotely control room temperatures and air conditioning. Figure 6.26 shows an overview of the architecture. In summary, Kaleidos consists of four layers, communicating with each other through standard interfaces. Each layer is independent from the other, but it is possible to set from the upper application layer, through the whole layer chain, all the relevant parameters of the sensor nodes lying in the lowest wireless sensor network layer or to upgrade their firmware.

In the bottom layer, a low cost set of sensor nodes is deployed in each switching plant. Each node (battery powered), includes radio communication capability, compliant to ZigBee Standard and one or more sensors integrated in the same board. Sensor nodes do not communicate with each other directly, but through a mesh network of routers, constituting of the self-organised wireless routers that are 'always on' and should be powered from the grid. The Gateway (generally one for each plant), acts as a data sink for the sensor nodes and connects the distributed sensing platform to the service platform through an Internet connection (LAN or GSM-GPRS) and a proper API.

The service layer represents the core of Kaleidos and is able to manage and control the sensing platform to collect and store data, and to make this available to applications via a web service; the platform was fully developed by TI (Telecom Italia). In essence, this layer is an ICT platform for the management and control of the distributed sensing platform which keeps the complexity of wireless sensor protocols hidden from applications by providing an open web-service interface; This layer offers common functionalities to the different applications and services like archiving, logging and localisation by means of a direct communication with the gateways via a standard interface (ZigBee standard). The idea is to enable applications to use multiple distributed sensing platforms as well as one sensing platform to be used by multiple applications. This isolation by means of a service oriented application (SOA) makes opaque the underline network and can even communicate with non-ZigBee devices (e.g. TinyOS). Finally, on the top of the service layer, any potential application can be deployed in order to implement energy management, healthcare, homeland security, and so on.

The first trial started in April 2007 in a plant situated in Turin and demonstrated an energy saving in the order of 10–15%. It is possible by simply optimising the air conditioning system's working parameters. In 2008, TI decided to deploy the system into its 300 larger plants. A typical site consists of 10–15 rooms, 2000–6000 m^2, 1–5 GWh/year energy consumption and it is equipped with a sensor network of 20–30 nodes and 20–25 environmental nodes. Kaleidos is used not only to monitor and control the energy and thermal behaviour of sites, but it has been shown to be effective in supporting the design of fine-grained energy saving actions and in real-time evaluation of those actions.

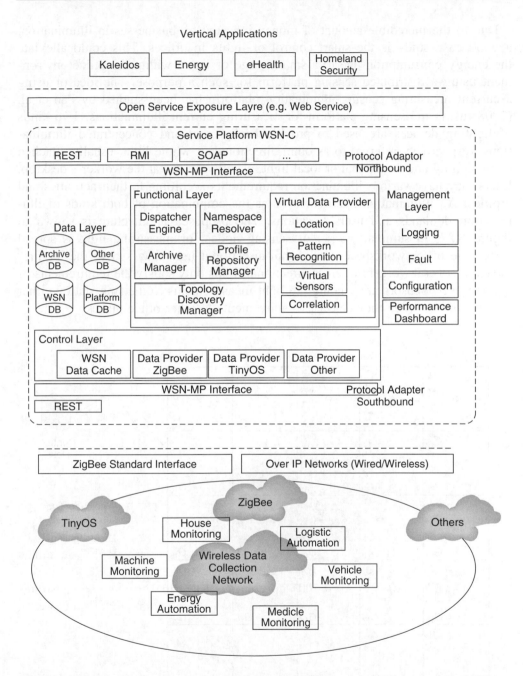

Figure 6.26 Architecture overview proposed in Kaleidos framework (Genova et al., 2009).

Due to the incredible amount of money invested by businesses in illumination, a good case study is the smart control of lights in offices. This could alleviate the energy consumption due to a smart usage of lights within the office environment using a distributed sensing platform for such a purpose. The idea of using a current measuring sensing illumination platform has been studied by Pan et al. (2008) using the sensing platform for measuring current illuminations. Two kinds of lighting devices are used to provide background and concentrated illuminations, respectively referred to as omni-directional or whole light, usually attached to the ceiling and directional or local light, usually located at the worker's desktop. Users may have various illumination requirements according to their activities and profiles. An illumination requirement is as the combination of both kinds of illumination demands and users' locations. The proposed architecture is shown in Figure 6.27. In summary, a set of sensing nodes for measuring light is spread across the office workspace. Each omni-directional light may satisfy the requirements of several workers whereas directional lights only satisfy the requirements of a given user. Sensors are sending the light measurements to the sink station either periodically or when they detect user movements to other office places.

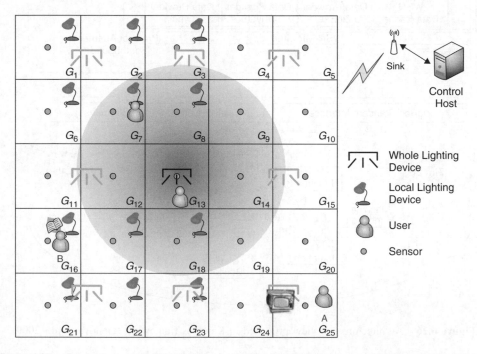

Figure 6.27 Scenario proposed by (Pan et al., 2008) in which there is smart management of lights in the office. © 2008 IEEE. Reprinted, with permission, from Pan et al. (2008). A WSN-based Intelligent Light Control System Considering User Activities and Profiles.

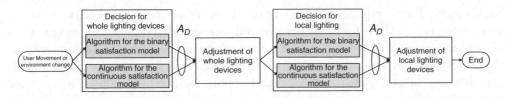

Figure 6.28 Decision algorithm proposed by (Pan et al., 2008) for deciding the lights to be activated in the office. © 2008 IEEE. Reprinted, with permission, from Pan et al. (2008). A WSN-based Intelligent Light Control System Considering User Activities and Profiles.

The information is received at the sink node and processed. In essence, two different satisfaction models are used, binary and continuous. In the former, a user who is satisfied returns a satisfaction value of one; otherwise, a zero is returned. In the latter model, a satisfaction value that is a function of the specified illumination interval and the sensed light intensity is returned. For the binary model, the goal is to satisfy all users such that the total power consumption is minimised. For the continuous model, the goal is to satisfy all users such that the total satisfaction value is maximised. However, in both models, it may not be possible to satisfy all users simultaneously. In this case, users' illumination intervals are gradually relaxed until all users are satisfied to a certain degree. These two satisfying models are used in a two-step algorithm for calculating the light adjustment over a given scenario. The first step calculates the onmi-directional lights whereas the second step used this information to calculate the status and intensity of direct lights. Figure 6.28 shows the algorithm proposed.

Note that information reported by sensors is accumulated as values; these values then need to be converted to the actual adjustment amounts. If the actual amounts do not match the target amounts, a binary search technique is adopted to gradually approach these amounts. The binary search procedure can be explained by the following example. Suppose that a device's current on-level is 40% with contribution 300 lux and the adjustment has to be done to 200 lux. The control host will first adjust the on-level to 20%. After first iteration, the control host will collect sensors' reports to compute thus calculating the new adjustment. The next guess will be an on-level of 10% or 30%. The similar trial is done for all whole and local lighting devices. In practice, the on-levels of dimmers are discrete and have finite levels. The termination conditions of the above binary search can be controlled by threshold where the system controls the lights providing the most efficient way of satisfying all the workers' requirements while it optimises as a local maximum the energy consumption spent in lighting.

Deployment of a large population of sensors for sophisticated sensing and control in the industrial and commercial infrastructures is a challenging research area. Unlike the office networks, the industrial environment for wireless sensor networks is harsher due to the unpredictable variations in temperature, pressure, humidity, presence of heavy equipment and so on. Low et al. (2005) describe a set of possible case studies for smart industrial scenarios. Firstly, the usage of a distributed sensing platform for autonomous mobile robots is of great interest to the industry. They are very suited for applications that are delicate, risky, heavy, repetitive or labour-intensive tasks. Most robots used in the industries are fixed and dedicated such as machine loading, welding, assembly/disassembly and so on. The use of intelligent mobile robots and wireless distributed platforms can enable robots to take measurements at different locations to approximate the actual placement of sensors with better accuracy. Moreover, this enables robots to directly get environment information in order to provide context-aware information for performing industrial tasks.

For example, multi-robot task allocation proposed by (Batalin and Sukhatme, 2004) allocates tasks explicitly to robots by a pre-deployed static sensor network. In this way, the robots can navigate efficiently throughout the workspace using the deployed sensor network to explore their environment. Potential applications include moving stacks of containers between machine rooms and warehouse, industrial floor cleaning robots, patrolling robots in unsupervised areas or toxic waste clean up, and so on.

Another application scenario in an industrial environment can be real-time inventory management. Note that inventory management based on manual processes may cause out-of-stocks, expedited shipments, production slowdowns, excess buffer inventory, and billing delay situations. With a distributed sensing platform, the inventory and asset could be monitored in real time and information such as the arrival of the raw materials and so on. could be routed across a distance to the gateway for management decision and control. The BP petroleum factory is using a distributed sensing platform for remotely monitoring its industrial customers' LPG (Liquefied Petroleum Gas) tank fill levels (King, 2005). Battery-powered ultrasonic sensors are used and the information is transmitted by radio signal to a low earth orbit satellite, which relays the data for timely deliveries. With this real time inventory data, BP could accurately and efficiently deliver services to the customers for about 200 tanks in England. They reported that there is improved delivery efficiency over 33%.

Process and equipment monitoring is another direct application scenario in which distributed sensing platforms enable remote monitoring of the health of machinery without the infrastructure needs for cabling. Note the importance of continuously monitoring the temperature, pressure, vibrations and power usage for enabling the manufacturers to reduce unnecessary costs incurred due to failures or malfunctions in machinery. For example, General Motors (Hochmuth, 2005) applies a distributed sensing platform to monitor the manufacturing equipment such as the conveyer belts and other types of machinery. Measured data such as the vibration, heat and other factors are detected and transmitted to a computer via a wireless mesh. By periodically collecting information from the machinery, the technicians could predict a machine's failure, mean time to failure and perform pre-emptive maintenance. The collected data also facilitates future improvement and faster repair of equipment from the data collected.

Another interesting application of distributed sensing platforms for industry is the environmental monitoring. The sensing platform can be used to provide solutions in the industry for leakage detection, climate reporting, radiation check, intrusion notification and so on. Emergency alerts could be sent to the operating managers requesting immediate preventive actions. The presence or the movement of abnormalities such as toxic chemicals, biological, radioactive agents or unauthorised personnel can be tracked throughout the facility using the sensing platform.

7

Public Safety Applications

Public safety involves the detection, prevention and protection from events that could endanger the safety of the general public causing a significant danger, injury, or damage. Earthquakes, cyber attacks, terrorism, forest fires, health diseases caused by environmental dangers such as lethal toxins, virus or chemical components are some of the events which would put public safety at risk.

In general, public safety refers to events that potentially could affect significantly the health of the population. For this reason, governments give careful attention to research and development initiatives focused on the prevention and protection of this kind of event. Indeed, according to the official web page of Hennepin Country, Minnesota (County, 2011), chosen as a mere example, the public safety operating budget for 2011 is 16.3% out of the total country budget. These numbers provide us with an indication of the importance of the design and implementation of effective infrastructures and systems for ensuring citizen safety.

There are a significant number of applications and networks specifically designed for ensuring and protecting public safety nowadays. These systems are generally composed of a set of interconnected devices in charge of detecting, preventing, protecting and alerting of dangerous events as fast as possible.

A public safety network is divided into three zones clearly differentiated as can be seen in Figure 7.1: Reception, Distribution and Detection zones. A detection zone is composed of a device or a set of interconnected devices distributed in a warning zone in order to perform the detection of hazards therein. These devices can be potentially any hardware with sensorial capabilities such as computers, laptops, sensors, mobile phones or even a citizen notifying an alert by mobile phone. Once the dangerous event is detected, all details have to be quickly distributed as widely as possible in order to ensure that all the involved entities in the management and mitigation of the detected event are notified as quickly as possible. This is done by means of the devices located in the distribution zone using any wire or wireless

Distributed Sensor Systems: Practice and Applications, First Edition.
Habib F. Rashvand and Jose M. Alcaraz Calero.
© 2012 John Wiley & Sons, Ltd. Published 2012 by John Wiley & Sons, Ltd.

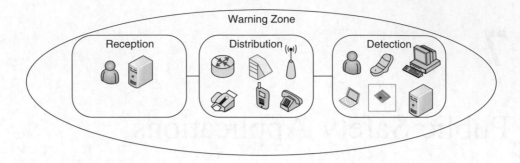

Figure 7.1   General IT public safety infrastructure.

technology to widely and quickly spread the information. Finally, the reception zone is composed of all the users and machines that may be notified about the dangerous events in order to manually or automatically start the management and coordination of the crisis situation.

Wireless sensor networks (WSN) play a crucial and critical role in the design of such infrastructures, especially in designing the detection zone for public safety infrastructures, being a really suitable approach due to its intrinsicity.

Due to the nature and purpose of public safety networks, the design of the detection zone has strict design requirements being a clear differentiating point with respect to other types of services and applications, essentially, due to the critical aspect of the information managed therein.

The detection zone designed for public safety requires a fast detection of dangerous events and a fast propagation of alerts once dangerous events have been detected in order to achieve an efficient system in which the up-to-date information arrives at the distribution zone on time. Note that a crisis situation always entails the necessity of continuous up-to-date information. This can be addressed from distributing sensing platforms in which a set of interconnected sensors is spread across the warning zone enabling the definition of an efficient monitoring area being able to route monitored information to the base station nodes connected to the distribution zone.

A reliable monitoring with fault tolerance on the devices and on the communications between devices is also a key aspect since it ensures the correct detection and delivery of the detected hazards. Redundancy and no central points of failures are also design requirements in order to avoid any denial of service for the critical detection and prevention services deployed in the detection zone. This can be achieved by mean of scenarios in which multiple paths are established between sensors and using a redundant deployment strategy in which more than one sensor is monitoring the same area and multiple base nodes are deployed. Reliable and accurate data gathered from sensors monitoring environmental variables are critical aspects in order to avoid false positive or false negative detections.

Secure sensor infrastructures are necessary for protection against malicious devices and for establishing secure communications between devices. It provides protection against cyber attacks and against the sensitive information exchanged between devices, which are directly related to the public safety. This security aspect can be addressed using a sensing platform by means of power-aware algorithms for providing efficient device authentication and encrypted communications.

Desourdis et al. (2002) have done an analysis of requirements in public safety wireless infrastructure. They emphasise that the authentication of the devices is crucial in this kind of infrastructure. Moreover, authors also indicate that the capability to prioritise messages according to their type of message is a suitable feature, providing a way, which quickly propagates high priority messages containing critical information.

One of the most differentiating features of the detection zones in public safety infrastructures is that the location of the warning areas are usually (but not necessarily) outdoor. Some outdoor examples are those related to the detection of natural and man-based disasters such as forest fires, earthquakes and air pollution monitoring. This requirement imposes several additional requirements in the design of these detection zones.

First, the sensors may provide autonomic management functions in order to perform automatic adaption of the sensors to the environment status with minimal human intervention. Second, the sensor has to be designed to be alive for a long-time period maximising the probability of detecting a hazard in case it happens. This requirement entails a cross-layer design of power-aware algorithms for all the optimisation of the functions provided by the sensors. Third, the sensor devices have to be adapted to outdoor environments. This requires a special design of the devices for isolating the electronic circuits from any weather inclemency and even from extreme conditions of temperature, humidity, pressure, and so on. Finally, the detection zone can be determined and modified at any time and this feature has to be flexible enough to not assume any infrastructure underneath for interconnecting sensors. In essence, this fact requires a pure ad-hoc sensor infrastructure with minimal connection assumptions.

Once the particular features of public safety infrastructures have been unveiled, the following subsections explain in detail the state-of-the-art in some common applications and scenarios which use distributed smart sensors as public safety infrastructures.

7.1 Monitoring Airborne Toxins

The monitoring of airborne toxins is a critical process to ensure public safety, especially, in big populations. The presence of any toxic substance in the town air may cause drastic diseases in the population. This is well known by military and government agents who are constantly monitoring the air searching for novice substances. There is a wide set of causes by which such substances could be in the air. The cyber-terrorists can potentially poison the air trying to cause potential victims, the great amount of chemical enterprises could have any critical problem in their installation poisoning the air, the nuclear central could have fissures in their structures, and so on. Moreover, traditionally, monitoring airborne diseases has been done using isolated specialised sensors, which have to be located in special areas in which the monitoring is done. In this context, the usage of distributed smart sensors could potentiate the detection of airborne hazards since they are monitoring many areas collaboratively augmenting the covered area sensed by a single isolated device. This infrastructure could also be smart enough for providing ad-hoc alerting platforms for quickly notifying about emergencies when they are detected. Thus, the usage of distributed smart devices enables low-cost and low-maintenance distributed monitoring systems, which can be used for detection of and protection from any hazard in the air.

A clear example is provided by Kijewski-Correa et al. (2009). The authors propose a distributed sensing platform for carrying out the detection of plumes of chemical agents such as sulphur hexafluoride (SF_6) in urban zones. Such a system is deployed in reality at the University of Notre Dame and in a major metropolitan area of South Bend, Indiana, USA.

Three different types of nodes are proposed for the detection zone, Instrumentation nodes, or 'I-nodes', which sense data from the environment continuously. These nodes store the monitored information locally and periodically and continuously send the stored information to the nearest Gateway node or 'G-node'. The G-node collects the data from the surrounding clusters of I-nodes over a predefined time period and then it uploads such information to a central database using a GPRS module installed in such nodes. Moreover, in case the distance between the I-node and the G-node is too great, a Repeater node, or 'R-node', can be added, providing a bridge between the two nodes. There is a slight difference between I-nodes and R-nodes since they can be the same type of physical device acting with a different role according to the routing protocol.

Two different types of 'I-nodes' are deployed according to the attached sensor. On the one hand 'I-nodes' are attached to an *InfraRan Specific Vapor Analyzer* sensor manufactured by *Wilks Enterprise, Inc.*, for the detection and measurement of the concentration of SF_6 in the air. On the other hand 'I-nodes' are attached to a *Vaisala WXT 510 Meteorological Station* with ultrasonic wind velocity measurement, as well as relative humidity, barometric pressure and temperature.

The proposed architecture is composed of heterogeneous nodes which entail the presence of diverse agents in the network performing different tasks. Thus, a basic in-network processing is done. However, there is no real data management in the network beyond the basic data aggregation technique based on time windows performed in the Gateway node.

The integration of the monitored information with the CT-Analyst software is a clear innovation introduced by authors. CT-Analyst is real-time software for calculating computational models in order to predict the hazard area based on the meteorological data and the level of airborne toxins. This integration enables not only the detection of potential chemical hazard in the airborne, but also the prediction of propagation models. This software is based on multiple pre-computed algorithms of computational fluid dynamics. The combination of the distributed sensing platform and the software enables the prediction on how the detected hazard can be propagated around the metropolitan area in real-time. This is a critical feature for the public safety infrastructure enabling a faster management of the crisis, producing effective evacuation and scheduling plans. The CT-Analyst software shows the information is in real-time enabling it to predict further hazard zones.

(Lim et al., 2011a) propose other distributed sensing platforms for monitoring biochemical reactions that occur in sewer pipes which produce a considerable amount of hydrogen sulphide gas (H_2S corrosive and poisonous), methane gas (CH_4 explosive and a major climate change contributor), carbon dioxide (CO_2 a major climate change contributor), and other volatile substances (collectively known as in-sewer gases). These toxic gases lead to contamination of the natural environment, sewer pipe corrosion, costly operational expense, public safety issues, and legal disputes. In order to prevent biochemical reactions and to maintain healthy sewer pipes, frequent inspections are vital. Various schemes are designed and developed to identify functional deficiencies in a Wastewater Collection System (WCS). Nevertheless, the current inspection techniques are not for mapping the sewer gas concentration. In addition, because of such a harsh and hazardous environment a comprehensive sewer gases inspection has been prohibitively expensive.

Authors propose *SewerSnort*, a low-cost, unmanned, fully automated in-sewer gas monitoring system. A sensor float is introduced at the upstream station and drifts down the sewer pipeline while the sensor float collects gas measurements along with location information of sampling points. This information is collected locally in the sensor following a *data mule* approach. This approach consists of the gathering of all the sensed information using the inherit drifting of the sensor along the pipelines for a later exchange of such information on milestone points.

The localisation of the sensor across the pipeline is achieved by means of the usage of stationary reference nodes, which are periodically sending beacon packages advertising their current localisation. This beacon of information is received by the mobile sensor updating its new localisation. Moreover, the stationary sensors placed across the path serve as milestone points in which the mobile node can transmit its locally stored information to the reference nodes, which also act as sink nodes.

The gas measurements are retrieved and used to generate gas concentrations, which are used to determine whether a maintenance or repair is necessary in the pipelines. Figure 7.2 depicts the scenario proposed. Note the smart usage of drifting water and floating sensors as an innovative idea to reduce power consumption

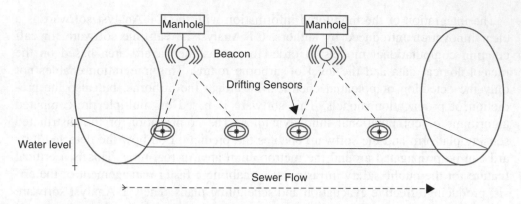

Figure 7.2 Monitoring scenario with nodes drifting along the sewer flow (Lim et al., 2011a).

associated with the movement of sensors by means of any engine. In fact, it may be potentially extrapolated to other scenarios. Authors carefully describe how the floating sensors and the different base stations must be built.

Another example for carrying out an airborne monitoring is proposed by (Kim et al., 2009). They design a wireless electronic nose network (WENn) for monitoring real-time gas mixture, NH_3 and H_2S, the main malodours in various environments.

The motes use micro-gas sensors with Sn02-CuO and Sn02-Pt sensing films for detecting the presence of target gases. Each node in the network performs a periodical sensing of the gases obtaining the isolated concentration of such gases. Then, each sensor performs a classification and concentration estimation of the binary gas mixtures in real-time. To do so, the sensors firstly use the fuzzy ARTMAP neuronal network (NN). This NN uses a set of weight values which have to be previously calculated and inserted in the motes a priori. Then, the NN uses the senses values and such weights in order to determine what kind of gas is being monitored. Then, when the type of gas has been identified, the fuzzy values are defuzzified in order to achieve both the gas species and its concentration. Figure 7.3 depicts the complete algorithm sequence proposed by the authors.

Note that the weights of subcategories of gases in a fuzzy ARTMAP neural network and the parameters of the membership functions have to be trained in order to achieve an accurate prediction method.

Note that some of these gases are not dangerous if isolated but they could become dangerous when they are binary mixed. Thus, each time the node gathers new up-to-date information about such a concentration, it estimates such gas concentration.

The proposed architecture is an event-based network and thus only when a new event is generated in a given sensor, is it transmitted to the sink node. Then, the algorithm assumes high and low concentration thresholds, which may produce new events in the distributed sensing platform.

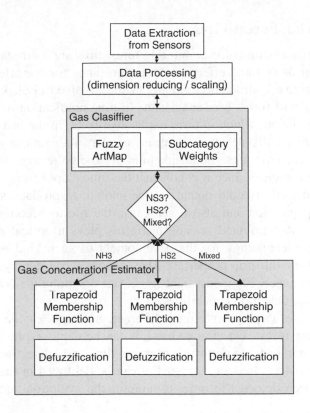

Figure 7.3   Algorithm sequence used for building a Wireless Electronic Nose (Kim et al., 2009).

Tsujita et al. (2005) propose another example composed of mobile nodes equipped with heterogeneous gas detectors (e.g. metal oxide sensors, electrochemical sensors and quartz crystal microbalance sensors). This architecture is deployed in a detection zone for air pollution monitoring. Authors are focused on the auto-calibration of the sensors, which is an important issue for those sensors specially designed for monitoring gases. Gas sensor outputs generally show drift over time and frequent recalibration of a number of sensors in the network is a laborious task. To solve this problem, authors propose an auto-calibration method for which a given sensor requests periodically the sensing information provided by the direct neighbour sensors.

This information is received in the sensor being recalibrated, which aggregates such information calculating the average value for the gas concentration, which is considered as an estimation of the true gas concentration. Then this value is used as a calibration parameter for the entire set of its neighbours. Note the added value of the in-network data processing in order to achieve free-maintenance gas sensors by means of a collaboration auto-calibration method.

7.2 Monitoring Forest Fires

Public safety infrastructures for monitoring forest fires are a critical component for the development of fast and effective evaluation plans for the affected population and for the effective coordination among all the involved civil protection agents. An early detection of forest fires enables the in-time notification of the alert among the affected population, which in turn, might save lives, in the last instance. Traditionally, fires are usually detected by means of alert notifications of people which have physically seen the fires and decide to notify the policeman and fireman. This has several disadvantages since it entails that the time elapse from the fire initiation and the initial detection could potentially be enough to produce serial damage in structures and population and also it can cause the fire to become uncontrollable. Thus, the usage of distributed sensing platforms plays a critical role in this context as a suitable technology for the development of such kind of infrastructures due mainly to their intrinsic properties. They enable a fast deployment of a set of wirelessly interconnected low-cost lightweight sensors along the potential detection zone for early fire detection capabilities.

According to the deployment time, different architectures may be potentially created across the forest. On the one hand, it can be deployed in the forest before any fire detection in order to perform a near real-time early detection of such natural or man-made disaster. On the other hand, it can be deployed in the forest once the forest fire has been detected in order to monitor the evolution and propagation of the fire.

Several examples of these scenarios have been provided during the last years for carrying out the detection of forest fires. For example, Hefeeda and Bagheri (2009) propose an architecture where the nodes are deployed uniformly at random in the forest. Since the lifetime of nodes in active mode is much shorter than even a fraction of one forest fire season, node deployment is assumed to be relatively dense such that each sensor is active only during a short period of time and the monitoring task is rotated among all nodes to achieve the target network lifetime. Moreover, active nodes are not continuously monitoring the area. Rather, they periodically (e.g. every 30 minutes) perform the sensing task. Therefore, nodes in the active mode are further divided into *active-sense* and *active-listen* modes. In order to determine which are the nodes being active and in sleeping mode for a given time, authors model the forest fire detection as a coverage problem over a dense sensor network. A k-coverage problem provides finding the minimal set of sensor nodes which need to be awake in order to ensure that at least k sensor nodes are monitoring all the sensing points of the interested area. Note that at least two nodes could be suitable to cover a given area since the detection of fires may imply the auto-destruction of the sensor node. The proposed system is an event-based system and it only transmits notifications to the sink node when an event is detected, that is, a fire.

Wenning et al. (2008) propose a routing algorithm to take into account the realistic fact that a deployment of a distributed sensing platform in forest fire scenarios can destroy the sensors devices in any moment due to fire detection. This has direct implications for the network lifetime, performance and robustness. Authors focused on node failures caused by the sensed phenomenon itself proposing a routing method that is aware of the node's destruction threat and adapts the routes accordingly, before node failure results in broken routes, delay and power consuming route rediscovery.

The key aspect of this novel routing protocol based on AODV is the usage of the node health, affected by the sensed phenomenon, as the most relevant routing criterion. This introduces a critical factor in the selection of the best suitable routes for routing the sensing information to the base station. In essence, this routing strategy used a best neighbour selection algorithm based on the node health, hop count and RSSI values. This contribution has an advantage in the field since it enables the usage of most robust environment routing protocols for this kind of critical scenario in which the reliability of a measure data can be a decisive factor in the crisis management or in evacuation plans. Moreover, note how smart data processing of environmental factors is directly utilised at the routing layer for determining the most suitable routers for notifying fire detections.

Regarding in-network data processing, (Yu et al., 2005) and (Soliman et al., 2010) propose similar architecture for real-time forest fire detection in which a large number of sensors are deployed in a forest. Sensor nodes collect measured data (that is, temperature, relative humidity) and send such information to their respective cluster head nodes that collaboratively collect and process the data. The cluster nodes use a *Neural Network* to take the measured data as input to produce a weather index, which measures the likelihood of the weather causing a fire. Cluster headers send the weather indexes to the base station which concludes the forest fire danger rate.

Authors are focused on the development of an efficient neuronal network which produces an accurate detection method for forest fires based on the gathered humidity and temperature obtained from all the sensors, but they do not cover any networking protocol or sensor deployment strategy. Similarly, (Sathik et al., 2010) propose an analogous architecture which uses *Support Vector Machines* rather than *Neuronal Network* for carrying out the classification of forest fire detection using two input metrics: temperature and humidity.

Other research fostered by (Kosucu et al., 2009) has shown that most of the studies in forest fires choose simulating their proposed solutions instead of doing experiments in real testbed environments, as this kind of setup exposes additional difficulties. Thus, they have designed a simulation environment that precisely models an outdoor testbed environment. To this end, they combine the best of both simulation and testbed worlds. In essence, they physically construct a number of outdoor testbeds with randomly deployed unattained sensor nodes, and collect actual sensor readings from this multi-hop sensor network. Moreover, the authors use a well-known fire simulator to burn a forest area with similar canopy characteristics, and collect the temperature readings that are mapped to the actual sensor positions in the testbed configurations. Finally, they merge the fire simulation results into outdoor testbeds to test the fire detection algorithms as if there is a real fire event occurring in the monitored region.

In essence, a real distributed sensing platform is deployed into the forest. Then, temperature values are simulated and merged with the real deployment providing such information to the final user.

From this outdoor testbed, authors proved that their proposed forest fire detection algorithm provided a real success ratio ranging from 30% to 80% when node failures occur due to the fire destruction and ranging from 80% and 100% when link failures occur due to lack of connectivity. These ranges vary mainly due to the ignition point in which the fire has started and its relationship with the network topology established. Authors also have analysed different factors such as topology and physical configuration changes, wind direction, ignition point position and sampling period variations.

The previous described works base the forest fire detection on sensing metrics such as temperature, humidity and smoke levels associated with already started fires. (Somov et al., 2010) realised this issue and focus on the development of an architecture trying to detect fires in more early states. They proposed a distributed sensing platform for detecting fires by means of gas sensors used for monitoring the pyrolysis effect which happens in a fire before even smoke starts therein. The different states for a fire are depicted in Figure 7.4. Authors are focused on the earliest stage of the fire producing a more efficient architecture in terms of response time in which the gap between fire ignition and forest fire detection event is around 1,359 seconds on average according to the authors' results.

Once a fire has been detected, a distributed sensing platform can be deployed in the forest in order to monitor the evolution and propagation of the fire. This is the case of *FireWxNet*, the system proposed by (Hartung et al., 2006). They proposed a portable wireless system for monitoring weather conditions in rugged wild land fire environments. This is a complete developed system used in the *Selway-Salmon Complex Fires* of 2005. *FireWxNet* provides the fire fighting community with the

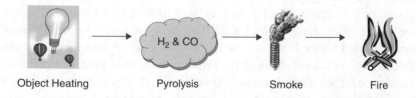

Object Heating   Pyrolysis   Smoke   Fire

Figure 7.4   Chemical states of the fire.

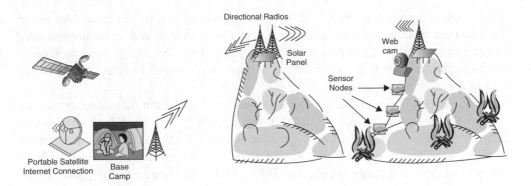

Figure 7.5 Monitoring forest fire architecture (Hartung et al., 2006).

ability to safely and easily measure and view fire and weather conditions over a wide range of locations and elevations within forest fires. This previously unattainable information allows fire behaviour analysts to better predict fire behaviour, heightening safety considerations. The architecture for the proposed system is depicted in Figure 7.5.

The opportunistic base camp is equipped with a portable satellite connection provided by *Skycasters* manufacturer (128 Kbps up, 512 Kbps down) in order to provide Internet connectivity. For the main links in the backhaul network, authors used two different types of radio made by *TrangoBroadband Wireless*: The *Trango Access5830* and the *Trango M900S Access Point/Subscriber Module Radios* (AP/SU). The primary radios were the *Access5830s*. These radios were strictly point-to-point directional radios that used polarised directional antennas to achieve a range of roughly 50 kilometres. They operated at 10 megabits per second in the frequency range of 900 MHz to 930 MHz. For shorter links authors used the *M900S* AP/SU connection to an external *Yogi* antenna to increase the connection range to just over 32 kilometres. The devices used in the backhaul connection are self-powered using solar panels and large batteries to power the equipments. The base station which acts as a connection point between the backhaul and the sensors is designed as can be seen in Figure 7.6. In essence, it has an Ethernet

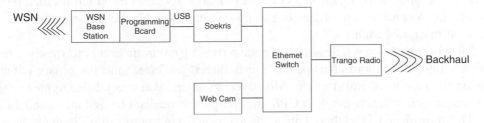

Figure 7.6 Logical base station structure proposed by (Hartung et al., 2006).

switch connection to a *Trango* Radio providing connection to the base. Moreover, the Ethernet is also used to directly connect a webcam and a mini-embedded device running Linux (*Soekris*) into the network. This *Soekris* is used to receive the information from the sensor acting as base station and to send the collected data to the distribution zone.

For the sensor nodes, *Mica2* devices are modified using the 52-pin external connection for using a *Chipcon CC1000* radio operating at *900 MHz*, and a set of external sensors. Concretely, they provide a more accurate temperature sensor, humidity sensor and anemometer sensor than that originally provided by Mica2 devices. Authors originally attached a GPS module but Mica2 are only powered by two AA power devices and they decided to remove this power hungry GPS device and to manually localise sensors (since they are immobile) during the deployment of them into the forest. Authors also carefully explain how they do the sensor deployment. In essence, the average distance between nodes was 138m with the longest link nearly 393m (usually, the deployments are done in a 30 meter range to ensure connectivity). They achieve such large distances by exploiting a phenomenon called Fresnel Zones which is the cause of multipath interference.

Authors propose a 15 minute period where the nodes would sleep for 14 minutes, wake up and send packets for 1 minute. However, note that nodes sleeping cannot forward packages and then time synchronisation is required in order to ensure that nodes wake up during the same minute within the period. For this purpose, they used a modified light version of the well-known FTSP time synchronisation protocol. During such minutes, the base station sends beacon packages every four seconds. These packages are used to perform route discovery, implement fault tolerance, and time synchronisation.

The beacons are propagated using a particular in-network data processing for which it was decided to use a smart re-forwarding of the beacons. In essence, a directed flooding algorithm is used which only retransmits the packets when the distance to base (DTB) of the originating node is less (closer to base) than the given node. To this end, the distance to the base is calculated since the nodes know their location which is pre-programmed at deployment phrase.

A different application for fire detection is proposed by the *FireNet* (Sha et al., 2006). This architecture is theoretical architecture which may be used for fire rescue applications. Authors define an outdoor scenario in which firemen need to be interconnected in order to get up-to-date co-ordinated information about critical rescue information.

Each fireman has a wireless sensor node providing environmental information, fire fighter information and also special events directly inferred into the sensor device using the monitored information. Moreover, there are also special fire fighter vehicles equipped with sensors with GPS providing their localisation and measured data.

The information is gathered in a PC with dual communication channels being the access point to the distribution zone. Authors use this scenario to identify

requirements and challengers on the WSN, especially related to networking protocols. A pure ad-hoc network with real-time self-organisation capabilities is a clear
requirement providing efficient mechanism for doing autonomic configuration and
optimisation capabilities. Moreover, due to the critical nature of the information
exchanged between nodes, fault-tolerance routing is a challenge in this scenario.
Another key on the design of this architecture is the service differentiation and message prioritisation in order to give priority to the critical messages to react as fast
as they can to the destination. Moreover, real-time and mobile localisation plays a
critical role in this scenario and it has to be carefully addressed since it could save
the life of fire fighters enabling their localisation and orientation.

7.3 Monitoring Structural Health

There are diverse reasons for which structural health monitoring (henceforth referred
to as SHM) is used as a critical process for ensuring public safety nowadays. All
the materials used for structures such as bridges, buildings and roads have inherent defects. Moreover, these materials can be potentially damaged by any external
agents such as natural and man-made disasters, weather conditions, traffic collisions, and so on. An early detection of such kind of defects and damages in the
structure may be really effective to enable the prevention of more serious dangers
in the structure, reducing significantly the structure maintenance cost, and in the
last instance, preventing bigger disasters such as evacuations and demolitions. Public safely plays an important role in these scenarios since structures are frequently
transited by people and the detection of hazards in the infrastructure can produce
an effective evacuation plan on-time. The usage of a distributed sensing platform
for SHM can provide accurate information provided by a set of smart interconnected nodes, which are able to correlate information to get a real-time status of
the monitored structure. This approach could be better than the traditional approach
in which humans are doing periodic inspections of the building status, and thus,
health information cannot be done continuously monitoring at real-time.

There are some specific features and challenges related to SHM scenarios which
may be an important issue if they are not addressed properly. Firstly, there are
important bandwidth, memory and processing capability limitations in the motes.
Moreover, this is emphasised in SHM scenarios in which sampling rates are relatively high in order to achieve accurate monitoring and this causes high data
ratings. For this reason, algorithms designed for SHM might be efficient enough
to deal with these requirements. Secondly, this kind of scenario may potentially
suffer a high rate of packet loss due to alteration in the connectivity, sleeping times
for motes, weather conditions, obstacles, multi-path effects, and so on. Then, best-
effort reliable transport protocol may be considered. Finally, time synchronisation
is an important feature in these scenarios in which data collection is usually done
synchronously in order to enable efficient data correlation for the monitored data.

Several suitable architectures have been proposed during the last few years for carrying out the monitoring of structural health. One of first real-world examples has been provided by (Chintalapudi et al., 2006) in the *Wisden* architecture. This architecture is deployed in buildings to detect and localise damage and to collect an analysis of structural responses to ambient and forced excitation. The system is based on *Mica2* motes using the *TinyOs* operating system. A *CXL02LF3* tri-axial accelerometer card and a *MDA400CA* vibration card have been incorporated in each mote. The logical topology is a one-to-one routing scheme in which a hybrid error recovery method is established in order to achieve reliable transmission. This error recovery method combines hop-by-hop and end-to-end error strategies.

In the hop-by-hop error recovery strategy, when a package is lost during data communications, the affected node starts the error recovery mode. This strategy assumes that nodes maintain a buffer of already sent nodes for a window of time and consequently the affected node initialises a message exchange (see Figure 7.7) requesting a copy of the lost packet to be immediately sent to the previous node.

In the end-to-end error recovery strategy, when a package is lost during data communications, the base station detects this fact and starts the error recovery mode. This strategy assumes that the base station knows how many packages are yet to arrive and in case one does not arrive after a window of time, it assumes that the package has been lost and initiates a protocol for requesting the lost package to be sent to the source node. Note that the source side also required a buffer for already sent packages in order to implement this strategy.

On the one hand, hop-by-hop error recovery strategy is almost necessary for performance optimisations in which distributed sensor deployments with lost links of up to 30% are not uncommon. Otherwise, an important degradation in bandwidth

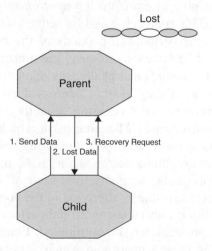

Figure 7.7   Hop-by-hop recovery protocol.

and data processing appears in the network performance mainly due to signalling and overheads caused by global retransmission. On the other hand, end-to-end error recovery can deal with topology changes, which could cause loss of missing packet list information. In fact, heavy packet losses can lead to large missing packet lists that might exceed the memory of the motes and it could also be addressed using this scheme. This is the main reason for which authors have decided to use both complementary smart in-network alternatives in the SHM deployment in order to achieve a reliable data package interchange.

Regarding data sampling, authors decided to reduce the messages sent through the WSN for only those in which samples exceed a certain threshold. In essence, the architecture is continuously sampling data and storing it locally during quiescent periods doing a local compression of the monitored information. When sensors have enough data to be submitted, the package is sent into the network. However, in case an event occurs in the structure, the packages have to arrive as fast as possible to the base station and for this reason, in this case, they are not compressed. It addresses the high data rates by means of the usage of data compression of the information reducing significantly the bandwidth required in data exchange.

Regarding time synchronisation, time stamping is clearly required in order to distinguish responses in a synchronous sampling. *Wisden* architecture relies on a light-weight time stamping process done consistently in the base station. While it requests some extra information in the package size, it does not entail global clock synchronisation. Each node calculates the amount of time spent by a sample at that particular node using its local clock. This amount is added to a *residence time* field attached to a packet. Thus, the delay from the time of generation of the sample to the time it is received by the base station (or any node) can be calculated. This time is stored in the packet as the sample travels through different nodes in the network. In summary, this is the time the packet resides in the network. The base station (or any node) can thus calculate the time of generation of the sample by subtracting the residence time. The main limitation of this approach is that it is impacted by clock drifts especially when residence times are long where timestamp can be significantly skewed. Moreover, clock drift can change the sample clocking, that is, individual samples may not be exactly 10 ms apart when sampling at 100 Hz. Authors analysed the impact of clock drifting in the proposed architecture. In essence, they stressed a system with 'artificial' sensing data in order to determine the average clock drifting in real testbeds. As a result, they found that the time drifts with *residence time* is approximately 10 ppm (packages per minute). If it is assumed that 10 ppm is a reasonable drift for the Mica-2s, this means that a sample can stay in the network for at most 1000 seconds (about 15 minutes) before its timestamp accumulates more than 10 ms error. This means that the buffering of messages done in the sensors in quiescence period encoding must not accumulate silence periods longer than a few minutes providing very reasonable results in the accuracy of the time synchronisation.

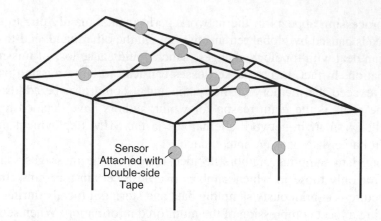

Figure 7.8   Ceiling structure used for the deployment done in Wisden.

(Chintalapudi et al., 2006) have deployed *Wisdem* architecture in two different real-world scenarios. The first deployed consists of 25 motes spread across all the floors of a three-floor building. The intention of this testbed was the reliability of the architecture proposed. Fifteen motes were programmed to generate artificial traffic in the network, the other 10 for routing and forwarding. Then, the 15 nodes programmed to generate traffic did so at one of the following rates: 0.1 packet/sec, 0.2 packet/sec, 0.25 packet/sec, 0.5 packet/sec and 1 packet/sec. Each packet was 80 bytes long and carried 18 samples. For context, a 1 packet/sec rate can enable the network to support readings from a structure that vibrates about 18% of the time (assuming no packet losses). In each run, every node sent out 200 packets, resulting in a total of 3000 packets per run. In all experiments, we achieved 100% reliability. After that, they also deployed a 10-node *Wisden* system on a test structure. This structure resembles the frame of a hospital ceiling, about 40 ft long and 20 ft wide as can be seen in Figure 7.8.

When completed, this structure had a shaker to impart forced vibrations, but it does not really suffer from the shaking. They instrumented this structure stressing by affixing the accelerometers with heavy-duty double-sided tape (on the advice of a local structural engineer), and wrapped the rest of the assembly with gaffer tape. They then repeatedly hit the structure for 20 seconds. The 10 motes formed a multi-hop network and transmitted all of the recorded vibration data back to base station within 5 minutes. The average residence time incurred by a packet in our experiment was 142 seconds; some of the delay can be attributed to the sustained excitation, and some to packet loss.

Another clear example of real-world deployment for SHM is shown in Figure 7.9. It depicts the distribution of the distributed sensing platform in the Golden Gate Bridge carried out by Kim et al. (2007). This infrastructure monitors the vibration along the whole infrastructure real to determine the response of the structure to

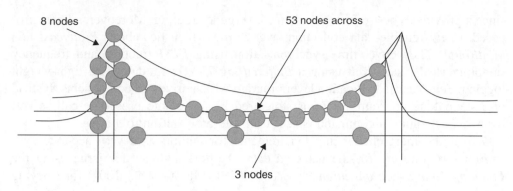

Figure 7.9   Sensor distribution in the Golden Gate Bridge to perform Structure Health Monitoring.

both ambient and extreme conditions. It enables the detection of possible structural damage due to earthquakes, seaquakes, or strong winds.

Sixty-four stationary *MicaZ* sensor nodes were deployed across the 4200 ft of the bridge collecting ambient vibrations synchronously at 1 kHz rate, with less than 10 µs jitter, and with an accuracy of 30 µG. Authors claim that it has been the largest WSN for SHM deployed until 2007.

The topology, due to the structure shape, shows that some sensors need more than 64-hops in order to research the base station. The current deployment achieves a 441 B/s bandwidth for the farthest nodes (those at 64 hops). Note that 1 kHz rate sampling is a strong requirement for the design of the architecture. To cope with this requirement, authors proposed the architecture depicted in Figure 7.10.

Figure 7.10 exposes a *TinyOs* architectural diagram in which a *Best-Effort Single-Hop* strategy for communications between nodes is assumed. The low latency dissemination service is achieved using *TinyOs Broadcast* module, which provides an unreliable dissemination service, with repeated broadcast, 100% eventual reliability can be achieved in practice. *MintRoute* module was used for information reply

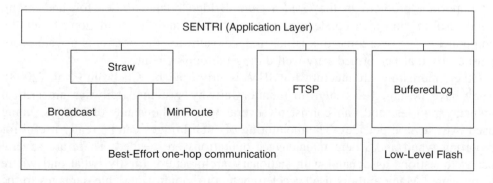

Figure 7.10   Logical stack used for monitoring the Golden Gate Bridge.

since it provides a best-effort multi-hop convergence routing. Moreover, authors proposed a new reliable data collection layer *Straw* which lies above *Broadcast* and *MintRoute*. They cover time synchronisation using *FTSP* module and frequency sampling is also achieved using a *BufferedLog TinyOs* module with light-weight logging. *Sentri* is an application layer program which drives all components. Instead of a standalone program, *Sentri* is structured like a remote procedure call server: for every operation a command is sent from the base station to a node.

One of the most challenging requirements for authors is a very accurate synchronous monitoring for the gathered data. Authors address this issue using the *Flooding Time Synchronization Protocol (FTSP)* (Maróti et al., 2004) for carrying out this synchronisation.

This is a high precise time synchronisation protocol robust to failures and flexible for topology changes using low bandwidth. This is a combination of other previous existing solutions for time synchronisation. It uses *Reference Broadcast Synchronization (RBS)* (Elson et al., 2002) at transport level (OSI level 3) which incorporates signalling messages in which timestamp is embedded in the messages. The main problem associated with this protocol is that it does not behave well in multi-hop networks. *FTSP* also uses *Timing-sync Protocol for Sensor Network (TPSN)* (Ganeriwal et al., 2003) which is a MAC layer (OSI level 2) time stamping using a spanning tree which requires knowledge about the parent neighbour. This fact makes this protocol not robust or flexible for topology changes. Then, *FTSP* joins both protocols using the TPSN MAC layer time stamping protocol and the RBS protocol for carrying out the skew compensation with linear regression providing the best of both protocols together. Finally, the dynamic topology management is achieved by means of a periodic flooding of messages.

In fact, Kim et al. (2007) require an even more precise time synchronisation than that achieved with *FTSP*. To do so, they modified FTSP in order to also deal with both spatial and time jitter providing a significant increase in the time synchronisation accuracy near to 10 µs jitter. Moreover, the authors provide a wide overview of issues related to a real-world deployment. For example, they found an important problem in the *Golden Gate* Bridge related to sea fog and strong wind resulting in quick condensation of salty water and fast oxidation of metallic components. To overcome this issue, the enclosure for the boards is a waterproof plastic box that performed very well during the deployment.

Other interesting architecture for SHM is proposed by Chebrolu et al. (2008). They have designed a system to monitor railway systems. Railways are critical in many regions, and can consist of several tens of thousands of bridges, being used over several decades. The monitoring of such bridges can be really useful for reporting when and where maintenance operations are needed. There are several design questions associated with stationary architecture. Firstly, what and where to measure? Many authors used accelerometers to monitor possible damages to the structure. This proposal also decides to use this kind of sensor. Moreover, the sensor

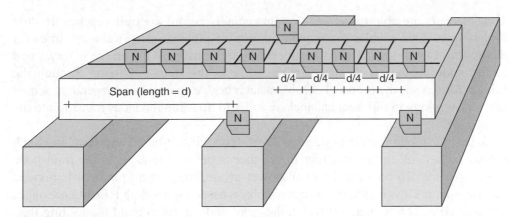

Figure 7.11 Deployment diagram of sensors proposed by (Chebrolu et al., 2008) to monitor railway bridges.

deployment is done according to the diagram depicted in Figure 7.11. A bridge span is defined as the longitudinal fragment of the road that holds over two transversal columns. Authors promote the distribution of four equidistant sensors along each span. Moreover, an extra sensors in the opposite part of the bridge in addition to extra sensors on the columns which gather a complete set of structural information about external events.

Secondly, when and how long to collect the data? *Forced vibrations* are defined to those causes by the train passing the infrastructure and *free vibrations* to those caused by its own structure once the train has already passed. Several research works have empirically determined that the frequency components of interest for these structures are in the range of about 0.25 Hz to 20 Hz. For 0.25 Hz, authors decide to monitor five time periods, equivalent to 20 seconds. The total data collection duration is thus about 40 seconds (20 seconds each for forced and free vibrations).

Thirdly, the time synchronisation is important to correlate the events monitored by the sensors. This proposal determines an accuracy of around 5 ms for time drifting (at maximum frequency, 20 Hz), so this is not an important issue since many synchronisation protocols already solved this accuracy.

Regarding quality of the measured data, authors used an accelerometer over-sampling at 400 Hz and the sampled information is averaged in order to eliminate noise in the samples. This accelerometer is sampling at 12-bit. So, assuming six sensor nodes per bridge span, the total amount of transferred data is: 3 channels $(x, y, z) \times 12$ bit $\times 20$ Hz $\times 40$ sec $= 692$ Kbits.

A notable and interesting fact of this particular architecture is the placement of a node at the train. This node is in charge of sending periodical advertisement beacons. The nodes at the bridge are *TMote sky* sensor nodes equipped with an external antenna. This fact enables an achievement of a significant radio range.

Thus, sensors are able to detect the train seconds before the train reaches the first bridge span. This enables the implementation of efficient sleep/wake up times for the nodes in the bridge. The data gathered by the sensors in the bridge is forwarded to the node located at the front of the train as the train is passing through the bridge. Moreover, the proposal considers each bridge span as an independent sensor network working in different channels in order to do a simultaneous transition to the train node.

Regarding routing, authors decided to go for a cluster-based approach in which a node gathers all the information from other nodes and sends it to the train node when possible. To do so and remain power-aware, the routing protocol is divided into two-phases, the neighbour discovery phase based on flooding HELLO messages and tree construction phase in which the head node starts to build the routing tree. The key factor here is that these phases are started only when any node in the network receives an advertisement message sent from the train node.

The architecture includes another head node located before the train reaches the bridge span; this enables fault tolerance against head node failures. One of the reasons to go for a cluster-based routing protocol is to reduce the transmission time to the train node. Note that the train is going fast and the connectivity time between train node and head node is very limited. In the empirical experiments, the proposed routing algorithm has enough performance for sending the information on time. The result of these experiments shows that 60 meters of connectivity are required at 30 Km/h against 140 meters required at 60 Km/h. To cope with higher transfer rates, the train node is equipped with an external antenna which enables a wide radio area and note that all the transmissions are done 360 meters before to react with the bridge span at maximum speed. So, this is a very prominent result which enables trains to travel at high velocity since connectivity is before and after the bridge span.

Ceriotti et al. (2009) propose another architecture for monitoring a historical building deployed in *Torre Aquila*, a medieval tower in Trento (Italy).This deployment entails specific features: i) heterogeneous sensors are available in the network, ii) the time of the phenomena of interest requires monitoring to span months or even years. In contrast, the systems found in the literature typically operate for at most a few weeks. iii) the ability to change the behaviour of the sensing infrastructure based on external inputs.

Authors use 3Mates! Motes developed by *Tretec*. *Environmental nodes* extended with simple analogue temperature, relative humidity and light sensors. *Deformation nodes* are extended with a dedicated *Fiber Optic Sensor (FOS)* providing a minimally-invasive solution with very high precision. *Acceleration nodes* are extended with an analogue ultra-compact tri-axial acceleration.

The base station requires a computing device with enough storage space and processing power to collect and store the monitored information by all the other devices. Moreover, this device must double as a gateway to interconnect with the

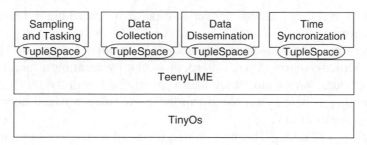

Figure 7.12 High-level overview for the architecture proposed by (Ceriotti et al., 2009) used to monitor the structure health of Torre Aquila, a historical building.

front-end, allowing remote users to interact with the system. The authors used *Gumstix*[1] device, easily customisable embedded PCs with a very small form factor.

Figure 7.12 shows the proposed architecture by (Ceriotti et al., 2009). It uses the framework *TinyOS* as a base for their development. *TeenyLINE* (Costa et al., 2007) is a free available abstraction for sharing memory (blackboard) between all nodes into a 1-hop distance where data is read and written as tuples, sequences of typed fields. *TeenyLINE* is used for carrying out the sampling and data collection of the monitored information which is stored in the shared memory.

The heterogeneity of the deployed sensors entails a great impact in the sampling requirements. Then, the usage of *TeenyLINE* acts also as homogenisation of the sampled data and also to the routing protocols in which reliability plays a critical role.

The authors identify two different kinds of traffic into the network. The *high-rate* traffic data and *strong reliability* requirements and the *low-rate* traffic with *weak reliability* requirements. The former is generally coming from the acceleration nodes. Note that this information has to be really accurate up-to-date information since it is critical for identifying damage in the infrastructure. For this kind of traffic, a local buffer is used to store information sending all this information in a *burst* after sampling. Authors use (Huffman, 1952) compression to carry out this send. Concretely, this is a differentiating point with *Wisden* approach because, this time (Ceriotti et al., 2009) use a loss-less compression algorithm to keep all the semantic richness of the vibration data. The latter generally come from environmental and deformation nodes. For this case, even if one sample is occasionally lost, a meaningful data analysis can still be carried out.

The routing protocol used by (Ceriotti et al., 2009) builds a tree topology rooted at the sink node. The tree is periodically rebuilt to account for connectivity changes. The process is performed by flooding a special control tuple. Each node re-propagates the tuple by writing a copy of it in the tuple space of every node within communication range. Moreover, since tuples are shared with 1-hop nodes,

---

[1] www.gumstix.com.

a hop-by-hop recovery scheme is very suitable for reaching reliability in the communications.

Regarding time synchronisation, authors use a modified version of TPSN protocol. The modified version of such protocol works by creating a hierarchy among the network nodes, whose clocks are then synchronised with the root's clock. Synchronisation then is based on a round-trip tuple exchange between nodes at level i and $i - 1$ in the hierarchy.

They deployed 17 nodes. This base station is placed on the top floor, the only spot guaranteeing access to the external Wi-Fi network. The sensor position is chosen to detect early symptoms of deterioration of the structure. The joint between the ancient parts of the tower and the more recent ones is today perfectly visible, but the degree of structural connection of this joint is still a major point of uncertainty. This is the reason for which authors decided to place the nodes in strategic places following this visual join of the building.

7.4 Monitoring Traffic

There is a continuous increase in the number of vehicles available in the cities. In fact, major congestion of cars is a daily reality for a vast majority of cities. Any problem is a vehicle on the road that may entail risks for either pedestrians or other vehicles. Thus, public safety architectures for monitoring vehicular traffic are critical scenarios in which several applications can be deployed. Vehicle speed monitoring and vehicle localisation monitoring, for example, can contribute to the provision of traffic information to the driver, to the police or even to other vehicles notifying dangers, accidents, and so on. Other applications are the adaptive control of traffic lights or the deviance of traffic to avoid congestion and to optimise the traffic queues or even the monitoring of the road conditions in order to warn about dangers. Building new roads is not a feasible solution for many cities since they have important space limitations and it entails a high cost of demolition for the old ones. Then, a solution comes with the use of distributed sensing systems in order to reduce congestion.

In this context, different architectures and technologies can be utilised to carry out the traffic monitoring. A much utilised architecture for these purposes is the vehicular ad-hoc network (VANET). Kiess et al. (2007) describe the main requirements associated with VANET with respect to other types of networks such as traditional WSN and Mobile ad-hoc networks (MANET). In essence, the following features are the differentiating keys of VANETS with respect to WSN.

Firstly, the communication pattern in safety applications and cooperative driving systems does not match that which is provided in WSN: the data needs to be distributed among all vehicles in a certain region rather than to one or more dedicated sinks. Secondly, although the primary purpose of WSN is collecting information, they are usually designed to transmit this information to one static sink, not to a large

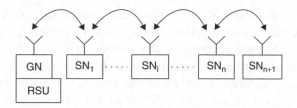

Figure 7.13 WSN architecture proposed in SAFESPOT for traffic monitoring (Franceschinis et al., 2009).

number of mobile nodes. Thirdly, Internet access is an application for vehicular networks. At least one gateway between the wireless and the fixed network is required. Capacity limitations in wireless multi-hop networks as well as latency bounds in fact enforce a large number of gateways. This leads to an infrastructure-based solution. Even if it were technically possible, there are no economical reasons to duplicate the functionality of existing (and already deployed) solutions such as UMTS by means of MANET or WSN. Finally, point-to-point communication is another possible scenario for chatting or file exchange that requires a connection between specific cars. Clearly such communication reaches far beyond the traditional WSN paradigm where unicast within the sensor network is not supported efficiently.

Despite the clear differentiating features between these technologies, this section deals with several scenarios related to traffic monitoring either using only WSN technologies or in combination with VANET networks.

The SAFESPOT (Franceschinis et al., 2009) project proposes a WSN architecture used for automatically generating safety warnings at black spots along the road network in order to *avoid collisions* preserving the public safety. The architecture is shown in Figure 7.13. It consists of one Gateway Node (GN) and n smart sensors (SNs) deployed along the roadside according to an approximately linear topology. Each SN is able to monitor a section of road, collecting parameters such as vehicle count, speed and direction. Data from the SNs is collected by the GN and delivered to a Road Side Unit (RSU) responsible for fusing it with traffic-related data generated by alternative sources. Note that the density of the number of sensors is directly related to the routing reliability since more neighbours means more routing alternatives in case of remote node failures or bad connectivity.

The GN behaves as a master node in a reactive network, polling the desired SN information with requests, while SNs play the role of slaves replying to the GN through data packets or acknowledgement (ACK) packets. The messages used in the network are classified as: *control messages* used infrequently and asynchronously to access and change SN configuration parameters and *data request messages* used in a pseudo-periodic fashion to query the node for traffic information.

Each SN is equipped with two sensors, an Anisotropic Magneto-Resistive (AMR) sensor for measuring the Earth's magnetic field and, consequently, its variation when vehicles pass by and a Pyroelectric (Pyro) sensor providing accurate temperature measurements; hence, a sudden variation witnesses the presence of vehicles as well as of human beings. Sensors are periodically sensing in order to detect vehicles in the road and collect this information, and then send it to the GN, when it is requested.

Although this architecture is merely reactive and there is no important in-network processing, it is a good example of WSN road side deployment. Moreover, it is worth mentioning that the routing algorithm used tries to forward a packet, first selecting the farthest reachable one-hop neighbour along the path to the destination. In case of a link failure, the next-hop destination address is decremented by one and so on. This protocol exploits data packets to keep the best one-hop neighbour as updated as possible. Then, the information is forwarded to the RSU, which can process the information and relay it to the distribution zone.

A smarter architecture composed of a hybrid WSN-VANET architecture is proposed by (Weingärtner and Kargl, 2007). The authors propose a road side distributed sensing platform deployment with constant availability and dense coverage across the road and also a VANET architecture which might have only sparse coverage. The distributed sensing platform constantly senses the road conditions and vehicle presences in order to avoid collision and unexpected race conditions. Then, the sensing platform communicates the sensed data to the vehicles driving on the road, delivering them with accurate and up-to-date sensor information. Vehicles communicate between them in order to disseminate this information over comparatively long distances. Such hybrid architecture is suited for all applications where a stationary node collects sensor data that is disseminated only on a small scale within the network, and then delivered to vehicles which transfer it to other regions by multi-hop routing or vehicle movement.

Authors proposed the scenario depicted in Figure 7.14. In essence, first, a smart sensor node detects ice on the road (road conditions) and shares this information with one-hop neighbouring motes (see 1) inside a small region. When a vehicle A enters this region, the stationary nodes trigger a vehicle-present event and the information is transmitted to the car (see 2). The driver of A would receive a short-term warning and can react accordingly. Vehicle A forwards the information via the long-range VANET to vehicle B (see 3). As the road splits at position B, the icy road information would otherwise become unavailable to vehicles approaching from the lower road. So B transfers the information back to the stationary nodes (see 4) and the nodes distribute it again in a small one-hop neighbour area for redundancy purposes (see 5). A and B leave this road segment (see 6) and are out of communication range, when vehicle C approaches the intersection and still receives the warning information from the stationary nodes (see 7). The vehicle displays a warning to the driver who has plenty of time to adapt his driving style to the approaching danger.

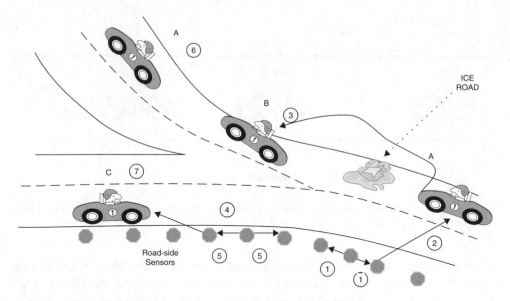

Figure 7.14 Scenario proposed by (Weingärtner and Kargl, 2007) showing how to use the sensing devices to avoid traffic collisions.

Note that Vehicle C is never connected to either vehicle A or B and also there is only a limited number of package interchanges between nodes in comparison to architectures with multi-hop routing protocols.

Festag et al. (2008) propose an analogous alternative to the (Weingärtner and Kargl, 2007) proposal. In this case, the distributed sensing platform uses a multi-hop routing algorithm. Each sensor monitors environmental data, stores the collected information and processes such information inferring if there is an alert situation in the road such as to stop a car, ice in the road, wild animals, and so on. The sensed information is stored with timestamp and geo-information. Then, the architecture only communicates the data to passing vehicles when an alert is inferred; otherwise sensors are sleeping in order to save power. When an alert is detected, this one is distributed across the network in order to prepare the entire network to set alert messages when vehicles are detected. The communication between vehicles using VANET is analogous to the (Weingärtner and Kargl, 2007) proposal.

Biswas et al. (2006) propose a V2V DSS for highway cooperative collision avoidance (CCA), primarily focusing on the Medium Access Control (MAC) and the routing layer. Authors describe a scenario in which three cars are driving in the same direction as shown in Figure 7.15. The first car has an accident. Then, the idea of the proposed architecture is to notify the other cars as fast as possible about the event in order to stop on time and avoid an imminent collision.

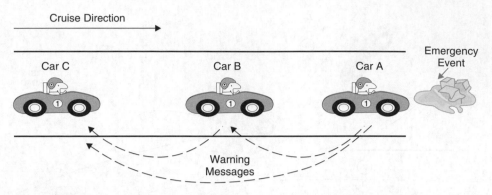

Figure 7.15 Car collision scenario in which a distributed sensing platform is used to avoid collisions (Biswas et al., 2006).

From the proposed scenario, two conclusions can be made. First, using a high-speed DSS, it is possible to design CCA systems that can improve highway safety by avoiding chain collisions. Second, reliable and fast warning message delivery is a crucial requirement for such CCA systems to be able to give leverage to the underlying networking infrastructure.

Then, when a vehicle meets an emergency situation, it needs to send an alert message to all cars behind. Since the identities of those prospective receivers may not be known a priori, classical unicast and multicast routing will not work. Authors propose an alternative routing approach in which the broadcast of the message is done periodically and then all its recipients selectively forward the message based on its direction-of-arrival. In case a recipient decides to forward the message again, it starts to do it periodically. Concretely, only vehicles receiving the package from the front of the vehicle start to decelerate immediately and to relay the package to other vehicles. This is called direction-aware broadcast forwarding. This mechanism ensures that the message will be delivered eventually to all the vehicles in collision range. Moreover, as there is a nonstop condition for the propagation message, it will be received by all the cars if somehow indirectly connected to the affected vehicle.

Note how this routing exposes excessive message forwarding, which leads to message collisions for MAC at layer 2. High MAC collisions reduce the message-delivery rate, and also increase the delivery latency, because successful delivery after message drops will have to rely on the periodic retransmissions from the event-detecting vehicle. To avoid these, authors introduce an implicit acknowledgement-based message generation and transmission strategy at the MAC Layer.

In summary, after starting the periodic broadcast, if an event-detecting vehicle receives the same message from behind, it infers that at least one vehicle in the back has received that message and will be responsible for propagating it along the rest of vehicles. In that case, the vehicle stops the broadcast avoiding unnecessary messages. In the case of such multiple receptions, a vehicle acts only upon the first reception. Moreover, authors assume that the DSS can be used for other purposes and then they assume the prioritisation of data from safety-related ITS applications over low-priority ITS applications.

(Piran and Murthy, 2010) describe an architecture optimised for carrying out traffic monitoring using a vehicular to roadside architecture. Each mobile sensor node is able to cover 500 meters which is a reasonable distance nowadays with current devices when power is not a critical issue since they are able to be powered by external power suppliers such as a car battery, solar panel, and so on. The roadside sensors do not have this possibility and they have to be designed to be more power efficient.

The routing protocol established an advertisement message periodically sent by the mobile node (vehicle) containing vehicle ID, driver license ID, vehicle speed and emergency status. This information is received by the roadside sensors once the vehicle is on the radio range. Only when the vehicle is exceeding the max velocity allowed or the emergency status indicates an emergency, the roadside node relay the message to other nodes in the multi-hop roadside network reaching the base station.

Authors keep some nodes in the network sleep mode to save power. The decision of whether to stay awake or in sleep mode is decided by a probability decided by the following factors: remaining energy in the sensor, a generated random number, the previous state of the node and the importance of the message. After some simulations and experiments, authors determined some application scenario in which the routing protocol can be suitable: vehicle velocity monitoring, vehicle positioning information and incident and accident reporting.

Another clear example of traffic monitoring is the adaptive control of traffic lights. Tubaishat et al. (2007) provide a distributed sensing platform for this purpose. They use Tiny OS-based sensors, each one equipped with two HMC1051Z magnetic sensors based on anisotropic magneto-resistive (AMR) sensor technology. The sensor receives one magnetometer sample for 0.9msec in order to perform the detection of vehicles in the land. Such magnetometers are turned off between samples for energy conservation.

Since the magnetic distortion depends on the ferrous material, its size and orientation, a magnetic signature is induced corresponding to the vehicle's shape and configuration. For detecting the presence of a vehicle, measurements of the (vertical) z-axis is a better choice as it is more localised and the signal from vehicles on adjacent lanes can be neglected. This way of detection entails that sensor nodes have to be placed vertically over the road.

The data at the sensor nodes is periodically transferred to a distinguished node called access point (AP) for purposes of control. The sensor nodes have limited (transmit) power and energy. Consequently, communication from nodes must travel over several hops to reach the access point and for this sensor and for a scheduling algorithm this 'distinguished' role is assumed homogeneously by the different sensors.

The authors have designed the architecture using a power-aware Medium Access Control (OSI Layer 2) called PEDAMACS (Sing Yiu Cheung, 2006). This MAC layer has been designed to meet both delay and energy requirements of traffic applications by exploiting the special characteristics of sensor networks. This is based on a scheme that discovers the topology of the network and keeps the nodes synchronised to validate the execution of a TDMA schedule. PEDAMACS protocol operates in four phases: the topology learning phase, the topology collection phase, the scheduling phase and the adjustment phase.

In the topology learning phase, each node identifies its topology information and its parent node in the routing tree rooted at the AP. In the topology collection phase, each node sends this topology information to the AP so, at the end of this phase, the AP knows the full network topology. In the scheduling phase, the AP broadcasts a schedule. Each node then follows the schedule. In particular, the node sleeps when it is not scheduled either to transmit a packet or to listen for one. The adjustment phase is included if necessary to learn the local topology information that was not discovered in the topology learning phase or that changed, depending on the application and the number of successfully scheduled nodes in scheduling phase. The determination of the schedule based on the topology of the network at the AP is performed according to the PEDAMACS scheduling algorithm.

Figure 7.16 shows a deployment done in a road intersection. Authors assume some restrictions to carry out the traffic light control. They propose four traffic

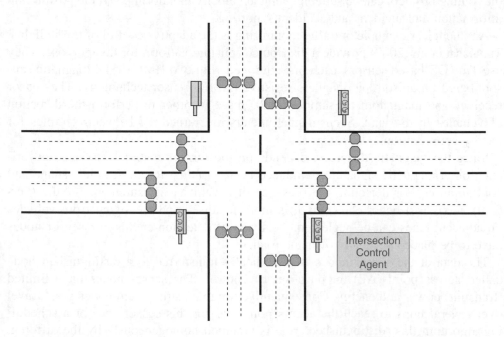

Figure 7.16   Road intersection configuration proposed by (Tubaishat et al., 2007).

lights in each intersection, each one responsible for controlling traffic in three lanes. Moreover, the right lane turns right only, the centre lane goes straight or left and the left lane goes left only. Note how the deployment is able to monitor vehicles before and after passing the road intersection. Moreover, the architecture also contains an Intersection Control Agent (ICA) in charge of controlling the four traffic lights.

Sensors send location, lane number and number of vehicles passed within t time to the AP. In fact, sensors located before the interception are positioned at a distance far enough from the traffic light to allow for the data to multi-hop to the intersection control agent, analyse the information and then send the result to the targeted traffic light. The following types of messages are shared between nodes.

- Message Sensor-ICA: Sensor nodes count number of vehicles approaching an intersection. Every node monitors one lane. The message sent from the sensor nodes to the ICA includes number of vehicles, time duration of the collected data, and lane number.
- Message ICA-ICA: ICAs can exchange information between them to improve the flow of vehicles in a wider area.

Regarding V2R architectures, Figure 7.17, there are mobile nodes available, WITS system (Chen et al., 2006) is an alternative for traffic monitoring architecture deployed at road intersections. This architecture is composed of three different types of nodes: the vehicle unit installed in every vehicle and monitoring the vehicle parameters; the roadside unit deployed along both sides of road (every 50~200 m

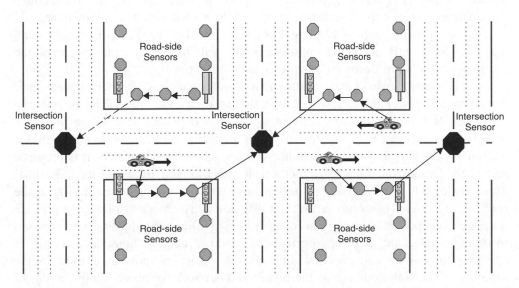

Figure 7.17 Deployment proposed by WITS system to monitor car traffic in cities (Chen et al., 2006).

according to the wireless cover range) and gathers the information of the vehicles around. Each unit gathers information about vehicular circulating in one sense. Then, they transfer it to the intersection unit, but an aggregation of the data is done before transferring them; and the intersection unit located in the centre of the intersection, which receives and analyses the information from other units, and passes them to the strategy sub-system.

The roadside units broadcast messages every second. The message includes the node ID, node location at the intersection. Normally, the vehicle unit is in the listening state. When a vehicle comes into the broadcast range of the roadside units and receives the broadcasted message, the vehicle unit switches to the active mode. Then, if a vehicle unit receives messages from more than three nodes, it can calculate the location (x, y) and velocity v (Stark and Davis, 2004). After that, the vehicle unit sends the information (x, y, v) to the roadside unit nearby. Then, the roadside unit collects and aggregates the data and sends it to the intersection unit which uses this information to adapt traffic lights accordingly.

The time-synchronisation done between nodes relies on TPSN protocol. Note that TPSN has a hierarchy of lower-level nodes and higher-level nodes, which is quite fit for the hierarchy of WITS nodes.

7.5 Case Study: Sink Location

Data gathering is a fundamental task of distributed sensing platforms. It aims to collect sensor readings from sensory fields at pre-defined sinks (without aggregating at intermediate nodes) for analysis and processing. Traditionally, the architectures are designed as reactive multi-hop networks in which sensors are in charge of sensing information and of sending the monitored data to the base station (sink node). Usually, only one static sink node is available in the network design and the routing protocols are designed over such an assumption.

Several research works have shown that sensors near a sink node deplete their battery power faster than those far apart due to their heavy overhead of relaying messages. Non-uniform energy consumption causes degraded network performance and shortens network lifetime. Recently, sink mobility has been exploited to reduce and balance energy expenditure among sensors. The effectiveness of this proposal has been demonstrated both by theoretical analysis and by experimental study. This fact entails the necessity to design new network architectures in which the role of 'sink node' is mobile across the different sensor nodes in order to provide homogeneous power consumption for the nodes. Different algorithms are proposed in order to select the most appropriate sink node to achieve homogeneous power consumption within the network. In case the reader is more interested in the mobility of the sink node across the distributed network for power optimisation, Li et al. (2010) provide a taxonomy and a comprehensive survey of state of the art on the topic.

This case study tackles architectures in which there are multiple possible sink nodes and the most appropriate has to be determined. This scenario often occurs when a moving user is awaiting some information to be delivered, for example, about hazards in the road. In this case, the selection of the most suitable roadside sensor node to act as sink node is determined based on the geographical deployment of the nodes along the road and the speed and direction of the mobile node in order to minimise the delay in the notification.

Figure 7.18 depicts the proposed scenario. The car equipped with a mote is driving across the road (MS–Mobile Sensor). There is a linear roadside deployment of sensors. These sensors are labelled as VS (Vice Sink) sensors since they can be good candidates for acting as sink nodes notifying the road conditions to the car.

Let us suppose that a maximum speed for a car in the road is around 120 Km/h. This means that the car travels 100 meters in 3 seconds. Note that the connection time between MS and VS is very short. Thus, the information interchange between VS and MS has to be sorted quickly. Moreover, the VSs have to be sparse along wide area zones and for this reason it cannot be assumed that there is a direct connection between VSs due to their separation. Moreover, the MS can be frequently out of any sensor range. Let's assume a few radio ranges separated from each VS.

Moreover, there is another set of sensors labelled as SN. They are sensors conceptually equal to VS but without direct communication (one-hop) to MSs. Note that VSs and SNs compose a platform in charge of sensing environmental information and processing it in order to detect hazards in the road. This platform is a multi-hop network.

In this scenario, two different possibilities arise. The first possibility is that the road side system has detected a hazard and it has to notify about imminent alerts

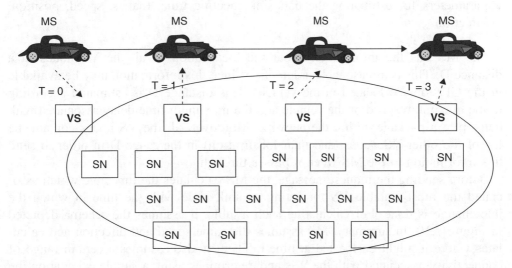

Figure 7.18 Proposed scenario for determining the sink node in a mobile sink node scenario.

to MSs, that is, an Infrastructure-to-Vehicle communication. In this case, once the alarm notification is generated in a given SN, this information can be flooded through the network notifying all the sensor nodes about this fact, especially the VS nodes. Then, VSs keep such information and as far as the MS is detected in the first VS, this sensor can directly notify such information to the MS really quickly.

A more sophisticated scenario arises when the MS is the originator of the information exchange, for example, requesting information about the road condition, whether there are traffic congestions, and so on. This would be, that is, a Vehicular-to-Infrastructure communication. In such a case, at least two message interchanges between MS and VS have to be done in a request-response fashion. In fact, an indeterminate elapse of time is required to gather the requested information from its source and deliver it to the MS again. Thus, it cannot be assumed that both request and response messages are exchanged with the same VS, mainly, due to the inherit velocity of the car. In this case, an in-network procedure to calculate the suitable sink node for storing the requested information and lately sending such information to the MS is required.

(Chen et al., 2009) propose an algorithm for calculating the suitable VS and for performing an optimised geographical routing of the request and responses messages in order to minimise the delivery latency.

To explain this approach, let us assume that at time $T = 0$, the MS is in the range of the VS_1 and send a request message. This message may include the following information:

- types of requested information (temperature, weather conditions);
- the timetable that maps the time slots and the VS;
- parameters for estimating the data-sink meeting time, that is, speed, position, direction.

The MS has the information about the localisation of all the VS nodes in a distance D. This is a reasonable assumption since this information may be available in the GPS device located in the vehicle. This distance D is estimated according to the maximum speed of the vehicle and the maximum time delay associated with delivering a package within the network. Moreover, all the VS nodes contain the list of the other VS nodes (and their localisation) in the range D in order to send this information to the MS when it passes through.

Before sending the request message, the MS calculates the timetable which associated the most suitable VS for acting as sink node with the time in which the interchange is done. For calculating such a table, it assumes the schema depicted in Figure 7.19. In summary, MS assumes the same speed and direction and calculates where it might be located at time t. Then, it also assumes a certain range of connectivity associated with the VSs and determines using a simple calculation the times in which the given VSs are reachable.

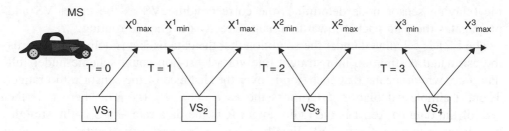

Figure 7.19   Schema used for calculating the timetable associated with VSs.

Thus, this timetable may contain entries in which a range of time is specified and the associated VS for such a range. For the example shown in Figure 7.19, the associated table is the following one:

$$(T_0, T_1, VS_1)$$
$$(T_1, T_2, VS_2)$$
$$(T_2, T_3, VS_3)$$
$$(T_3, T_4, VS_4)$$

Note that this architecture requires a time synchronisation protocol available in the network nodes. Any periodically synchronisation protocol can be used for this purpose such as FTSP.

The VS node receiving the message request needs to know where the source of the information requested is. For this purpose, several strategies can be employed, including dynamic service discovering, preloading service lists or relying on the routing algorithm for sharing this information along the network. This is out of the scope of this case study and from our point of view the source node ID can be achieved directly from the type of requested information.

The VS directly send this request message to the associated source node. When the source node receives the request, it is processed at the same time as gathering the information. In cases where the requested information requires more than one sensor interaction, the VS will determine the closest node to the geographical centre of the area which covers all the involved nodes in the request. Such a node is designated as a cluster node. The cluster node is the receiver of the request (analogous to the source node) and also it is the rendezvous point at which all the information provided by the rest of the sensors is gathered and aggregated if necessary.

Once all the information is achieved, the source node (or cluster node) knows the time elapsed from the original incoming of the request and the current moment. Then, before forwarding the packet, each relaying sensor calculates the moment of time when the data packet is estimated to meet the MS by adding the estimated propagation time (EPT) to the current time. According to the time frame calculated, the relaying sensor node determines the corresponding VS as the target VS and propagates the data packet towards it using greedy geographic routing.

The EPT is calculated under the condition that the data packet is delivered towards the estimated destination in a straight line via a least-hop count scheme under full data rate, which means that each hop is over the distance of the sensor radio range. Then, a simple calculation determines the $EPT = (d \cdot L)/(r \cdot R)$, where L is the size of the data packet; r is the radio range; R is the data rate and d is the straight line distance to the estimated VS. Finally, when the request arrives at the associated VS, it will be immediately sent to the MS as long as it is within the coverage range, receiving the requested information up-to-date.

The change of the *sink node* role between motes may require a complete re-initialisation of the WSN routing protocol or at least a re-adaptation phase. This re-adaptation phase can be considered as a sink location process in which motes need to discover a new sink node and recalculate the routing information. Note that sink mobility can cause unexpected changes of the network topology and routing paths. For a sensing architecture with mobile sinks, one of the challenges is to design scalable routing protocols to accommodate the sink mobility.

Several strategies can be proposed in order to perform the update scheme of the sink location. In this context, some scalable routing protocols have already been proposed in the literature. In case the reader is more interested, Al-Karaki and Kamal (2004a) provide a comprehensible revision of the topic.

7.6 Case Study: Congestion Avoidance

The congestion avoidance is another common technique that enables an intelligent routing of a time-critical message to ensure it reaches its destination with a low risk of being trapped in any congested part of the network. This is especially important for public safety scenarios in which network reliability is crucial to ensure that critical packages reach the destination. Data success rate of these critical networks is equally important as the energy efficiency and congestion in the network causes packet drop which reduces the reliability of data transmission.

There is a wide range of research work done in congestion avoidance in WSN. In fact, Flora et al. (2011) provide a comprehensible and complete overview of the current congestion avoidance protocols in WSN.

Usually, a distributed sensor platform requires information exchange between sink node and regular nodes. The downstream traffic from the sink to the sensor nodes is usually a one-to-many multicast package whereas the upstream traffic from sensor nodes to the sink is a many-to-one communication. Regarding the upstream traffic, it could be catalogued, according to Wang et al. (2007), in event-based, continuous, query-based and hybrid communications. The event-based communication occurs when any event is monitored in the sensor. In the continuous communication, the sensor sends the monitored information to the sink node periodically. The query-based communication is reactive, and thus, the sensor only sends information when it is requested. Finally, a combination of them can be done simultaneously.

Due to the convergent nature of upstream traffic, congestion is more probable in the upstream direction. Two types of congestion could occur. The first type is node-level congestion caused by a buffer overflow in the node, which can lead to packet loss, and increased packet delay. The second type is the link-level congestion caused by multiple active sensor nodes trying to seize the channel at the same time. This kind of congestion increases packet service time and decreases both link utilisation and overall throughput. Both node-level and link-level congestions have direct impact on energy efficiency and QoS of the network.

There are two general approaches to control congestion: network resource management and traffic control. The first approach tries to increase network resource to mitigate congestion when it occurs. The second approach controls the traffic in the network to mitigate the congestion through adjusting traffic rate at source nodes or intermediate nodes. According to their control behaviour, there are two general methods for traffic control: end-to-end and hop-by-hop.

The first one determines an exact rate adjustment at each source node and simplifies the design at intermediate nodes; however, it results in a slow response and relies heavily on the round-trip time. The hop-by-hop congestion control has a faster response, but it is usually difficult to adjust the packet-forwarding rate at intermediate nodes.

This case study is focused on the Congestion Detection and Avoidance (CODA) protocol, designed for avoiding congestion in upstream traffic. This is congestion control protocol composed of three different modules. Firstly, a *congestion detection* module is in charge of providing an accurate congestion status of the network. To this end, *queue management* is often used in traditional data networks for congestion detection, however, when link-layer does not manage acknowledgements (not necessary for many smart sensor scenarios) buffer occupancy or queue length cannot be used as an accurate indication of congestion. Moreover, in CSMA networks it is imperative to listen to the channel, hence, a straightforward method is to trace the channel busy time and calculate the local channel loading conditions. This *channel loading* gives accurate information about how busy the surrounding network is, but it has a limited effect in detecting large-scale congestion caused by data impulses from sparsely located sources that generate high-rate traffic. Finally, another way for detecting congestion is the *fidelity measurement*. In essence, the sinks expect a certain sampling rate or reporting rate coming from the sources. This rate is highly application-specific, and can be seen as an indication of event fidelity. When a sink consistently receives a less than desired reporting rate, it can be inferred that packets are being dropped along the path, most probably due to congestion. The usage of three metrics provides accurate information about the congestion status and they are used in the other modules of this protocol to provide efficient congestion avoidance.

The second module is the *hop-by-hop congestion avoidance*. In essence, each node is measuring the congestion metrics and thus detecting congestion in the network. A node broadcasts a suppression message when it detects congestion based on channel loading and buffer occupancy. This node will continue broadcasting the message up to a certain maximum number of times with minimum separation as long as congestion persists. The message is propagated upstream toward the source to the sink node. Then, nodes receiving such signals regulate their sending rates and also determine whether or not the backpressure signal continues to be propagated (to the next hop), according to the network conditions.

This *hop-by-hop congestion avoidance* module uses a buffer queue size metric since it does not entail any overhead and complex computation. However, channel condition monitoring is a challenge since it entails that the radio is always at least in receiving mode and it is against the principles of power saving. Thus, authors propose the usage of this metric only when a node is receiving a packet and needs to forward it. In this situation, there is something to send and listening to the media does not imply any overhead since it is already an imperative fact. Moreover, authors define the term *depth of congestion* to indicate the number of hops that the backpressure message has traversed before a non-congested node is encountered. This metric can be used as a metric for congestion-aware (dynamic) routing protocols or also to determine smart routing like dropping advisement routing packages, and so on., to prioritise the important data packages whereas the congestion is minimised.

Note that this *hop-by-hop congestion avoidance* module is the primary fast time scale control mechanism when congestion occurs. However, there is a necessity of control congestion over multiple sources from a single sink when congestion is persistent. Moreover, the sink plays an important role as a 1-to-N controller over multiple sources. Thus, authors provide a third *multi-source regulation* module. In normal situations, sources would regulate themselves at predefined rates without the intervention of sink regulation and consequently this module is deactivated. However, when the source event rate exceeds a certain threshold which is a fraction of the theoretical maximum throughput, then this source is more likely to contribute to congestion and therefore this *multi-source regulation* module is triggered. In essence, a source requires constant feedback from the sink (ACK) to maintain its rate. A source triggers sink regulation by setting a regulate bit in the event packets it forwards toward the sink. Reception of packets with the regulate bit set force the sink to send ACKs to regulate all sources associated with a particular data event. ACKs can be sent either in an application-specific manner, for example only along paths it wants to reinforce or globally in order to do a global congestion control. The reception of ACKs at sources serves as a self-clocking mechanism allowing the sources to maintain the current event rate. Finally, when the source reduces the value of event rate below the threshold established, this module is deactivated going to normal operations again.

ACKs can be lost when forcing sources to drop their event rate, according to some rate decrease function. Moreover, the sink can stop sending ACKs based on its view of network conditions. Note that a sink is capable of measuring its own local channel loading and determining if it wants to stop sending ACKs to sources. This module used the *fidelity measurement* metric at sink node in order to determine if a certain reporting rate is consistently less than the desired reporting rate, inferring that packets are being dropped along the path due to persistent congestion. This control avoidance module enables a multi-source regulation in order to provide long-path congestion control.

As a result, a complete congestion control is done providing a combination of hop-by-hop and end-to-end congestion protocols. This protocol produces a fast and efficient congestion protocol based on broadcasting congestion status and is the first wall of defence. Moreover, the automatic regulation of all source nodes when necessary with selective regulation acting only over those nodes is more likely to contribute to congestion in the network.

7.7   Case Study: Target Tracking and Surveillance

Target tracking is a non trivial application of wireless sensor network which is set up in the areas of field surveillance, habitat monitoring, indoor buildings and intruder tracking. Tracking an object moving through a field of sensors has received considerable attention in recent years and intended solutions can be mainly classified, according to (Bhatti and Xu, 2009) into five schemes, which are: tree-based tracking, cluster-based tracking, prediction-based tracking, mobicast message-based tracking and hybrid methods. If the reader is more interested in target tracking protocols for DSS, Bhatti and Xu (2009) provide a taxonomy and a complete survey.

Moreover, target tracking can be classified as attending to different angles. On the one hand, there are target tracking schemes specifically designed for tracking of individual targets, for example people, animals, vehicles, and so on. On the other hand, there are target tracking schemes specifically designed for tracking the spread of continuous objects or phenomena, such as wild fires, toxic gases, and so on.

Continuous objects differ from multiple individual targets in that they are continuously distributed across a region and usually cover a large area rather than discretely distributed over the area of interest. Moreover, they have a fundamental characteristic since they tend to increase in size, change in shape, or even split into multiple relatively smaller continuous objects over time.

In this case study, a cluster-based DSS target tracking algorithm specifically designed to monitor continuous objects has been described. This algorithm is named Continuous Object Detection and Tracking Algorithm (CODA) and it is proposed by (Chang et al., 2008). Static clusters are those initially established which are fixed in size, covered area, members, and so on. being not suitable for tracking objects since it cannot be adapted to this highly dynamic scenario, whereas dynamic clusters continuously change such parameters in order to use the cluster shape to track the object. However, dynamic clusters can incur high communication costs if the tracked object moves rapidly across the sensing field since the sensors are required to exchange messages more frequently in order to construct appropriate dynamic clusters. Thus, CODA applies a hybrid static/dynamic clustering scheme to achieve continuous object detection and tracking with low communication overheads.

Regarding the management of the static clusters, CODA designated a number of static clusters which are constructed during the initial network deployment stage. Authors do not explain the way in which such static clusters are generated leaving it open to the usage of any reactive or proactive cluster-based protocol such as LEACH, PEGASIS, and so on. When sensors are joining a given cluster (CH), a message will be exchanged between them. First, the sensor sends its localisation to the CH and the CH replies with the cluster ID establishing the association of the sensor to such a cluster. Then, when the CH do not receive a new register request, it calculates those sensors which are located in the boundary area of the cluster. To do so, authors use the Graham scan (O'Rourke, 1998) algorithm used to solve the convex haul problem, that is, to find a set of n points $S = [P_1, P_2, \ldots, P_n]$ so that it is defined as the smallest convex set of points in the space which contains S.

Firstly, the Graham scan determines P_0 as the point with the lowest y-coordinate available in the set of locations of sensors (if more than one point exists with the same minimal y-coordinate, the point with the lowest x-coordinate out of the candidates should be chosen). Then, the set of points must be sorted in increasing order of the angle they and the point P make with the x-axis.

Sequentially for each point in the set it is determined whether moving from the two previously considered points to this point is a 'left turn' or a 'right turn'. Let us suppose a 2-D space in which the first three points are defined as $P_0 = (x_0, y_0) P_1 = (x_1, y_1)$, $P_2 = (x_2, x_2)$. Then, if $(x_1 - x_0)(y_2 - y_0) - (y_1 - y_0)(x_2 - x_0)$ is positive, P_2 is a 'left turn', otherwise it is a 'right turn'. If it is a 'right turn', this means that the second-to-last point (P_1 is the previous example) is not part of the convex hull and should be removed from consideration.

Once the set of boundary sensors are determined, they are notified about this fact status. Figure 7.20 depicts a running scenario in which five clusters are statistically determined and the nodes notified as boundary sensors are represented in a different colour from those non-boundary sensors and from those defined as cluster heads. Moreover, the same picture represents the target object being monitored, for example, a fire, which in this case is spread along two different clusters: 3 and 4.

Two control messages are used to transmit the detection information to the CH whenever the object is detected. A *SENSE* message, used only by the non-boundary nodes, is sent to the CH when the sensor detects the target object. The reception of this message implies that the CH will request a *REPORT* message from the boundary nodes.

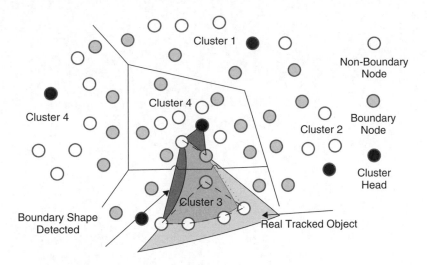

Figure 7.20   Sample scenario for tracking an object using a cluster-based approach.

A *REPORT* message, used only by the boundary nodes, is sent either if it is requested by the CH (due to a target detection initiated by a non-boundary sensor) or when the boundary node detects the target. This is composed of a bitmap in which the nth bit is set to 1 if the cluster neighbours of such a boundary node also detect the target. The boundary node communicates with all of its one-hop neighbours in other clusters to query their object detection information and fill such a bitmap.

Then, when the information of all the report MESSAGES is sent to the CH, it is able to determine all of the static clusters within which the target object has spread (at least all directly connected to the given cluster). After that, the CH applies again a Graham scan considering only the set of points associated with those sensors available in the cluster detected the target. This result provides the boundary shape of the traced object within the cluster zone. In Figure 7.20, the detected shape performed by each cluster using discontinuous points to identify such shapes are shown.

Note that this algorithm generated a *dynamic* sub-cluster composed of the boundary sensors which detect the target object. *Dynamic* in this context is referred to as the possibility of changing the state of being a boundary sensor or not for a given target object.

Finally, all the CH send both the list of adjacent clusters in which the object has been detected and the list of sensors (and their locations) within the cluster which constitutes the shape of the tracked object inside of the sink node in a compressed way. Then, by compiling the integrated boundary information received from each of the CHs in the network, the sink is able to determine the entire boundary of the continuous object. This inferred boundary shape is also represented in Figure 7.20 in order to enable the reader to decide the difference between the real object and the detected shape.

8

Geographical Applications

There are a vast number of applications of distributed sensing platforms in scenarios located outside of buildings in which there is not any prebuilt networking infrastructure at all. This chapter covers many of these applications and scenarios. Firstly, the application of distributed sensing platforms into the farming industry is described covering scenarios such as vineyards, sugar farms and orchard monitoring and management. After that, the special requirements and designs of sensing platforms in order to work correctly in holistic scenarios such as the coal mining industry are also presented. Transportation is another direct application scenario covered in this chapter in which different applications such as maintenance of cold chain and vehicular health monitoring are exposed. The application in earth resource monitoring is also addressed, covering scenarios like volcano monitoring or dynamic spatial fields. Finally, the usage of distributed sensing platforms for underwater applications and its challenges associated directly to the transmission of signal in this medium is also described.

8.1 Farming Industry

The importance of using sensors in agriculture was well understood a long time ago. Even the application of primitive humidity sensors could have been facilitating the work of a farmer by directing his efforts to particular areas of the plantation that are in need of water. More sophisticated sensor devices with today's capabilities of chemical and biological sensing, not only help in the field maintenance process, but also drastically increase the quality of the final product. Several processes can be improved in the farming industry due to the usage of a distributed sensing platform. Good examples come from *precision agriculture*. Traditionally, a large farm has been considered as a homogeneous field in terms of resource distribution. This includes seeding, water resources, herbicides, pesticides, fertilizers, and so on. Then,

Distributed Sensor Systems: Practice and Applications, First Edition.
Habib F. Rashvand and Jose M. Alcaraz Calero.
© 2012 John Wiley & Sons, Ltd. Published 2012 by John Wiley & Sons, Ltd.

treating it as a uniform field can potentially cause an inefficient use of resources and consequent loss of productivity. Thus, *precision agriculture* is informally defined as a method of farm management that enables farmers to produce more efficiently through a frugal usage of resources. This type of farm management requires a set of technologies for helping in the following processes: yield monitors, yield mapping, variable rate fertilisers, weed mapping, variable spraying, topography and boundaries, salinity mapping, guidance systems, and so on.

One good example is the application of distributed sensing platforms in *vineyard yield monitoring*. Temperature is the predominant parameter of *vineyards* since it directly affects the quality as well as the quantity of the grapes. Grapes see no real growth until the temperature goes above 10 °C and also different grapes have different requirements for heat units. Then, a distributed sensor deployment could aim to measure the temperature over a 10 °C baseline that a site accumulates over the growing season. (Burrell et al., 2004) deploy a distributed sensing platform to monitor and characterise variation in the temperature of a wine vineyard.

This application scenario provides some differentiating features which may require the design of a specific sensing platform. A vineyard is accessible and farmers and even family dogs are usually moving up and down across the rows of the yield. Thus, authors realise that any of these moving bodies (even the dogs) could serve as a *data mule* by carrying a small device that invisibly gathers data wirelessly from the static, distributed motes. This design principle does save on equipment costs because motes can be distributed more sparsely throughout the vineyard, as they don't need to communicate with neighbouring motes.

Moreover, vineyard managers do not need to calculate heat units immediately and a latency of a few hours or a day is suitable. This requirement relaxes the need of live-data and emphasises the usage of in-network data processing. Note that architectures based on a pure *data mule* model do not provide any power efficiency since it requires motes still in listening mode to detect *data mule* motes. The motes in listening mode consume a considerably higheramount of energy than sleep time and this *data mule* approach requires the minimisation of the usage of sleep mode in order to avoid disconnection times in which the data mole cannot get the data even if it is in range. To minimise the power hunger of this design, Burrell et al. decide to sense the temperature information every minute in all the motes and store the high and low temperatures of the day. Then, the only information transmitted via RF is these values which can be transferred via infrequent activation of the RF listening motes to detect any data mule in range.

The network is physically deployed in a real vineyard composed of 65 nodes deployed in a grid-like pattern covering about two acres. The deployment of the sensors was relatively easily done by one person in a day. The reasons are directly associated with the self-configuration nature of the distributed sensor network together with the inherent structured layout of vineyard fields. The main requirements identified for this application scenario are the placement of nodes in an area of viticulture interest.

Another important factor in *yield monitoring* is the water quality. The water quality has a direct impact on the productivity of the fields. This fact is especially important for the cases in which the water used is extracted from underground. Thus, the deployment of a structure for monitoring the underground water salinity can be a direct application scenario. Hu et al. (2010) deploy a platform for such a purpose in a Sugar Farm located at the Lower Burdekin region of Queensland, Australia. For this deployment, authors decide to use Fleck3 motes (Sikka et al., 2004). These are motes using radio devices working in 915 MHz (Industrial, Scientific and Medical band) with bitrates of around 100 Kbps. The main reason for this choice is the imperative necessity to provide a system for sensing wide coverage areas and these motes are able to perform communication links up to 1000 meters using standard unity-gain quarter-wavelength antennas. The sensors used for water quality monitoring attached to the Fleck3 motes are: a) Sensorex TCS1000 for monitoring the salinity; b) Tyco PS100 pressure sensors for monitoring the water level; and, c) Krohne electromagnetic, a flow meter sensor.

Hu et al. (2010) use TinyOS for implementing the distributed smart processing in the sensor. At the transport layer (OSI Layer 4), authors decide to use an end-to-end Negative Acknowledgement (NACK) with aggregated positive Acknowledgement (ACK) mechanism. The reason is that there is a sparse (not dense) distribution of the sensors across the monitored area. This fact can cause packet drops. Thus, this transport protocol enables the base station to detect missing packets.

Regarding network layer, authors chose the well known sensor network routing protocol MintRoute (Woo et al., 2003). MintRoute is a reliable multi-hop routing protocol that uses link quality as its routing metric. The performance of MintRoute has been shown to be superior to other routing protocols (including shortest-path, DSDV, AODV) in unreliable wireless environment. The main problem associated with the usage of this protocol is that Fleck3 sensors do not provide link quality or even RSS (Received Strength Signal) information. Then, authors decide to estimate link quality from the statistics of received beacon messages and snooping.

Regarding the MAC layer, the long communications links are exacerbated and lead to high packet-loss rates. Thus, authors decide to use hop-to-hop packet recovery which has been demonstrated to greatly increase end-to-end reliability in sensor networks (Woo et al., 2003). Then, the authors implemented a Carrier Sense Multiple Access (CSMA) style Medium Access Control (MAC) with acknowledgement.

Eight sensors were physically deployed in the land following the real topology shown in Figure 8.1.

Most of the sensing nodes are AC powered since this is available at the pumping sites to operate the pump. The rest of the sensors use a large solar panel and car battery to operate. From Figure 8.1, we can see that the average distance between nodes is around 800 meters. The black arrows depict the sense in which data is flooded

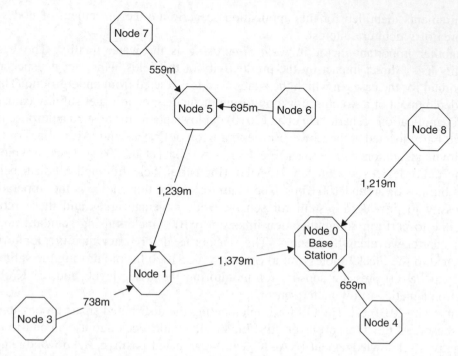

Figure 8.1 Real deployment done by (Hu et al., 2010) used for monitoring underground water salinity.

to the base station. Gathering empirical results, the authors determine that the network keeps this topology more than 70% of the time using MintRoute protocol.

Note how almost all of the nodes choose the geographically closest node as their parent node. Moreover, the authors observed the receipt of NACKs with substantial delay in our deployment. Note that almost all the protocols such as *MintRoute* assume that data is routed in one way. Consequently, nodes only store upstream paths in their routing table and use broadcast for downstream traffic. Therefore, while the upstream traffic can be delivered efficiently, it is very inefficient to deliver downstream traffic, for example ACKs and NACKs.

Yang et al. (2011) have also proposed another application scenario in precision agriculture. They proposed a distributed sensing architecture for monitoring agro-meteorological parameters which plays an important role in orchard management. The architecture is designed and developed for remote real-time monitoring and collecting of soil and atmospheric data. The proposed architecture is divided into several wireless sensing clusters and one routing cluster. Functionally, each sensing cluster is a wireless sensor sub-network which is made up of sensing nodes and one sink node. Each sensing node is equipped with sensors for monitoring air temperature, wind speed and direction, precipitation, atmospheric pressure, air humidity and global solar radiation. Note that the more sensing nodes in a sensing

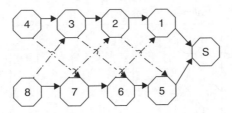

Figure 8.2 Double-chain network topology.

cluster, the faster the energy consumption of sensing nodes closer to the sink node. Thus, the authors decide to use a 4-bit encoding scheme so that a sensing cluster includes 16 nodes at most.

Sensors within the cluster have to be logically grouped according to a routing protocol. In this case, authors decide to use a double chain network in which each node maintains a double chain topology with their neighbours. An example of the double chain network topology can be observed in Figure 8.2 in which S is the sink node and numbers labelled from 1 to 8 are sensing nodes within the cluster.

This architecture is fabricated in the laboratory and tested in Yang's campus orchard. A sensing cluster is made up of eight sensing nodes encoded by $1 \sim 8$ and one sink node by 0. Base station connects with a sink node directly through RS232. The testing result shows that the proposed scheme has the capability of robust data transmission and high sensing accuracy of micro-meteorological parameters.

The *habitant monitoring* is another direct application of the farming industry in order to monitor animal behaviour and to increase the productivity of the farm. For example, an application scenario is monitoring in poultry farms. Distributed sensing platforms can provide important benefits since the most vital climatic factors for the productivity of a poultry farm are temperature and humidity and their maintenance. For example, a one-day-old chicken requires 33 °C at a relative air humidity of 50%. If the outside temperature is 24 °C and the air is headed straight into the zone occupied by the nascent birds without being heated first, then the weather would be too hard to resist. Then, heating solutions are usually provided along with appropriate ventilation. In this context, Murad et al. (2009) deployed an architecture for automating the monitoring of these climatic parameters. The favourable poultry farm climate adjustment, especially during the summer and winter seasons, can help to improve the productivity of the farm and economise on the energy usage.

The architecture uses a WSN only for monitoring the values, which are sent to a central control unit in which they are analysed to determinine the different actions to be performed over the heating thermostat and ventilators available in the room. Authors use CrossBow and TelosB motes which can sensor both temperature and humidity. The deployed network uses the multi-hop distance vector routing protocol Collection Tree Protocol (CTP) for performing the routing across the motes. The sensors were physically deployed in the N-W.F.P. Agricultural University's research

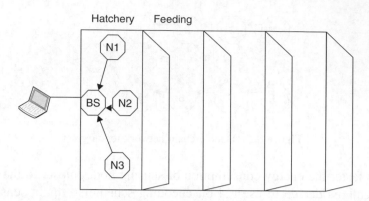

Figure 8.3 Sensing platform deployed in a poultry farm proposed by (Murad et al., 2009).

poultry farm in Peshawar in the North-Western Frontier Province of Pakistan. The size of the poultry farm is 30×90 meters. It is divided into four zones (see Figure 8.3): hatchery, feeding, egg production and multiplier breeder flocks zone which provides broiler hatching eggs.

The motes are deployed in a star topology in each of the four different zones of the poultry farm. Basically, three motes are deployed in each zone, measuring climate variants and communicating directly with the base station node for data relaying. One base station mote acted as a coordinator and received the measured data from the sensor motes. Authors decide that the wake time for a given mote is 10 seconds and sleep time is 600 seconds (10 min). Then, only one of the four motes equipped with temperature and humidity sensors is reading data from the sensors and transmitting it to the base station mote. This technique avoids packet collisions and distributes the power consumption homogeneously across the motes.

Another application scenario is the habitat monitoring of *hog farms*. Many hog farmers are suffering from high pig mortality rates due to various wasting diseases and increased breeding costs. A systematic and scientific pig production technology can increase productivity and produce high quality pork. Hwang and Yoe (2010) propose a distributed sensing system to collect environmental and image information not only in monitoring the hog farm but also to facilitate the control of hog farm facilities in remote locations.

The proposed scenario entails particular requirements to be analysed in order to provide an effective solution. The temperature in the hog farm should be set at a level the pigs find comfortable. It entails methods for comparing monitored values with those provided by external thermometers, methods for mitigating temperature variations controlling the speed of the ventilation fans. The sensory temperature of the pigs is more important than the absolute temperature available at the hog farm, and in turn, it varies according to absolute temperature, floor material, humidity, air flow speed and radiation from heating lamps.

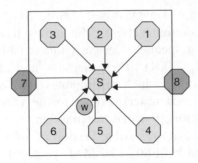

Figure 8.4 Hog farm sensor distribution proposed by (Hwang and Yoe, 2010).

Then, the authors designed a platform composed of motes with the SHT71 temperature and humidity sensor and the GL5547 luminance sensor attached and a TEKCELL 5.6 battery. The mote is provided with a CC2420 RF chip which enables ZigBee protocol stack. The CO_2 level is measured externally by a wired sensor already placed in the hog farm.

The architecture is deployed in a hog farm located at Seo-myeon, Suncheon-si, Jeollanam-do, Korea. This hog farm is composed of two different hog farms. Only one of these hog farms is monitored using the WSN architecture together with processing software, which is in charge of automating the fan speeds and air conditioner temperature. Figure 8.4 shows the sensors distribution of the distributed sensor system in the hog farm. Note that sensors 7 and 8 are located outside the hog farm monitoring external values whereas sensors from 3 to 8 are located inside the farm. Moreover, the wired sensor and the base station are localised in the centre of the farm trying to do a star deployment topology.

After the deployment, authors analysed the data and the user maintained hog farm is in very good condition from 06:00am until 20:00pm because the user directly controlled the facility, but a sudden variation in temperature and humidity inside the hog farm occurred from 20:00 until 05:00 the next day when the user did not control the facility directly himself. This sudden variation of temperature and humidity in the hog farm will produce stress on the pigs, if the level of the stress is serious, then it may lead to the pigs dying or cause a disease such as a cold. On the other hand, the hog farm with the WSN is always kept in the normal boundary levels for humidity and temperature since it is automatically regulated from the constantly monitored information.

Habitant monitoring is not only limited to in-land monitoring but also can be useful at in-water scenarios such as *fish farms*. Traditionally, a very specialised staff measures periodically the living conditions of the fish farm (pH, temperature, NH4+) with a high accuracy in order to keep the fishes' living environment in the best condition. However, this periodical test cannot detect immediate changes. Thus, López et al. (2010) propose a architecture for carrying out this monitoring process automatically.

The motes used are the TinyOS-based *TMote Sky*, a compliant device for IEEE802.15.4 wireless communications. The motes have attached a pH and a NH4+ sensor based on a specially designed ion sensitive field effect transistor (ISFET) (Errachid et al., 2004). These sensors are cached using a specially designed underwater module, which incorporates the sensors. The transmission of sensed data from the underwater modules to the wireless node is based on a proprietary polling asynchronous wired transmission.

Authors decided to use a power-aware version of AODV distance vector-based routing algorithm over the IEEE802.15.4 MAC protocol. Note that AODV works as a reactive routing algorithm. The notes are stationary and deployed at different places in the fish farm, so once the route has been established, the rest of the transmissions for a given source node will follow the same path, and then, the discovery route should only be done when the network wakes up for the first time and in exceptional circumstances when a node falls due to battery problems.

The deployment done in the fish farm is composed of 60 nodes homogeneously spread across the farm. The maximum number of hops to the base station is 6. The base station requests the monitored data every 5 minutes. Notes are able to sleep during idle communications periods. The proposed architecture is able to periodically monitor automatically the pH, NH4+ and temperature available in the different places of the fish farm and to send it to the processing software.

Due to uneven natural distribution of rainwater it is crucial for farmers to monitor and control the equal distribution of water to all crops in the whole farm or as per the requirement of the crop. The usage of distributed sensing platforms for automatic environment monitoring and controlling the parameters of the greenhouse may deal with a significant increase of the product quality. Chaudhary et al. (2011) propose a platform for carrying out the monitoring of greenhouses. They propose the usage of three different types of sensors: outdoor climatic monitoring, indoor climatic monitoring and soil sensors. Figure 8.5 shows the proposed layout of such sensors. Two hundred nodes are sufficient if the size of greenhouse is 35 m × 200 m. This is the physical size of the targeted area. A maximum of two nodes is more than enough for outside climatic monitoring. Regarding inside climatic monitoring, sensors can be placed at a distance of 10 to 15 meters in diameter, to capture precise environmental condition. Soil sensors, which are recommended for use, as per the layout plan of the crop plantation. They can also control waterflow of an irrigation system used in greenhouse.

Outside sensors sense wind flow, wind direction, ambient light, temperature, ambient pressure, humidity and percentage of CO_2 whereas inside climatic sensors monitor ambient light, temperature, ambient pressure, and humidity and CO_2 percentage from the inside of the greenhouse. Moreover, soil sensor nodes are specially designed to monitor the soil conditions like humidity of soil, temperature, pH value, and electric conductivity of a soil.

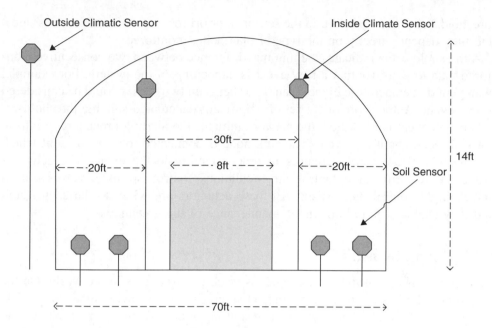

Figure 8.5 Greenhouse sensor deployment proposed by (Chaudhary et al., 2011).

Note that growth of plants depends on the photosynthesis process, which is a measure of photo synthetically active radiation. It is observed that proper temperature level influences the speed of sugar production by photosynthesis radiation. Thus, temperature has to be controlled since it is a critical component for the product quality. Water vapour inside the greenhouse is one of the most significant variables affecting the crop growth. High humidity may increase the probability of diseases and decrease transpiration whereas low humidity may cause hydria stress, closing the stomata and thus it may lower down the process of photosynthesis, which depends on the CO_2 assimilation. Moreover, soil water also affects the crop growth. Consequently, the monitoring of soil condition has a specific interest, because good condition of a soil may produce the proper yield.

The architecture proposed assumes that sensors are measuring periodically the monitored parameters storing such information locally. In fact, each sensor also calculates the maximum and minimum monitored values measured within the whole day. There is no information interchange between sensors and sink node until an event is identified in the greenhouse, that is, it is an event-based network design. The key aspect is the way in which an event is identified. Authors propose the continuous and periodical monitoring of the monitoring parameters. Then, they establish a threshold value per each monitored parameter. This threshold value is used to determine if an event is being generated in the network or not. These thresholds have to be inserted in the sensors a priori (or via configuration messages) and they depend directly on the type of plan being monitored.

A threshold value indicates the minimal difference between two consecutive monitored data to consider that a new event is happening in the system. For example, change in the temperature higher than $0.36°$ between two senses could be considered a new event. Authors provide a set of empirically calculated suitable thresholds.

Once an event is identified, the network initiates the alerting procedure to inform the sink node about the event. The sink node is connected to a central unit which is able to activate some actuators in order to mitigate the event. The actuators proposed are: drift valve control, air circulations control, sprinkler valves control and daylight control mechanism. All these actuators are wired to the central unit and they enable a partial automatic maintenance of the greenhouse.

8.2   Mining Industry

Coal mining is one of the most dangerous work environments in the world. On the 3rd August 2007, six miners were trapped at the Crandall Canyon mine Utah, USA. A shaft 6.4 cms and 26 cms was drilled and an omni-directional microphone and video camera were lowered down and an air sample was taken. The measurements were 20% O_2, little CO_2 and no CH_4. The amount of O_2 was sufficient enough to sustain life for some additional days, the absence of methane gave hope that there would be no immediate danger of explosion and the absence of CO_2 and the evidence from the camera and the microphone undermined the expectation of finding the lost persons alive. It took six days just to gather this evidence, which was absolutely essential for coordinating the rescue tasks. Moreover, this time the source of the disaster was a seismic shift but this is not the only cause. Explosions sparked by methane gas and coal dust which mean that CO cannot disperse into the air producing poisoning gases are other causes.

Distributed Sensing Systems can be applied to *coal mining* to monitor a variety of conditions completely below ground, ranging from *location of individuals* and *collapsed holes* to the measurement and forecast of *seismic shifts* and *concentration of gases*.

Misra et al. (2010) describe the communication channel characteristics available in a coalmine as a high-stress communication environment affected by the following environmental conditions. Firstly, the location of mine walls and faces may alter continuously as a result of mining operations. Secondly, the usage of collapse zones where there are no supports and the faces are allowed to collapse as mining operations proceed, or in the event of seismic activity. Thirdly, there are restrictions on communication arising from the normal mine arrangement of long orthogonal tunnels, support pillars, tunnel blockages and floor undulations. This cause imposes significant excess attenuation. Moreover, note that fires generate ionised air, which can act as plasma and disturb EM propagation in mines. Additionally, the relative humidity in mines is high, typically greater than 90% and the ambient temperature is commonly around 28 °C. And finally, the main component of the flammable gases that leak from coal seams is methane. When the concentration of methane exceeds a critical threshold, an explosive mixture is formed with a risk of gas blasts.

This application scenario entails several challenges due to the extremely hostile environment for radio communications available in the mine. The turns and twists of underground tunnels make it impossible to maintain a line-of-sight communication link. Moreover signals are highly reflected, refracted and scattered. This is emphasised with the high percentage of humidity available therein which directly causes signal absorption and attenuation is extremely high.

Misra et al., (2010) also describe several requirements that need to be addressed in WSN deployment in a coal mine:

- to make WSN nodes usable in the mining environment and able to provide the desired data;
- long sensor node life through use of both batteries high in energy density or rechargeable using available energy sources, and techniques to minimise node power consumption;
- physical protection of the WSN nodes and sensors to prevent damage or faulty operation in normal and post-accident circumstances without adversely affecting communications;
- network protocols to store, exchange and retrieve information reliably under harsh operating conditions;
- system health monitoring to establish and report the functional status of the system during normal conditions as well as after the occurrence of a mine accident;
- system maintainability, that is, the effort required to keep the system operational in both normal and emergency conditions.

Regarding *collapsed holes*, (Li and Liu, 2009) provide an architecture designed with the following principles: detect rapidly the collapsed area and report to the sink node; maintain the system integrity when the sensor network structure is altered; and provide a robust mechanism for efficiently handling queries over the sensor network under unstable circumstances. They did a real deployment in the D. L. coal mine. It is one of the most automated coal mines, yielding the second largest production of coal worldwide. The D. L. coal mine is a typical slope mine. This structure of this mine is depicted in Figure 8.6.

A slightly sloped 14-kilometer long main tunnel starts from the entrance above the ground surface and goes 200 meters deep underground to the working bed. The main tunnel is the primary passage for miners and equipment. To monitor the underground environment in a coal mine, authors designed and deployed the Structure-Aware Self-Adaptive (SASA) system along the main tunnel and working spaces.

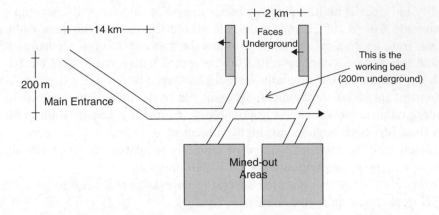

Figure 8.6 D. L. Coal Mine structure provided by (Li and Liu, 2009).

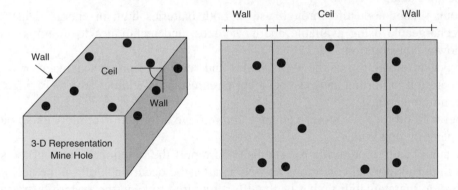

Figure 8.7 2D localisation plane of the 3D deployment proposed by (Li and Liu, 2009).

The proposed architecture is composed of stationary sensor nodes, which are deployed on the walls and ceiling of tunnels, in a mesh network fashion. To facilitate hole detection, SASA unfolds the two walls of the tunnel and builds a 2-D representation of the 3-D deployment on the inner surface of the tunnel, as depicted in Figure 8.7. The preconfigured location in each node is a 2-D location coordinate on the 2-D surface. Note that Figure 8.7(c) shows the distance between any two nodes in the 3-D real environment. This distance is always less than or equal to the distance between the pair in the unfolded 2-D view. Thus, the real connectivity of the sensor network is no less than shown in the 2-D representation this representation being a pessimistic estimator.

Regarding the mote employed, authors use the Mica2 mote platform. They decided to use a MPR400 radio device. This device is an 868/916 MHz tuneable Chipcon CC1000 multichannel transceiver with a 38.4 Kbps transmission rate with a 500-foot outdoor range. A sensor board is connected to the Mica2 mote performing environmental data collection. However, while sensors are more important for other purposes in coal mining, the monitoring of collapsed holes does not entail the usage of special sensing devices, according to the detection algorithm provided by the authors.

In essence, when a collapse occurs, the motes in the accident region are moved or removed and a hole of sensor nodes emerges. In a reasonable density of sensor node deployment, the sensor hole should reflect the actual collapsed hole. When this hole emerges, the nodes on the hole edge will have a loss of neighbour nodes, and these nodes outline the hole if their localisation is known. The basic idea in detecting a hole is to let sensor nodes maintain a set of their neighbours (and their localisations). When a node suddenly finds that a subset of its neighbours has disappeared, it should be aware that it is now likely to be an edge node of a hole.

Different routing protocols can be used for this periodical detection of neighbours. On the one hand, nodes periodically can probe their neighbours. This approach is costly in terms of traffic overhead. Then, on the other hand, each node can actively report its existence. When an edge node detects loss in the density of neighbours higher than one node, it sends this alert information to the sink. This simultaneous notification creates a traffic peak and increases the collision domain. Thus, *Randomized Forward Latency* can be inserted in order to reduce significantly simultaneous sends into the network.

After the alert notification, the platform has to start its reconfiguration in order to maintain the system validity as soon as possible. In the reconfiguration, both centralised and decentralised mechanisms are employed to achieve short detection latency. In the centralised mechanisms, when the sink receives the alert messages with the edge nodes locations and approximate the hole region, it sends to every node in this zone a reconfiguration message to start detecting its surroundings and check its location from a beacon message. This mechanism is combined with a decentralised approach, which is independent of the distance to the sink, and is based on the self-identification of the node status and acts in consequence. Note that edge nodes that lose neighbours but themselves do not move do not need to be reconfigured since their location is correct. Moreover, edge nodes that fall into an area where no normal node exists do not need to be reconfigured either. And also, edge nodes that fall into other normal node ranges have to stop beaconing. The combination of these two approaches is efficient and reliable in various situations and they can always be turned off and reconfigured to conform to their new positions after an important event such as a collapsed hole-detection.

The routing protocol used is based on the concept of *synopsis diffusion* (*Nathy* et al., 2004), this protocol achieves significantly more accurate and reliable answers by combining energy-efficient multi-path routing schemes with techniques that avoid double-counting in data aggregation. Synopsis diffusion avoids double-counting through the use of *order- and duplicate-insensitive (ODI) synopses* that compactly summarise intermediate results during in-network aggregation.

Another application in coal mining is *gas detection*. In this sense, (Li-min et al., 2008) propose a ZigBee WSN for monitoring temperature, humidity and methane values underground of a coal mine. Authors monitor cyclically the sensors attached to the devices and store the monitored data locally. Then, only when there is any measured value, which causes any safety danger for the miner's health, the routing protocol is in charge of forwarding the information to the sink node.

Authors proposed the usage of the tree-based AODV routing protocol together with a beaconing protocol for broadcasting routing update messages. The power hungry associated to the beaconing protocols is controlled with the inclusion of several techniques. Firstly, they restrict the routing updates to the root node. Secondly, they added a limitation of the number of nodes added to a cluster-head node. Thirdly, they added routing maintenance functions to repair local routing. Finally, the adjustment of the node retains energy to make more rational use of it.

Li-min et al. (2008) propose a deployment based on the division of the tunnel in four different regions on which different clusters of sensors are deployed. Each region contains a cluster-node with a longer communication radio and a set of sensing nodes with more limited radio coverage. Moreover, in order to avoid frequency overlapping between clusters the distance of the cluster should be established at the at 2/3 * (communication radio) of the cluster node (Li and Liu, 2007).

Another distributed sensing system for monitoring the concentration of gases and seismic monitoring in the coalmines is proposed by Tan et al. (2007). Authors provide hybrid architecture composed of stationary and mobile nodes. The mobile nodes are placed at the miner pocket and they are constantly collecting the gas concentration within its region and transmitting this information to the stationary nodes, in a data mule approach. Then, these stationary nodes are in charge of distributing the monitored values across the network to the base station where it is processed and an alert message is notified if some dangerous situation arises.

Regarding the MAC layer, authors propose the usage of S-MAC protocol. This protocol turns nodes into sleep mode during broadcasting if they are not the receiver and the dispatcher. Moreover, they also propose an energy-aware multi-hop routing protocol using energy consumption as a metric for selecting the most appropriate route to reach the destination.

8.3 Transportation

Transportation opens new research horizons for the usage of wireless sensor devices in such application. The status monitoring of the transported loads, automatic vehicle guidance, fleet tracking and in-vehicle monitoring are some direct examples of the application of distributed sensing platforms in the transportation.

A clear example for transportation is provided by Carullo et al. (2009). They propose architecture for monitoring the cold chain inside of a refrigerated vehicle. This is intended for monitoring temperature-sensitive products during their

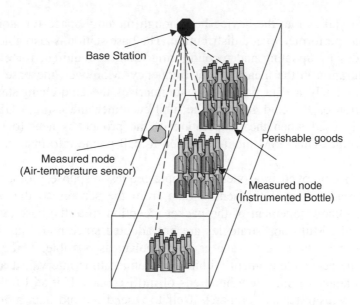

Figure 8.8 Deployment proposed by (Carullo et al., 2009) for cold chain monitoring.

distribution with the aim of conforming to the cold-chain assurance requirements. There are many products which require strict cold-chain maintenance. For example, chilled products need to be stored near 0 °C, where a rise in temperature of just a few degrees could cause microbial growth. The situation is more serious in the case of pharmaceutical products since an uncontrolled change in the temperature, even for short time intervals, could cancel the effectiveness of the product or even make it dangerous. The change in the internal temperature of the vehicle can fluctuate due to by several factors such as loading and unloading of the goods, changes in the air flux, temporary interruptions of the refrigeration function, doors opening, and so on.

Authors proposed a new method composed of two different types of sensors: air-node and bottle-node. Air-node sensors are placed inside the lorry in order to monitor the inside air-temperature of the goods whereas bottle-node sensors are dummy goods inserted into the packaging among the real products for monitoring the product temperature. For example, the nodes may be placed inside a bottle of milk and placed into the pallet of milks for monitoring the temperature of such a pallet. Figure 8.8 shows a diagram of the architecture proposed.

Traditional general purpose sensor devices may not be suitable for monitoring temperatures in this scenario since they are rather big to be instrumented inside a bottle of milk or any similar product. Then, the authors decided to select *Mica2Dot* system, which is a wireless node of a circular shape with a diameter of 2.5 cm, very appropriate for this kind of environment.

The base station is designed to be completely autonomous (without a main controller) and it is in charge of collecting the measurements from all the measuring nodes. This requires that the device has enough memory space for storing all the measurements performed over a distribution. The base station is also able to interact with the nodes by updating and reconfiguring them as required, for example, for the reconfiguration in the measurement frequency. Moreover, the base station can provide periodically a data packet that summarises the cold-chain status providing information on the cold chain to the final customer and using a GSM module which is only used when the temperature of the product is near to reaching an impossible value. This GSM module is used to call phone information about such a situation.

The routing employed is a one-hop simple request-response protocol based on periodical broadcasting of the sensed data from the sensors to the base station and unicast acknowledgement of the packet reception from the base station to the associated node. Multi-hop strategies are not adopted since nodes are in sleeping mode for most of the time to save as much power as possible. If the node does not receive the acknowledgement within a predefined time interval, it attempts the transmission again within a few hundreds of milliseconds. If it still fails, the node turns off the transmitter and prepares itself to resend the old data along with the new one during the next scheduled radio transmission.

Another application for transformation is the structure health monitoring of vehicles. Health monitoring of tyres is a clear scenario in which distributed sensing can be valuable. Mishra et al. (2006) propose an architecture for providing intelligent tyres into the vehicles. Tyres that can measure strain within themselves are known as intelligent tyres. A vehicle equipped with intelligent tyres is considered far more reliable than the one which has regular tyres. Besides measuring tyre pressures, tyre manufactures are also examining ways to measure strain and temperature as well as enhancing overall safety of a vehicle.

In essence, authors propose to insert sensor nodes either inside or outdoor of the tyres, directly attached to them. The former provide better signal strength and it is easily accessible for recharging batteries if necessary whereas the latter implies that the sensor unit need power sources to keep running the sensor for a long period due to their hard accessibility. However, the inside placement of the sensors provides more accurate and reliable strain and temperature information.

In fact, the in-tyre placement of sensors may require alternatives to the power consumption which can be the usage of a simple battery charger up to the transceivers' batteries and place a plug on the tyre's rim where the user can hook up a power source to the charger. Moreover, signals inside the tyre are subject to multi-path fading due to the high density of metallic components surrounding the transceiver. Several solutions are proposed to deal with such a problem, such as to make the antenna weave through the tyre's rubber, Stick the antenna out of the tyre's valve or small hole on the rim, applying a coating to the inner rim's surface to prevent signals from bouncing off and the use of a specialised container to protect the unit from temperature and shock variations.

The routing protocol used for communication in-tyre sensors and base station may rely in a one-hop or in a multi-hop routing protocol and being the latest one is an alternative to be explored to solve the problem of weak signal strength. The base station acts as the command and control centre and is used to monitor the information provided by the in-tyre sensors. A reliable data transport mechanism is used sending ACK hop-by-hop rather than end-to-end due to the high packet loss which can be expected in this scenario.

The strain measured by each node is monitored periodically but will not be transmitted till the base station requests such information. Thus, in-tier nodes act as storage agents which will transmit the sensed information when requested.

Transportation, by definition, also implies the movement of objects between two geographical points. Traditionally, the driving and control of the vehicle is directly performed by humans, however, recently there a new trend to focus the research attention in providing automatic guidance systems has appeared. The use of collaborative agents between vehicles is crucial to achieve safety, reliable and effective automatic guidance of vehicles. A team of aerial vehicles can be used for exploration, detection, precise localisation, monitoring and measuring the evolution of natural disasters, such as forest fires. Thus, the use of smart sensors to establish highly dynamic ad-hoc networks for performing such collaborative driving is a clear application scenario in which distributed sensing networks are valuable. Maza et al. (2010) propose an architecture for carrying out a distributed cooperation and control system for multiple Unmanned Aerial Vehicles (UAVs) with sensing and actuation capabilities.They develop a system in which multi-UAVs collaboratively transport load transportation. Note the intensive in-air distributed balance of the charge to be done between UAVs to keep the load horizontal without risks of falls. This system is developed under the project AWARE whose general objective is the design, development and experimentation of a platform providing the middleware and the functionalities required for the cooperation among UAVs and a ground sensor-actuator wireless network, including mobile nodes. Figure 8.9 shows the proposed architecture.

Each UAV is a UAV helicopter composed of two main layers: the On-board Deliberative Layer (ODL) and the proprietary Executive Layer. The former deals with high-level distributed decision-making whereas the latter is in charge of the execution of the tasks. In the interface between both layers, the ODL sends elementary task requests and receives the execution state of each task. Moreover, for distributed decision-making purposes, interactions among the ODLs of different UAVs are required in order to exchange contextual information and collaborative tasks. The Human-Machine Interface software allows the user to specify the missions and tasks to be executed by the platform, and also to monitor the execution state of the tasks and the status of the different sub-systems of AWARE.

The way in which the data is processed to achieve the collaborative decision which achieves the mission is based on the following scheme. A mission M specified a

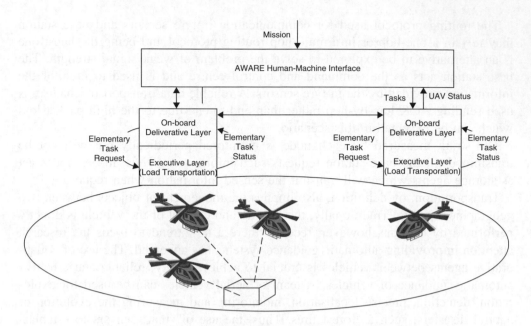

Figure 8.9 Architecture proposed by (Maza et al., 2010) for controlling UAV devices.

set of partially ordered tasks T. The mission is specified by the user of the system using the Human Machine Interface. These tasks can be allocated to the UAVs manually (statically) from the HMI or autonomously in a distributed way. The latter approach is based on the usage of the Contract Net Protocol (Smith, 1980). The CNP specifies the interaction between agents for fully automated competitive negotiation through the use of contracts. Thus, it manages the distributed task allocation process among the different UAVs.

A task is modelled as the tuple *Task (Type, Pre-conditions, Post-conditions, Status, Parameters)*. A given set of task types are available and well-known by the UAVs, for example, TAKE-OFF, LAND, GOTO, WAIT, and so on. Each task type has an associated set of parameters in order to enable the associated task execution. Moreover, a given task has an associated given status, for example SCHEDULED, RUNNING, MERGING, ABORTED, and so on. Then a task is allocated into a given UAV, it is processed at the ODL layer, and divided into simpler tasks, called elementary tasks that are finally sent to the executive layer of the UAV.

The division of the tasks in elementary tasks is done by the *Plan Builder/Optimiser* which can operate in both offline mode just using the predefined mission provided by the user and in on-line mode using a plan builder programmed to solve the specific planning problems of the UAVs in the AWARE platform. The set of elementary tasks are then scheduled to be executed in the Execution Layer using the *Task Manager* module. For each task, this module produces an insertion or an abort of the next task into the Execution Layer. The *synchronisation manager* is in charge of keeping the task dependencies coherent among the different tasks in the current plan of the UAV (preconditions and post-conditions fulfilment) and, also with the tasks of other UAVs (which are achieved by means of message exchanges with the other UAVs).

Another important issue when considering the different plans of the UAVs is that the main resource they share is the airspace. Therefore, a plan merging module has been included in the architecture to detect potential conflicts among the different trajectories and also to follow a policy in order to avoid them. This module interacts with the plan builder module and also with other UAVs to interchange the different 4D trajectories.

As a result, a collaborative communication of the UAVs and a constant update of the current tasks status enable the collaborative execution of the mission across all the UAVs involved achieving efficient load transportation. This deployment has been implemented, tested and validated. In fact, in case the reader is interested, authors provide multimedia resources available by means of the Internet in which the real helicopters are recorded doing the collaborative load transportation.

8.4 Remote Sensing and Imaging

Remote sensing is the acquisition of information about an object or phenomenon, without making physical contact with the object. The most prominent and traditional application of remote sensing is the usage of aerial sensor technologies to detect and classify objects on Earth such as lasers, sonars, radiometers, and so on. Analogously, remote imaging is a special case of remote sensing in which images about the monitored object are retrieved. The usage of distributed sensing platforms for implementing distributed remote sensing technologies has an important application in military and public safety scenarios. Note that traditional remote monitoring systems are centralised and the usage distributed approaches can augment significantly the reliability, accuracy and precision of the information sensed.

A clear example of remote sensing is provided by Schiewe (2005). He exposes an overview of the kinds of digital airborne sensor systems for carrying out remote sensing. This review includes large format camera systems and multi-sensor systems (acquiring image and laser scanning data). This kind of sensor enables the remote monitoring of pollution, toxins and topographical parameters. An early attempt for the usage of these sensors from remote sensing is provided by (Grüner et al., 1991) which monitors maritime pollution and the guidance of oil combating operations. The proposed architecture also allows measuring the volume of films on the sea surface, and classifying the substance type and parameters which are relevant for the evaluation of ecological conditions as, for example, the occurrence of plankton blooms.

The severe natural conditions and complex topology make it difficult to apply precise localisation in underground mines. Pei et al. (2009) propose a coal mine deployed in underground mines to provide a localisation method for mobile miners based on the remote sensing. The proposed architecture is tested through simulations with 100 nodes, and also with outdoor experiments using 15 ZigBee physical nodes, and with experiments in the mine gas explosion laboratory with 12 ZigBee nodes.

Localisation algorithms can be divided into two classes: anchor-based algorithms and anchor-free algorithms. Anchor-based algorithms assume that all reference nodes are anchor nodes or nodes whose real position coordinates are known in advance. Anchor-free localisation algorithms only require a few anchor nodes and the rest are able to discover their localisation coordinating with the anchored nodes. Typical anchor-free localisation algorithms estimate the coordinates of the reference nodes and complete precise localisation for mobile targets based on reference nodes, usually based on the Received Signal Strength (RSSI). Note that RSSI is a very prominent way for performing remote monitoring of the object.

Pei et al. propose an anchor-free localisation using Cicada devices for carrying out the distributed sensing platform deployment. There are six types of Cicada nodes including methane sensors, oxygen sensors, carbon monoxide sensors, smoke sensors, temperature-humidity sensors and voice sensors. All of them are based on a ZigBee protocol stack. These sensor nodes join the network, acquire the environment information on a fixed time cycle and transmit sensing data to the ZigBee gateway. ZigBee static router nodes are previously deployed to construct the ZigBee backbone network. They are also reference nodes for mobile targets. Voice sensor nodes are installed on miner's helmets. Miners are the mobile targets for localisation using such sensors.

The localisation algorithm proposed is based on a non-metric multidimensional scaling (MDS) algorithm. This type of algorithm receives as input a relationship matrix that specifies distances between every pair of nodes, and the output is a coordinate set of all the nodes. Thus, authors calculate the RSSI (Received Signal Strength: RSSI) matrix as the input for the algorithm. In essence, they can measure the RSSI between two adjacent nodes and use the shortest path algorithm to estimate the RSSI between adjacent nodes. Then, the localisation algorithm is composed of two phases. The first one is in charge of determining the localisation of the reference nodes using the non-metric MDS algorithm. The steps of this phase are as follows:

Step 1: All reference nodes broadcast a one-hop RSSI request message. The neighbour nodes measure the RSSI value between them and report the response message to the sink node.

Step 2: The sink node starts up the Dijkstra's shortest path algorithm to construct the RSSI relationship matrix for every pair of nodes, which is the input to the non-metric MDS.

Step 3: Finish the non-metric MDS algorithm process to obtain the relative coordinates of all reference nodes.

Step 4: Compute the absolute coordinates through shifting, translating, rotating
and/or reversing with anchor nodes.

Authors assume that there are at least three reference nodes as a minimum for
enabling the miner localisation. Note: accuracy of the miner location is directly
affected by the number of reference nodes used in the localisation algorithm.

Once the absolute coordinates of the reference nodes are calculated, they are sent
to their associated reference nodes in the network. Then, the localisation of mobile
targets can be started using the N-best sequence-based localisation (SBL) algorithm.
This localisation follows the next steps:

Step 1: The mobile targets broadcast one-hop RSSI request messages at a fixed
time cycle. When reference nodes receive such messages, they calculate the RSSI
values between them and report them to the sink node.

Step 2: The sink node aggregates the coordinate information of the involved reference nodes and executes the N-best SBL algorithm. It obtains the position
coordinates of mobile node.

The sequence-based localisation method is a novel anchor-based distributed localisation technique, which was recently proposed by (Yedavalli and Krishnamachari,
2008). The idea consists of the division of the space into distinct regions by the
perpendicular bisectors of lines joining pairs of reference nodes. Each region is
uniquely identified by a rank sequence that represents the distance ranks of anchor
nodes to that region. In summary, this localisation method follows the next steps:

Step 1. Each reference node is able to determine the RSSI for all the other reference
nodes.

Step 2. Each reference node can determine all the feasible localisation regions in
the localisation space and the rank distance to the other reference nodes based
on the received RSSI information. This information is stored locally.

Step 3. Each reference node can determine the RSSI of the mobile node. The
reference nodes send such information to the sink node, which is in charge of
aggregating it obtaining the location sequence of the mobile node.

Step 4. The sink node can search in such information for the 'nearest' sequence to
the location sequence of the mobile node.

Step 5. Once the region is determined, the sink node can take the centroid of the
region, which is presented by the 'nearest' location sequence, as the position of
the mobile node.

Another clear and innovative example of the usage of distributed sensing platform
for supporting remote sensing is provided by (Chanboun and Raissouni, 2011). They

propose a real deployment done at Kasr-Seghir, a site located in northern Morocco at approximately 45 km away from Tangier. Authors use a system deployed in a grid shape, covering a geographical area to be monitored. The idea is to use such a system to perform simple in situ Land Surface Temperature (LST) monitoring. The aim pursued is to provide an alternative distributed sensor-based system as support for the LST information provided by the satellite remote sensing.

Authors provide geo-localisation (GPS coordinated) for all the sensors available in the network in order to correlate the measured values to the geographical position on the Earth. Each sensor is composed of a Mica2DOT sensor with a thermal infra-red sensor. This is a remote monitoring sensor which is able to monitor the earth surface temperature. The sensor applied in this study is non expensive and energy-efficient, based on an infra-couple radiometer, namely the OSM101 (portable IR to analogue converter module). The software uploaded to the sensors is the XMESH multi-hop mesh protocol with several customisations like low-power listening, time synchronisation, sleep modes, any-to-base and base-to-any routing. Then, sensors are able to monitor the LST of the geo-referenced sensors and to gather such information into the base station providing the LST of the whole geographical area is sensed. In order to validate the deployed system, authors directly compared the average value from all the monitored information with the information provided by the Moderate Resolution Imaging Spectro Radiometer (MODIS) sensor, obtained from the daily LST product developed by the MODIS NASA Science Team, on board the TERRA satellite. As a result, authors validate the proposal as a suitable alternative to the LST remote satellite monitoring since their measured values over the same geographical area match perfectly.

8.5 Earth Resources Observation

Usually, the observation of Earth resources requires the usage of expensive devices that are difficult to move or require external supply voltage. Moreover, the deployment and maintenance of these devices require vehicle or helicopter assistance. Data storage also must be retrieved on a periodic basis. A direct application of distributed sensing platforms for monitoring the earth's resources could provide solutions to many of these problems.

A clear example of application for earth resource monitoring is *volcano monitoring*. Note that volcanoes usually are hidden from view occurring on the ocean floor along spreading ridges and other difficult to access locations. Thus, a large number of small, cheap, and self-organising nodes can be deployed to cover a vast field. The main advantages of the usage of systems in active volcano monitoring are the fast and economical deployment, the possibility to achieve high spatial diversity and the low maintenance routines required.

(Werner-Allen et al., 2006) deployed proof of a concept distributed sensing platform in Volcano Tungurahua in central Ecuador in 2004. After that, the same authors deployed a complete distributed sensing platform in Volcano Reventador in northern Ecuador in 2005. The latest deployment is composed of 16 TMote Sky motes each equipped with a microphone and a seismometer sensors and an alkaline D cell battery. The motes are deployed over 3 km gathering volcano eruption information over three weeks. The base station is located at 4 km from the sensing platform attached to a laptop in a makeshift volcano observatory. *FreeWave* radio modems using a 9-dBi directional Yagi antenna are used to establish a long distance radio link between the sensor network and the observatory.

Each mote samples two or four channels of seismo-acoustic data at 100 Hz, storing the data in local flash memory. This means that, according to the TMote Sky flash memory specification, each note is able to monitor 20 minutes of data before it is filled. Fortunately, a differentiating aspect of volcano monitoring is that many interesting signals occur in less than 60 seconds and occurred several dozen times per day. Thus, when the memory is filled the oldest values can be safely removed in a circular fashion. Moreover, due to this fact, authors decide to design the network to capture time-limited events, rather than continuous signals, that is, an event-based network.

Nodes also transmit periodic status messages and perform time synchronisation. When a node detects an interesting event, it routes a message to the base station laptop. Then, if enough nodes report an event within a short time interval, the laptop initiates data collection, which proceeds in a round-robin fashion.

The laptop downloads between 30 and 60 seconds of data from each node using a reliable data collection protocol. When data collection is completed, nodes return to sampling and storing sensor data.

Regarding the transport protocol, the *Fand so onh* protocol is used to retrieve buffered data from each node over a multi-hop network. This protocol tags each block of data with sequence numbers and time stamps. The block is fragmented into several chunks, each of which is sent in a single radio message. The base-station laptop retrieves a block by flooding a request to the network. The request contains the target node ID, the block sequence number, and a bitmap identifying missing chunks in the block. Then, the target node replies by sending the requested chunks over a multi-hop path to the base station.

The routing protocol used is *MultiHopLQI*, a variant of the *MintRoute4* routing protocol modified to select routes based on the *CC2420* link quality indicator (LQI) metric. Link-layer acknowledgements and retransmissions at each hop improve reliability.

Regarding the time-synchronisation protocol, authors decide to use the Flooding Time Synchronisation Protocol (FTSP) to establish a global clock across our network. Moreover, one of the nodes used a Garmin GPS receiver to map the FTSP global time to GMT. Unfortunately, FTSP occasionally exhibited unexpected behaviour, in which nodes would report inaccurate global times, preventing some data from being correctly time stamped. What authors need to do is to develop a correcting technique based on the large amount of status messages logged from each node, which provide a mapping from the local clock to the FTSP global time.

Song et al. (2010) provide a similar architecture with slight differences deployed in the Mount St. Helens' active volcano. In this case, authors use IMote2 motes, equipped with both seismic and ultrasonic sensors. This architecture is focused in a high re-configurability of already deployed sensors. To this end, authors imported Deluge, which is a de facto network reprogramming protocol that provides an efficient method for disseminating a program binary over the wireless network and having each node program itself with a new image for the IMote2 platform using it as a software upgrade.

A differentiating aspect is that in this scenario all the motes are equipped with a GPS which is used for two purposes, localisation and time-synchronisation. Then, the time synchronisation is done each 10 seconds using the GPS time and only if the GPS signal disappears, the system will switch to FTSP mode. Although, authors already know that GPS is a very power hungry device, they decided to use it since node localisation is a requirement on their scenario and they decided to use additional power suppliers.

Regarding transport protocol, authors developed the Reliable Data Transfer (RDT) protocol. This transport protocol uses bitmap to represent multiple ACK/NAKs, namely ANK, to reduce feedback traffic. An ANK packet has three fields (*start SeqNo, validBits, ankBitmap*), where *startSeqNo* is the starting sequence number to ANK, and *validBits* indicates number of valid bits in *ankBitmap* (e.g. number of packets to ANK). In ankBitmap, bit-1 denotes a received packet while bit-0 denotes a lost packet.

Regarding MAC protocol, authors developed a new TDMA MAC protocol based on IEEE 802.15.4 called TreeMAC to regulate the channel access. The design of *TreeMAC* is based on the key observation of multi-hop data collection networks that the bandwidth allocation of any node shall be no less than that of its subtree, so that the nodes closer to the sink have enough bandwidth to forward data packets for the nodes that are further away. With TreeMAC the network can achieve a data throughput to the gateway of at least 1/3 of the optimum assuming reliable links.

Another application is *earthquake monitoring*. Earthquakes cause two basic kinds of seismic waves: Primary waves (P-waves) and Secondary-waves (S-waves). The harmless P-waves are almost twice as fast as the S-waves, which cause most of the destructive shaking. Therefore, the time interval between the detection of the fast P-waves and the arrival of the slow S-waves, commonly termed 'warning time'. Typical values of the 'warning time' are only a few seconds, for example 4 seconds for Istanbul. Although this short time is not enough for people to leave their houses, this can still be sufficient to mitigate secondary damages safely shutting down nuclear reactors or gas and power suppliers. To deal with early detection of earthquakes, (Nachtigall and Redlich, 2011) propose an on-site deployment of a WSN system. On-site deployments are deployed in the city itself and are different from front-site deployments, which are located near to the expected epicentre. The author's deployment consists of 20 self-engineered sensor nodes. Two of these nodes are base stations attached to a laptop with Internet connectivity. Each sensor node is equipped with a seismometer (for detecting shaking), a GPS receiver (for timing and localisation) and a wireless radio.

Whenever a node triggers on shaking, it broadcasts a trigger message. This message contains the node's ID, its position and trigger time, received from GPS device. Radio link quality is extremely variable in earthquakes and this is an important requirement to be addressed in the transport and routing protocols. Then, a receiving node processes this message, stores its trigger information and then rebroadcasts it, if it has not done so before, that is, a flooding protocol.

However, note that all the motes should detect the earthquake at the same time and this protocol is not suitable since it can cause many packet collisions. Then, authors insert a randomised delay (using Gaussian distribution) that is proportional to the distance between sending and receiving nodes and between the reception of the trigger message and rebroadcast of it. Moreover, radio link reliability is addressed using implicit acknowledgements in the modified flooding routing protocol. Lossless aggregation is applied by piggybacking other nodes' states.

Other research work for earthquake monitoring is provided by (Goldoni and Gamba, 2010). The authors create a prototype system called *W-TREMOR* which can provide 1 KHz sampling rate and real-time data delivery using nodes which cost less than 80 Euro each. The motes are composed of Kistler 8330A3 accelerometer sensors, which are a single axis sensing element for measuring static acceleration up to 3g or low-level, low-frequency vibrations at 2000 Hz. They used an AVR ATMega 168 microcontroller with a Digi XBee radio module, an IEEE 802.15.4-compliant which operates in the 2.4 GHz band.

The transmission of acquired data is based on the PRISM (Protocol for Real-time Synchronous Monitoring) protocol. PRISM uses a TDMA-like scheduling to avoid collisions among transmissions assigning non-overlapping time slots. Given the maximum sampling frequency of the sensor network, PRISM calculates off-line the optimal scheduling of transmissions. Authors have done an empirical study to determine PRISM efficiency providing prominent values of tens of nodes acquiring more than one hundred samples per second in a star and multi-hop topologies.

PRISM not only covers a TDMA-based MAC layer but also provides some routing routines. In essence, it enables start (one-hop) and multihop topologies and different synchronisation protocols are used in each case. For the start topologies, a centralised in-band synchronisation mechanism is used in order to compensate clock discrepancy among sensor nodes without requiring dedicated hardware. Either a reference node or the gateway itself periodically broadcasts a clock signal to all the sensor nodes during a dedicated time slot. This solution features low execution time, low traffic overhead, and also provides accurate synchronisation for most practical

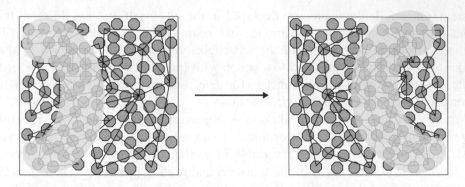

Figure 8.10   Dynamic spatial monitoring scenario proposed by (Duckham et al., 2005).

uses. For multi-hop topologies, authors foster the usage of Timing-sync Protocol for Sensor Networks (TPSN) protocol.

There is another application scenario related to earth resources monitoring where distributed sensing can play an important role. It is the *dynamic spatial fields monitoring*. This consists of the monitoring and querying of dynamic information about the geographic environment, in this case, using distributed sensing platforms. For example, this dynamic information includes mud flows, ash clouds, and gas concentrations, wildfires, flooding, detection and tracking of toxic spills or other types of environmental contamination, or monitoring the level of pesticides. (Duckham et al., 2005) propose a distributed design for monitoring dynamic spatial fields. The idea provided by the authors can be seen schematically in Figure 8.10. In essence, the black circles indicate the sensors that are currently turned on; the unfilled circles indicate sensors that are currently in sleep mode; a triangulation has been formed using as nodes the sensors that are currently active.

After the sensors have been deployed, the network is initialised. In the initialisation phase, a small proportion of sensors will be designated active, and active sensors will negotiate appropriate communication partners. The algorithm for building and maintaining the communication network is based on the central idea that can only add and remove sensors in such a way that individual triangles in the triangulation are added or removed. Thus, when adding or removing triangles, then a number of neighbouring nodes may need to communicate to coordinate their de/activation.

A sensor is activated itself either by detecting a significant change in its environment during periodic wake-up cycles, or receiving a request from a neighbouring sensor. In the active network phase, the sensor network is required to respond to qualitative changes in the dynamic field.

Note that the activation of the sensors may imply the deactivation of other sensors and thus the dynamic field is being tracked all the time activating high densities of sensors only in those important areas in which significant dynamic changes happen.

8.6 Underwater Sensing

The application for carrying out underwater sensing reveals many interesting and practical application scenarios. Underwater sensing systems which can be used for detecting hazards in the water such as shark detection, jellyfishes and water quality measurements, tracking shoals of fishes and individual fishes such as dolphins and whales, monitoring microorganisms and marine vegetables, and even aquiculture and soil monitoring.

Sensing underwater using wireless infrastructures is a very challenging scenario due to the intrinsic nature of the medium in which High Frequency (HF) radio waves are strongly attenuated. Thus, the available traditional radio modules which operate in MHz and GHz cannot be used underwater where, currently, radio signals attenuate rapid travelling to very short distances. Two main alternatives are proposed in order to provide efficient communications underwater, acoustic-based communications which are the most extended way for communications underwater and optical-based communications which are new emerging technologies and are focusing the attention of many research works.

A clear application of underwater monitoring is the monitoring of earth resources. In particular, Martinez and Hart (2010) propose a distributed sensing platform for understanding of sub glacial processes, especially to investigate their links with climate change, as well as developing the next generations of environmental sensor networks. They perform the monitoring of glaciers in the Arctic.

For this purpose, all the motes are custom sensor network hardware and software, mainly due to the lack of suitable off the shelf hardware. The motes are physically deployed in temperate valley glaciers in Norway and Iceland. Concretely, between 2003 and 2006, 30 nodes were deployed in Briksdalsbreen, Norway. Subsequently, a further eight nodes were deployed in Skalafellsjökull, Iceland in 2008. Figure 8.11 shows an overview of the deployment done in Iceland. Note that the base station is located outside of the water with the transceiver placed into the water and the rest of sensors are in the water.

The main geological objectives pursued are: i) to produce a long term record of water pressures changes in the ice and sub glacial sediment (till); ii) to produce the first ever record of three-dimensional probe movement; iii) to carry on the first ever study of till temperatures within the till, in order to examine heat balance and iv) to enable an investigation of the relationship between water pressure, till strength (case strain, resistance), and till temperature in order to understand till sedimentation. In particular, Table 8.1 shows some of the operation requirements associated with this scenario.

This application scenario entails many challenges for the design of the sensing platform due to the strong operational requirements imposed. The sampling rate of the sensors is fixed at once per four hours, partly because changes are expected to be slow but also to save power. Authors expected long radio-disconnection periods,

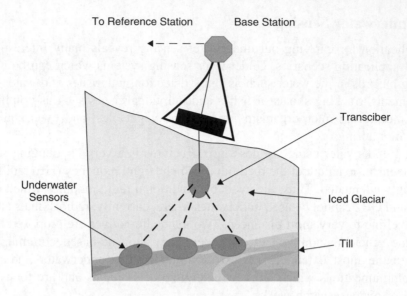

Figure 8.11   Sensor Deployment provided by (Martinez and Hart, 2010) to monitor glaciers.

Table 8.1   Operational Requirement for glacial monitoring provided by (Martinez and Hart, 2010)

Operating air temperatures	-20 to $+30\,°C$
Temperature in ice/till	-1 to $+1\,°C$
Radio frequencies required to transmit through 100 m temperate ice	Less than 500 MHz
Maximum diameter of probe case to fit borehole	Less than 0.1 m
Maximum depth of ice	100 m
Maximum ice pressure	900 kPa
Maximum water pressure	1,200 kPa
Battery life	4 years

but glaciologists wanted every data sample, even if this data is delivered months later. Therefore, a large ring buffer (6,000 readings) for the data is used on each mote in order to store data for up to a year. Note that this time the sensing platform acts as a remote sensing architecture in which all the monitored information is 'remotely' stored in the internal memory.

A GPS device is attached to the base station and a broadcasting of the GPS time to all motes is done daily as a way to keep synchronisation within seconds and save power by narrowing safety margins on wake-up scheduling. Power management is a key to satisfying the requirement for long-term system life. Since a daily data transfer is acceptable, the radios on every unit are completely off most of the time and limited time windows are given to those tasks that used them (e.g. 1–3 min per day, maximum 3 retries per packet).

When commercial nodes became common, authors still prefer to use their own hardware since they have power control and full sensor interfaces such as amplifiers and bridges (for strain and conductivity). Moreover, the radio systems available on commercial nodes are also not suitable since they typically use a frequency that is too high (e.g. 2.4 GHz) or have insufficient power (e.g. 10 mW). They used a power amplifier with a transceiver or a high power unit (e.g. 100 mW) since tests with 10mW showed them to be limited in range to tens of meters through ice. After several probes, authors decided to insert the transceiver of the base station into the water, because otherwise the loss of signal is significant and it is impossible to establish connections so far at 20 meters (testing several frequencies such as 868 MHz, 433 MHz, 172 MHz). This transceiver is standardised around the Radiometrix BiM unit tuned to 173.25 MHz but powered at 100mW rather than the default 10 mW.

The case designs for the motes are tested in a high-pressure water testing facility in the National Oceanographic Centre in Southampton for water leaks. The case is done in polyester due to its RF properties and low water absorption and strength. Each mote is equipped with a set of carefully selected sensors. First of all, the original temperature sensor does not provide enough resolution at 0 °C. Secondly, a MEMS accelerometer is needed to provide a 3-D visualisation of the till. Thirdly, a conductivity sensor is simply achieved by using a pair of stainless steel bolts, which is a more robust design than using off-the-shelf conductivity sensors. Fourth, a pressure gauge was selected that would operate at pressures up to 50% higher than the predicted maximum under the ice. Fifth, strain gauges used for measuring case strain (the least reliable of all the sensors with a 50% failure rate), presumably due to their delicate wiring. And, sixth, a light-reflecting sensor to provide more information about the nature of the material in which the mote is embedded. All of the sensors required calibration, mainly to determine their zero offset, especially at these temperatures.

Moreover, regarding routing, authors decided to implement a star network topology rather than a multi-hop since they consider it very risky for this kind of scenario. A TDMA-based ad-hoc protocol was designed around the fairly static environmental conditions for the MAC layer.

Regarding the base stations, it is implemented using the Gumstix device, a low power ARM-based boards running embedded Linux is used. The base stations only power-on units such as radios or GPS during their use. Thus, a microcontroller is used to control power to the peripherals. The latest base station deployed in Iceland used a custom MSP430 add-on PCB that also controlled power to the Gumstix device. Rather than using a standby mode of the ARM, the microcontroller powers-up the Gumstix, which, while using more energy in start-up, results in a low sleep power and hence longer life in this situation.

Moreover, the base station is constantly moving (about 15 cm/day) and melting (in summer this could be 20 cm/day). Then the authors decided to build a pyramid structure that cuts into the ice and needed no anchors. In practice, a rope anchored down a deep hole (15 m) is used as a backup.

Note that all the monitored information is kept in the distributed sensing system during the whole day or even more and only a few minutes a day are used to try to transmit the data gathered to the base station.

Another example of underwater monitoring is provided by (Zhang et al., 2004). They propose an underwater distributed sensing architecture for monitoring marine microorganisms. Microorganisms such as Phytoplankton are exceedingly small 2-3pm and distributed in the ocean at varying spatial scales. It is not practical to locate them by measuring their density everywhere. Thus, an alternative localisation could be done by means of establishing a correlation with chemical (e.g. nutrient concentration) and physical parameters (e.g. temperature, light intensity) in the marine environment.

There are two major factors that are important to the growth of microorganisms: light intensity and nutrients. In the ocean, the former comes from above (sunlight) and the latter comes from below. At a certain depth, there is a good balance between light intensity and nutrients, and the density of certain microorganisms may be expected to be high. Given the hypothesis that marine microorganisms bloom at a thermo-clime (a physically measurable phenomenon), authors propose a distributed approach to localise a thermo-clime using a wireless sensor network which implements a distributed binary search algorithm to find a local temperature gradient maxima.

The deployment is done using n nodes deployed in vertically where the topmost node is connected to the external world. Each node has the communication range limited being able to communicate only with its nearby nodes. Moreover, each node also has a pressure sensor, since the change of water pressure is linear in the change of depth. Hence, by measuring the pressure around it, the node is able to estimate its depth. Moreover authors assume that nodes can change their depth using an actuator to change the buoyancy of the node.

The nodes perform a distributed binary search implementation in order to determine the most suitable area for microorganisms. In summary, each node monitors its own temperature and pressure, defined as (t_i, p_i) and the ones provided by its neighbours, that is, the upper-most point and lower-most point. Such monitored data are defined as (t_u, p_u) and (t_b, p_b) respectively. Then, the node may move to the new location defined as the pressure $P_m = (p_u + p_b/2)$. Note that this is the equidistant distance between the given node and such boundary nodes and also P_m divides the search space in two regions. Then, the node sense again the data (t_i', p_i') and calculate the differences of temperature with the information provided by the upper and lower nodes, that is, $dT_t = |t'_i - t_t|$ and $dT_b = |t'_i - t_b|$, respectively. Then, if $(dT_t > dT_b)$, the node will discard the lower space and the remaining part of the search region is the new search region.

Authors define four different messages in the proposed architecture. An initialisation message used to indicate the sensor that they have to create to the routing tree and the registration message used to really build the routing tree. Moreover, another message for querying the maximum temperature gradient is sent from the user to the sensing platform and, finally, the most important message which is used to provide the result obtained. Such messages provide the information about the maximum gradient, its location (pos), the node who found it (id) and current resolution (posDiff).

Then, the proposal performs in-network data aggregation in order to calculate the response message distributed optimising the energy efficiency of the proposal. In summary, when a node receives a query message, it forwards the message to its children and, at the same time, it starts the local maximum gradient search. However, it just executes one step, and then waits for responses from its children.

After it receives replies from its children, the node compares the conclusions of its children and its own. As a result, the node may update the field available in the result message inserting any of these values or just relaying the message to its parent in the routing tree. In fact, the node always updates the maximum discarded temperature gradient field (available also in the report message) if necessary in order to suppress unnecessary local maximum search and messaging. At the end, the report message contains the global maximum temperature gradient and the information about the node, its position and who detected it.

Another application scenario underwater is the long-term monitoring of coral reefs and fisheries. In this sense, (Vasilescu et al., 2005) present a novel platform for underwater sensor networks to be used for this purpose. The distributed sensing platform consists of static and mobile underwater sensor nodes. Regarding the static nodes, they are Fleck CPU motes interfaced to a special optical communications board through 2 digital IO pins (IN and OUT). The Fleck is also interfaced with a sensor board. The boards are connected in a stack using stack-through connectors. The sensors and boards are contained in a yellow watertight *Otter box* which is guaranteed to be watertight up to a depth of 30 meters.

Each node has a high speed optical communication module with 2.2m of distance coverage within a cone range of 30 degrees and a maximum data rate of 320 kbits/s. Additionally, there is an acoustic communication module using 30 kHz FSK modulation with a range of 20m omni-directional, and a data rate of 50bit/s. Each node has a pressure sensor, temperature sensor, and a CMUCam camera capable of colour pictures with a 255×143 resolution.

The mobile node is an autonomous underwater vehicle (AUV) used to dock and transport stationary nodes and also to act as a data mule for retrieving information. This is an acrylic tube in which four different thrusters are located, two of its thrusters are aligned vertically and another two positioned horizontally to provide forward and backward movement in the horizontal plane as well as rotation. A pressure sensor is used to determine the deep and a magnetic compass used for orientation feedback. Note that this device is power hungry, thus is powered using a 140Wh lithium battery which provides around 4 hours of continuous working life whilst all sensors and systems are running. The bottom cap of the robot has a cone shaped cavity, designed for maximum mechanical reliability in docking and for optical communication. A robot can dock with sensor nodes in order to pick them up and transport them to a new location. This operation enables autonomous network deployment, reconfiguration and retrieval. Most of the electronics inside the robot, including the batteries, are placed in small Otter watertight cases.

Regarding data processing, authors use a data mule approach in which the AUV periodically visit the nodes to download the stored data. During the AUV-sensor communication, the AUV receives all the monitored information in the sensor and also uploads data to the static node, in essence, to adjust its clock and to change the data sampling rate. The key challenges for underwater data mulling are (1) locating the first node. Authors use the given GPS coordinates of the first sensor and the AUV can perform surface navigation guided by GPS to move toward the node; (2) locating the next node in the sequence. Authors use two complementary approaches: one based on image recognition and another one based on active beaconing of the sensors; (3) controlling the hover mode for the mobile node, which could also be controlled using active beaconing; (4) data transfer in a one-hop routing environment (point-to-point). This is done with a simple master-slave protocol in which the mobile node begins with a query about the available data and then the data is transmitted in 239 byte check summed packets. At the end of the data transmission, the mobile node asks the static node to reset and erase the data that was collected; and finally (5) synchronising clocks so that the data collected by the sensor network is time stamped in a consistent way. Authors use the first request-response package transmission to read the node's clock. Then, if this time is different from the current time on the mobile node's clock, all the data time stamps are adjusted and the current time of the node is reset to match the mobile node's time.

Appendix A

Further Details on Potential Devices and Systems

A.1   Accelerometers

In the last decade, the usage of accelerometers for elderly people has been very successful. Elderly people are expected to live longer and they have a choice to live at the place they prefer, while ensured a high quality of life, autonomy and security as well as simultaneously reducing their expenditures on in-patient care, which will have tremendous impacts in the very near future. This also includes assistance to carry out daily activities enhancing safety and security as well as getting access to social, medical and emergency systems. Receiving social and medical support in various innovative ways contributes to independent living and quality of life for many elderly and disabled people. All this can be summarised in the Ambient Assisted Living (AAL), being monitored as a patient for any vital signs of difficulty can be gathered which consequently can help and supervise from clinical data, such as health status, falling risk and, of course, the efficiency of rehabilitation. To this effect we look at (Dinh et al., 2009) new design work. They consider using accelerometers and sensors for medical applications. Those sensors have significant advantages, for example when used with micro- electro-mechanical system (MEMS) accelerometers one can monitor the movement of the subject while simultaneously not interrupting his or her daily routines. Also, more functional, lighter, more reliable optical tracking systems can be produced at a fraction of the cost of the conventional macro-scale accelerometer elements.

Distributed Sensor Systems: Practice and Applications, First Edition.
Habib F. Rashvand and Jose M. Alcaraz Calero.
© 2012 John Wiley & Sons, Ltd. Published 2012 by John Wiley & Sons, Ltd.

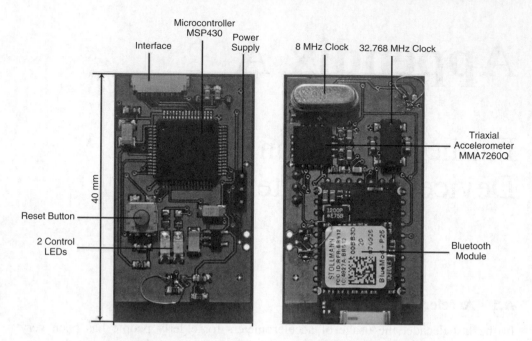

Figure A.1   MMA7260Q acceleration sensor unit with Bluetooth module.

Here a triaxial accelerometer MMA7260Q from a Freescale Semiconductor enhanced by a bluetooth module attached near the waist is used to capture the movements and to detect a fall (see Figure A.1). The sensitivity of the sensor is ±1.5 g, noise level of 4.7 mVrms, a sampling rate of 512 Hz and 12-bit resolution.

To clean the signal from noise and interferences filtering and further signal processing has been adopted. For example a simple one-sided moving average filter with an empirically determined window size of n = 5 samples lead to adequate results with the sensor module used in this work. Figure A.2 shows an illustrative example. As a consequence, the peak-to-peak noise level of the accelerometer could be reduced from 16 mg to 4 mg.

A testing benchmark included five test simulations of the fall scenarios for five times listed in Table A.1. Heavy falls result in a reliable detection rate. Nevertheless, the subjects tried not to fall too heavily. The results are achieved without considering the features' velocity and displacement.

Wearable gadgets are becoming an extremely popular part of recent sensor activities. Being attached to a shirt, on the neck or attached to a belt all these systems have been through many stages over the last decade. One good design is from (Hu et al., 2009) which includes ECG, EEG, RFID, and so on to make it a useful first aid medical care gadget. A few samples from the work are shown in

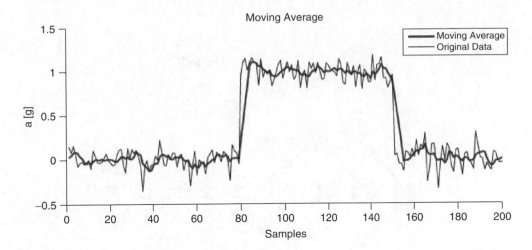

Figure A.2   Moving average filter with five samples window size applied to acceleration signal.

Table A.1   Results of sensitivity and specificity performing
different fall scenarios

Fall Scenario	Sensitivity	Specificity
Collapse	88%	99,76%
Forward	96%	99,64%
Backward	96%	99,40%
Sideward	96%	99,80%

Figure A.3 as a deployment of smart autonomous ECG sensors following a course of unstructured network.

In order to increase reliability of the EEG system and ensure safe warning of an abnormal heart beat or brain status, the health emergency warning and monitoring system is based on using predictable patterns. The mental disorder warning signals could be triggered in two points in the system. Figure A.4 shows the accurate and reliable design.

A.2   Equipment

We have also mentioned in various parts of the book that sensors and actuators, whether smart, autonomous or intelligent for any potential application, can be best used as an integrated part in a larger system. A pure set of homogenous DSS or its wireless version WSS interconnected together, therefore cannot provide a viable reality on its own.

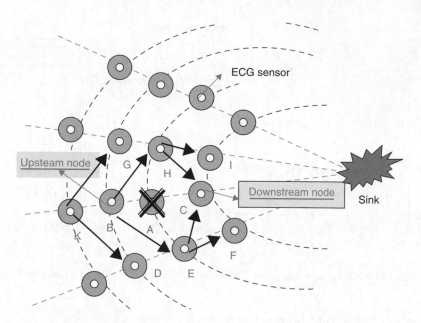

Figure A.3 Clustered sensing system using ripple-to-ripple loss recovery. © 2009 IEEE. Reprinted, with permission, from Hu et al (2009) Congestion-Aware, Loss-Resilient Bio-Monitoring Sensor Networking for Mobile Health Applications.

A.2.1 Tomography

The word tomography comes from building up an image from many complementary components, where each component carries a portion of the overall information. This case arises from the need to get an imaging knowledge of a region-of-interest (RoI), such as the human brain, by using some kind of radiation's properties (phenomena) such as reflection, refraction of non-destructive and harmless radiations. The theory highly depends on both the texture of RoI and the radiation and normally very complex, but basically relies on generating complementary components, orthogonal if possible, with minimum duplications of information required to build a complete image. This technique, though often related to medical images, has a wide range of applications in other fields and is expanding. It can use various waves and signal radiations such as atomic rays, electron rays, electric properties, magnetic, thermal radiations, neutrons, positron, optical, seismic waves, ultrasonic waves, X-ray, and so on.

The image processing part of the system usually demands very high computing power, and often comes with special architecture parallel processors to meet the required high volume of real time number crunching processing. For this there is equipment called computer tomography or computer-assisted tomography (CT).

Equipment, traditionally using X-ray, are bulky and very costly to maintain so that they have to be accommodated in a large dedicated room, so they are usually

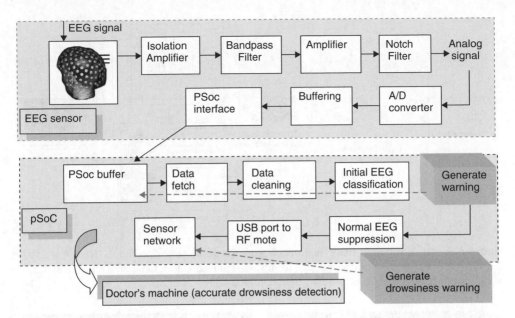

Figure A.4   EEG workflow design Drowsiness Warning System. © 2009 IEEE. Reprinted, with permission, from Hu et al (2009) Congestion-Aware, Loss-Resilient Bio-Monitoring Sensor Networking for Mobile Health Applications.

found in large hospitals and specialised institutions. With the rapid increase in demand and the cost of space, staff and energy many R&D activities are generating new innovative systems using new banks of sensor arrays for these and many other reasons such as:

- To make them intelligent and smarter by integrating them into new smart systems through two alternative strategies of (a) replacing those, if feasible, by miniaturised devices in the form of distributed systems or (b) enhance them to be interworking with other parts of smart systems.
- Alternative technologies for less radiations hazards, such as optical, ultrasonic.
- Alternative technologies for less energy, lighter and smaller for portability and reduced cost of maintenance.

Here, we briefly scan a few samples.

For example (Kim et al., 1997) work under 'Programmable Ultrasound Imaging Using Multimedia Technologies: A Next-Generation Ultrasound Machine', which brings in an enhanced algorithm for using new radiation of ultrasound technology. Their Programmable Ultrasound Image Processor (PUIP) facilitates a new ultrasound innovation for a multiple high-performance multimedia processor computing some 4 billion operations per second. Their algorithm provides the flexibility of making the system programmable so that it can be automatically configured to

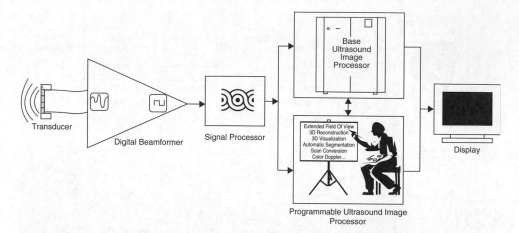

Figure A.5 Architecture of PUIP dual image processor in projected operation (Kim et al., 1997).
© 1997 IEEE. Reprinted, with permission, from Kim et al (1997) Programmable ultrasound imaging
using multimedia technologies: a next-generation ultra-sound machine.

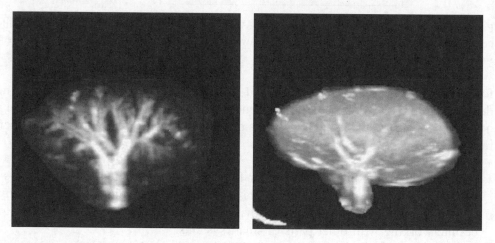

Figure A.6 Two different scans of 3-D-rendered images of the power mode images. © 1997
IEEE. Reprinted, with permission, from Kim et al (1997) Programmable ultrasound imaging using
multimedia technologies: a next-generation ultra-sound machine.

run optimally in various modes like B-mode, colour flow, cine and Doppler data.
Figure A.5 shows their projected system.

An example of the result of this image processor can be seen in Figure A.6.

Woolard et al. (2005) propose making use of a new unused frequency spectrum,
the Terahertz frequency sensing for imaging. As shown in Figure A.7, the THz
is an unused spectrum which means no interference and huge bandwidths need to
be utilised. The THz can yield at very small levels of power per frequency for
time-domain sampling, but naturally facilitate two-dimensional (2-D) and tomo-
graphic imaging. This also supports new developments for higher resolution EM

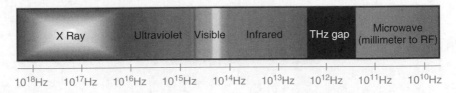

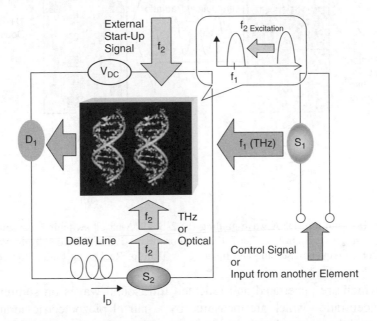

probing, analysis of semiconductor materials, non-invasive skin diagnostics, biomaterial, seismic and 3-D imaging.

An interesting adoption is the integrated architecture utilisation of combined electrical THz, and optical, called electro–THz–optical (ETO), which enables increased information-based THz spectral signature. This technique, which can be used for the proposed concept, is shown in Figure A.8, a tuneable DNA-based filter. The concept of the ETO-based architecture utilises two emission sources with 2D and 3D switching and DNA-based photonic band gap (PBG) shown in Figure A.9.

Another interesting extension to classic tomography is to make non-invasive use of ultrasonic and electrical properties of imaging systems for temperature and composition of the body that work simultaneously to build a multifunctional image. Kimoto and Shida (2008) achieve this multifunctional sensing using piezoelectric

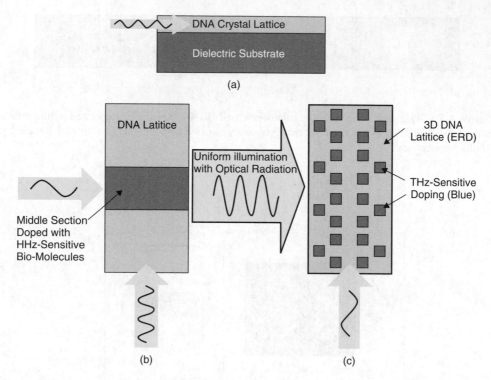

Figure A.9 (a) Thin-film DNA wave-guiding (b) 2-D DNA-based switching waveguide (c) 3-D DNA PBG with active control. © 2007 IEEE. Reprinted, with permission, from Woolard et al (2007) Terahertz Frequency Sensing and Imaging: A Time of Reckoning Future Applications?

ceramics which are penetrated and reflected ultrasound waves on sodium chloride (NaCl) concentration which are measured by a pair of piezoelectric ceramic transducers. Then, the electrodes on the surface of the transducers are measured to complement the process. Figure A.10 shows the multifunctional sensing method proposed and Figure A.11 shows the diagram for the measurement system.

For understanding the brain activities (Ohta et al., 2009) devise a new tomography-based imaging system to capture images from in the deep brain which claim to be very important for learning and memory. In this device they use a multi-modal CMOS-based device to simultaneously sense fluorescence and electrical potential in neural activities. Using its own previously developed devices integrated on one chip the captured image is not the whole image. Figures A.12, A.13 and A.14 show some operational details of the device.

A.2.2 Gadgets

As an alternative to the traditional electric shock to get a disordered heart back into its normal rhythms a self-activating device may help a patient to carry on his normal

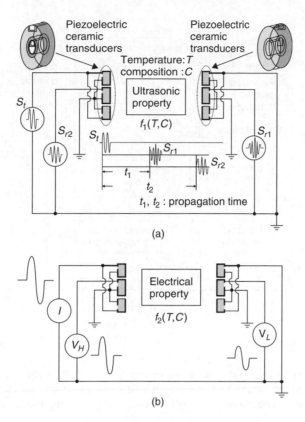

Figure A.10 Multifunctional sensing method that uses piezoelectric ceramic transducers (a) Measurement of the ultrasonic properties (b) Electrical measurement of the properties. © 2008 IEEE. Reprinted, with permission, from Kimoto and Shida (2008) A New Multifunctional Sensor Using Piezoelectric Ceramic Transducers for Simultaneous Measurements of Propagation Time and Electrical Conductance.

life. For most patients with severely disordered hearts an implanted device installed to act automatically is the best solution, so a gadget, normally called an implantable cardioverter defibrillator (ICD), shown in Figure A.15 is the best solution.

This is a battery-operated smart sensor-actuator device which keeps itself synchronised to the heartbeat, senses irregularities such as a missing pulse and generates an electric shock to the muscles to get the pulses back to normal. See Figure A.15.

Patients with cirrhosis often develop Porto systemic shunts to redistribute flow from the main portal vein in the liver. The patient needs to undergo liver transplantation, but the newly transplanted liver, if not controlled for the flow of blood could be damaged. For this problem (Jung et al., 2009) have designed a blood controlling gadget shown in Figure A.16. The gadget is a sensing tube integrated with a piezoelectric flow sensor to get all under control.

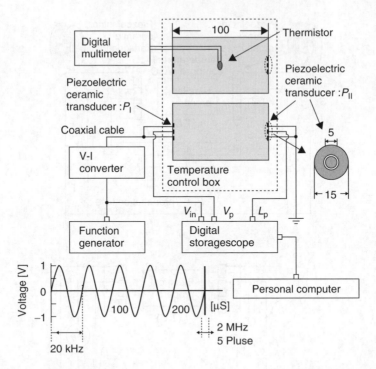

Figure A.11 Schematic diagram of the measurement system. © 2008 IEEE. Reprinted, with permission, from Kimoto and Shida (2008) A New Multifunctional Sensor Using Piezoelectric Ceramic Transducers for Simultaneous Measurements of Propagation Time and Electrical Conductance.

The device comes with two electrically isolated electrodes as input and output in Figure A.17. The electrodes are set on piezoelectric regions to improve the sensitivity of the sensor.

A.3 Smart Sensors Devices

This section describes the most common smart sensors used in the applications, scenarios and case studies described in this book. Only those general purpose sensors which are available on the market are exposed in this section leaving aside any custom-made sensor specifically developed by a particular scenario. The idea is not to provide a complete state-of-the-art topic but to supplement devices in the book in which the sensor devices used in the architectures and systems available are explained, giving the reader an idea of the internal hardware infrastructure.

A.3.1 Mica2 and Mica2Dot

The MICA2 and MICA2Dot Motes are used in a wide number of developments and applications throughout the book. They are the direct descendant of MICA Mote,

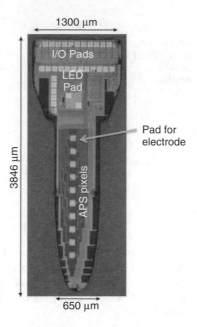

Figure A.12 Microphotograph of the sensor chip. © 2009 IEEE. Reprinted, with permission, from Ohta et al (2009) A multimodal sensing device for fluorescence imaging and electrical potential measurement of neural activities in a mouse deep brain.

developed by CrossBow Technology Inc and released in 2002 as the mote module used for enabling low-power, wireless, sensor networks.

The main difference between Mica 2 Dot and Mica 2 sensor nodes is shape and size of mote. The Mica 2 Dot has similar dimensions to a 2 Euro coin, and has a different pin arrangement of course. It has 24 solderless goldpins on its circumference. Mica 2 Dot has one LED on board as well as a temperature sensor. However, they share all the other specifications, thus both are going to be presented next. Figure A.18 shows an image of a Mica 2 Dot mote.

The MICA2/Mica2DOT Motes features several new improvements over the original MICA Mote. The following features make the MICA2 better suited to commercial deployment:

- 868/916 MHz, 433 or 315 MHz multi-channel transceiver with extended range;
- TinyOS (TOS) Distributed Software Operating System v1.0 with improved networking stack and improved debugging features;
- support for wireless remote reprogramming;
- wide range of sensor boards and data acquisition add-on boards.

They are based on the processor Atmel AVR ATmega 128L. The ATmega 128L is a low-power microcontroller which works at frequency 8 MHz. This processor possesses multiple assets which have given it an excellent reputation and have made it a

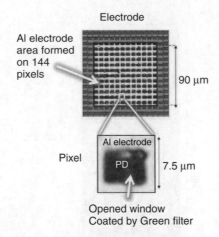

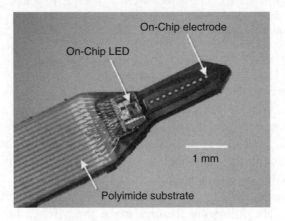

Figure A.13 Unit electrode area. Al was formed on the top of the pixels in the electrode area. The Al was removed only on the top of the PD. The PD surface was coated with Green filter to selectively transmitted fluorescence light. © 2009 IEEE. Reprinted, with permission, from Ohta et al (2009) A multimodal sensing device for fluorescence imaging and electrical potential measurement of neural activities in a mouse deep brain.

Figure A.14 Fabricated multimodal CMOS sensing device. © 2009 IEEE. Reprinted, with permission, from Ohta et al (2009) A multimodal sensing device for fluorescence imaging and electrical potential measurement of neural activities in a mouse deep brain.

perfect choice for a sensor node. It has 128K of program memory – an amount large enough for most applications; 4 KB SRAM for volatile data and 4 KB EEPROM for persistent data. Moreover ATMega 128L is characterised by extremely low power consumption in normal mode, aside from this it has six sleep modes. The best energy-saving mode processor only waits for interrupts while any other operations are stopped. Moreover it works at voltage range 2.7–5 V (mostly 3.3) which leads to a significant reduction of power needs compared to solutions, which operates at

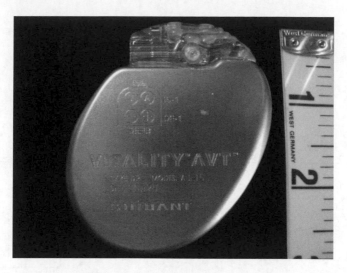

Figure A.15 Implantable cardioverter defibrillator. This photo is distributed under Creative Common License. Http://en.wikipedia.org.wiki/file:ICD.jpg.

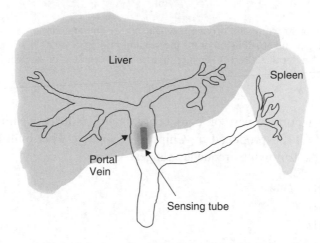

Figure A.16 Demonstration of the portal vein in place. © 2009 IEEE. Reprinted, with permission, from Jung et al (2009) A sensing tube with a integrated piozoelectric flow sensor for liver transplantation.

5 V. An important feature of ATMega 128L is the JTAG connection, which allows debugging 'on live', this means without a simulator, and SPI (Serial Programming Interface) which makes programming very simple. Moreover, it comes with a 10 bit ADC. The MICA2 51-pin expansion connector supports Analogue Inputs, Digital I/O, I2C, SPI, and UART interfaces. These interfaces make it easy to connect to a wide variety of external peripherals. Figure A.19 shows an image of a MICA2 mote.

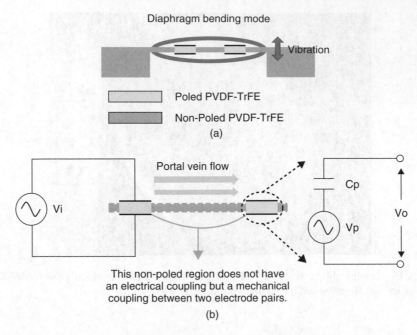

Diaphragm bending mode

Poled PVDF-TrFE

Non-Poled PVDF-TrFE

(a)

Portal vein flow

Cp

Vi

Vp

Vo

This non-poled region does not have
an electrical coupling but a mechanical
coupling between two electrode pairs.

(b)

Figure A.17 Sensor structure and driving mechanism: (a) The regions for input and output electrodes as a diaphragm and (b) Illustration of the sensing principle. © 2009 IEEE. Reprinted, with permission, from Jung et al (2009) A sensing tube with a integrated piozoelectric flow sensor for liver transplantation.

The second important part of core board is the radio transmitter. In Mica2/Mica2Dot the CC1100 produced by Chipcon is applied. This multichannel device can operate at the following frequency ranges:

- 300–348 MHz
- 400–464 MHz
- 800–928 MHz.

The operating range differs depending on the chosen frequency, the maximum value is 1000 ft (approx. 330–350 m). The transmission is performed at 38.4 kbaud speed, the data are encoded by means of Manchester code.

Various sensor and data acquisition boards are available from Crossbow. These boards connect to the MICA2 through a surface mount 51-pin connector. Crossbow supplies the several sensor boards extension for the incorporation of photocell, thermistors, microphone, magnetic and acceleration sensors, experiment board, and so on. Regarding power consumption, Mica2 is able to operate with two AA batteries ranging the voltage between 2.7 and 3.3 V. It enables autonomy for more than one year using the sleeping mode efficiently.

Figure A.18   Screenshot of Mica 2 Dot mote developed by Crossbow Inc. © 2010 IEEE. Reprinted, with permission, from Liao et al (2010) Human cognitive application by using Wearable Mobile Brain Comptuter Interface.

Figure A.19   Screenshot of MICA2 Mote developed by Crossbow Inc. © 2004 IEEE. Reprinted, with permission, from Lorincz et al (2004) Sensor Networks for emergency responses: Challenges and opportunities.

A.3.2   MicaZ

The MICAz is a 2.4 GHz Mote module used for enabling low-power, wireless sensor networks. Product features include:

- IEEE 802.15.4 compliant RF transceiver;
- 2.4 to 2.48 GHz, a globally compatible ISM band;

- direct sequence spread spectrum radio which is resistant to RF interference and provides inherent data security;
- 250 kbps data rate;
- plug and play with Crossbow's sensor boards, data acquisition boards, gateways, and software.

The MicaZ is based on the Atmel ATmega128L, analogously to Mica2/Mica2Dot. A single processor board can be configured to run a sensor application/processing and the network/radio communications stack simultaneously. The 51-pin expansion connector supports analogue Inputs, Digital I/O, I2C, SPI and UART interfaces. These interfaces make it easy to connect to a wide variety of external peripherals.

The main difference with Mica2 is the radio interface providing an IEEE 802.15.4 radio that offers both high speed (250 kbps) and hardware security (AES-128). The 51-pin expansion connect is analogous to that the provided in Mica2 motes. Figure A.20 shows a shoot of the MicaZ mote.

A.3.3   Telos and TMote Sky

Telos, developed by Moteiv Corp and released in 2004, is the predecessor of TMote Sky released in 2006, and developed by the same company has been intensively used for years in the development of wireless sensing platforms and many of the applications and scenarios described in this book use such motes. They share the same specifications and only minimal differences between them can be found. Thus, let us explain TMote Skype, which is an ultra low power wireless module for use in sensor networks, monitoring applications and rapid application prototyping. TMote Sky leverages industry standards like USB and IEEE 802.15.4 to interoperate seamlessly

Figure A.20   Screenshot of MicaZ mote developed by Crossbow Inc. © 2004 IEEE. Reprinted, with permission, from Lorincz et al (2004) Sensor Networks for emergency responses: Challenges and opportunities.

Figure A.21 Screenshot of TMote Sky mote developed by Moteiv Corp. © 2005 IEEE. Reprinted, with permission, from Polastre et al (2005) Telos: Enabling ultra low-power wireless research.

with other devices. By using industry standards, integrating humidity, temperature and light sensors, and providing flexible interconnection with peripherals, TMote Sky enables a wide range of mesh network applications. TMote Sky is a drop-in replacement for Moteiv's successful Telos design. Figure A.21 shows the TMote Sky mote developed by Moteiv Corp.

The key features of TMote Sky device are the following, extracted directly from the datasheet specification provided by the vendor:

- 250 kbps 2.4 GHz IEEE 802.15.4 Chipcon Wireless Transceiver;
- interoperability with other IEEE 802.15.4 devices;
- 8 MHz Texas Instruments MSP430 microcontroller (10k RAM, 48k Flash);
- integrated ADC, DAC, Supply Voltage Supervisor, and DMA Controller;
- integrated onboard antenna with 50 m range indoors/125 m range outdoors;
- integrated humidity, temperature, and light sensors;
- ultra low current consumption;
- fast wakeup from sleep ($< 6\,\mu s$);
- hardware link-layer encryption and authentication;
- programming and data collection via USB;
- 16-pin expansion support and optional SMA antenna connector;
- TinyOS support : mesh networking and communication implementation;
- complies with FCC Part 15 and Industry Canada regulations.

Regarding the power, TMote Sky may be powered by two AA batteries. The module was designed to fit the two AA battery form factor. AA cells may be used in the operating range of 2.1 to 3.6 V DC, however, the voltage must be at least 2.7 V when programming the microcontroller flash or external flash.

The low power operation of the TMote Sky module is due to the ultra low power Texas Instruments MSP430 F1611 microcontroller featuring 10 kB of RAM, 48 kB of flash, and 128B of information storage. This 16-bit RISC processor features extremely low active and sleep current consumption that permits TMote to run for years on a single pair of AA batteries. The MSP430 has an internal digitally controlled oscillator (DCO) that may operate up to 8 MHz. The MSP430 has eight external ADC ports and eight internal ADC ports. The ADC internal ports may be used to read the internal thermistor or monitor the battery voltage. A variety of peripherals are available including SPI, UART, digital I/O ports, Watchdog timer, and timers with capture and compare functionality. The F1611 also includes a two-port 12-bit DAC module, Supply Voltage Supervisor, and three-port DMA controller.

TMote Sky features the Chipcon CC2420 radio for wireless communications. The CC2420 is an IEEE 802.15.4 compliant radio providing the PHY and some MAC functions. The CC2420 is controlled by the TI MSP430 microcontroller through the SPI port and a series of digital I/O lines and interrupts. The radio may be shut off by the microcontroller for low power duty cycled operation. TMote Sky uses the ST M25P80 40 MHz serial code flash for external data and code storage. The flash holds 1024 kB of data and is decomposed into 16 segments, each 64 kB in size. The flash shares SPI communication lines with the CC2420 transceiver.

The humidity/temperature sensor is manufactured by Sensirion AG. The SHT11 and SHT15 models may be directly mounted on the TMote Sky module. The SHT11/SHT15 sensors are calibrated and produce a digital output. The calibration coefficients are stored in the sensor's onboard EEPROM. The difference between the SHT11 and SHT15 model is that the SHT15 produces higher accuracy readings. The sensor is produced using a CMOS process and is coupled with a 14-bit A/D converter. The low power relative humidity sensor is small in size and may be used for a variety of environmental monitoring applications.

TMote Sky has connections for two photodiodes. Moteiv currently uses photodiodes from Hamamatsu Corporation. The default diodes are the S1087 for sensing photo-synthetically active radiation and the S1087-01 for sensing the entire visible spectrum including infrared.

A.3.4 Fleck3 and FleckNano

Fleck motes are developed by CSIRO starting in 2003 with Fleck 1 and being Fleck 3 Rev.B and FleckNano the latest version available. Fleck3 mote is used in many applications described in this book. The Fleck devices incorporate a number of novel design features that set them apart from other devices on the market: superior radio for outdoor applications, superior power supply that is solar-capable, and an extensive range of sensors and sensor interfaces.

Fleck3 motes are based on the Atmel Atmega 1281 processor which has 128 kbyte program flash, 8 kbyte RAM and is able to run the TinyOS operating system smoothly. It comes with a wide set of built-in sensors including three LED indicators, temperature sensor, real-time clock which allows for very deep sleep mode, and 1 Megabyte external flash memory. Regarding radio interface, it relies on the Nordic 905 chipset for establishing a 433/915 MHz communication using GFSK modulation. It achieves an effective communication of 50 kbps covering a range of more than 1000 m in outdoor scenarios. The power supply ranges from 3.5–8 V and it enables a significant optimisation for sleeping modes in which only 33 uA is consumed. The supported power supplies are: 3xAA rechargeable batteries with overcharge protection, super capacitors and solar panel cells. The use of an onboard real-time clock means that the processor can completely shut down for minutes to hours at a time since it is freed from the burden of time keeping. Figure A.22 shows a snapshot of the Fleck3 mote.

Fleck Nano is designed for a variety of applications that require a small unobtrusive wireless sensor. Intended applications are contact logging for livestock research, indoor RF tracking and human movement analysis. The Fleck Nano can be interfaced to a variety of different sensors such as temperature, humidity, barometric and gyroscopic sensors. It is in essence an onboard 8051 microcontroller, EEPROM and a 3-axis accelerometer for motion detection. The Nano uses a Nordic 915 MHz RF transceiver which allows it to communicate with the Fleck3 platform, and the radio range is 7 m to 20 m indoors. The dimensions are 25 mm × 20 mm.

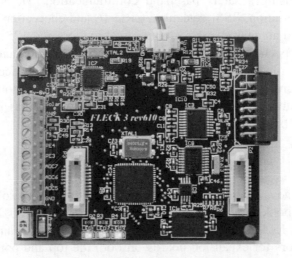

Figure A.22 Screenshot of Fleck3 mote developed by CSIRO. © 2007 IEEE. Reprinted, with permission, from Sitka et al (2007) Fleck - a platform for real world outside sensor network.

A.3.5   3Mate!

3Mate! is developed by TRETEC as a very low power sensor module featured for Wireless Sensor Networks. It's a small device developed mainly as a node for generic IEEE 802.15.4 wireless network. The aim of 3Mate! is to allow a rapid development of monitoring applications, as distributed environmental ones. The main components of 3Mate! are the Texas Instrument's MSP430F1611 microcontroller and the CC2420 radio transceiver: both chips on top of their product classes. They provide the principal calculation and computation features to the module and they are also specifically developed for low power consumption.

Moreover 3Mate! provides storage capabilities and a set of three useful expanding connectors for communication and analogue interfacing. Expanding connectors allow the module to be properly expanded with suitable daughter boards. In particular the daughter boards can support a wide range of sensors and different type of network interface standards like UART, USB, GPRS, Ethernet and Wi-Fi.

3Mate! supports TinyOS operating system out-of-the-box and it's fully interoperable with some other previous products like Telos or Telos rev.B. 3Mate! is small in size, low power, modular and customisable.

The 16-bit MSP430F1611 process provides a RISC architecture in which there is a 125 ns instruction cycle time. The process has 10 KB RAM, 48 KB Flash, General Purpose I/O lines, two internal timers and watchdog. Moreover, it exposes a 12-Bit A/D Converter with internal reference, sample-and-hold, and autoscan features. The mote come with 1 Mbyte flash memory, but it can be updated up to 4 Mbs. Regarding the network interface, it achieves 250 kbps using a 2.4 GHz CC2420 RF transceiver which provides a comfortable IEEE 802.15.4 communication interface using a safety baseband communication. Moreover, it provides On-chip synchronisation, RSSI detection, CRC-16 and encryption capabilities. The integrated onboard antenna enables a 50 m range indoors and 125 m range outdoors and the mote has been designed to support an optional external antenna. Expansion connectors in the mother board that give access to more advanced interfaces like USB, GPRS, Ethernet and Wi-Fi. One reset button, one user button and three LEDs for direct application control are also built-in. Moreover, it has been designed to work with 2xAA batteries or analogous.

A.3.6   IMote 2

The IMote2 is an advanced wireless sensor node platform used also in many systems described in this book. It is built around the low power PXA271 XScale CPU and integrates an 802.15.4 compliant radio. The design is modular and stackable with interface connectors for expansion boards on both the top and bottom sides.

The connectors provide a standard set of I/O signals for basic expansion boards and provide additional high-speed interfaces for application specific I/O. A battery board supplying a system power can be connected. Regarding the processor, the IMote2 contains the Intel PXA271 CPU. This processor can operate in a low voltage (0.85 V), low frequency (13 MHz) mode, hence enabling very low power operation.

The frequency can be scaled from 13 MHz to 416 MHz with Dynamic Voltage Scaling. The processor has a number of different low power modes such as sleep and deep sleep. The PXA271 is a multi-chip module that includes three chips in a single package, the CPU with 256 kB SRAM, 32 MB SDRAM and 32 MB of FLASH memory. It integrates many I/O options making it extremely flexible in supporting different sensors, A/Ds, radios, and so on. These I/O features include I2C, two synchronous serial ports (SPI) one of which is dedicated to the radio, three high speed UARTs, GPIOs, SDIO, USB client and host, AC97 and I2S audio codec interfaces, a fast infrared port, PWM, a camera interface and a high speed bus (mobile scalable link). The processor also supports numerous timers as well as a real time clock. The PXA271 includes a wireless MMX coprocessor to accelerate multimedia operations. It adds 30 new media processor (DSP) instructions, support for alignment and video operations and compatibility with Intel MMX and SSE integer instructions.

Regarding the radio antenna, the IMote2 uses the CC2420 IEEE 802.15.4 radio transceiver from Texas Instruments. The CC2420 supports a 250 kb/s data rate with 16 channels in the 2.4 GHz band. The IMote2 platform integrates a 2.4 GHz surface mount antenna which provides a nominal range of about 30 meters. For longer ranges a SMA connector can be soldered directly to the board to connect to an external antenna.

The primary battery is typically accomplished by attaching a Crossbow IMote2 battery board. Moreover, a rechargeable battery can be used, but it requires a specially configured battery board attached. The IMote2 has a built-in charger for Li-Ion or Li-Poly batteries. Another alternative is that IMote2 can be powered via the on-board mini-B USB connector. This mode can also be used to charge an attached battery. Figure A.23 shows an image of the IMote2 mote.

A.3.7 System-on-Chip CC2510

The CC2510 system-on-chip developed by Texas Instruments released in 2010 has been used in several scenarios proposed in this book. The CC2510Fx/CC2511Fx is a true low-cost 2.4 GHz system-on-chip (SoC) designed for low-power wireless applications. The CC2510Fx/CC2511Fx combines the excellent performance of the state-of-the-art RF transceiver CC2500 with an industry-standard enhanced 8051 MCU, up to 32 kB of in-system programmable flash memory and 4 kB of RAM, and many other powerful features. The small 6x6 mm package makes it very suited for applications with size limitations.

The CC2510Fx/CC2511Fx is highly suited for systems where very low power consumption is required. This is ensured by several advanced low-power operating

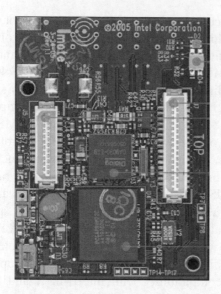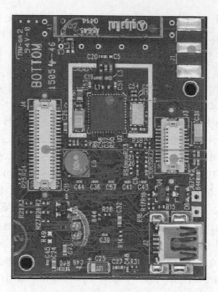

Figure A.23 Screenshot of IMote2 mote developed by CrossBow Inc. © 2008 IEEE. Reprinted, with permission, from Nachman et al (2008) Imote 2: Seriuous Computation at the edge.

modes. The CC2511Fx adds a full-speed USB controller to the feature set of the CC2510Fx. Interfacing to a PC using the USB interface is quick and easy, and the high data rate (12 Mbps) of the USB interface avoids the bottlenecks of RS-232 or low-speed USB interfaces.

Regarding the radio interface, it provides a high-performance RF transceiver based on the market-leading CC2500, excellent receiver selectivity and blocking performance working with a high sensitivity (-103 dBm at 2.4 kBaud). The radio frequency range is between 2400 and 2483.5 MHz and it provides extra added values like digital RSSI/LQI support.

The microcontroller employed is the high performance and low power 8051 microcontroller core. It is a 8/16/32 kB in-system programmable flash, and 1/2/4 kB RAM, with full-speed USB controller with 1 kB USB FIFO. Moreover, it offers I2S interface, 7–12 bit ADC with up to eight inputs, 128-bit AES security coprocessor, powerful DMA functionality, two UARTs, 16-bit timer with DSM mode, three 8-bit timers and hardware debug support apart from 19 general purpose I/O pins. The voltage ranges between 2.0 V and 3.6 V. It is important to emphasise that this mote does not support communications with ZeeBee-based motes even when they are working in the same 2.4 GHz radio frequency band just with the transceiver from the family CC25xx.

A.3.8 System-on-Chip CC2530

The CC2530, developed by Texas Instruments, is a true system-on-chip solution for 2.4-GHz IEEE 802.15.4 and ZigBee Applications. Note that it provides compatibility with ZigBee communication. The CC2530 has been released in 2010. Regarding the microcontroller, the CC2530 comes with a High-Performance and Low-Power 8051 microcontroller core with code prefh and so on. It is a 32-, 64-, 128-, or 256-KB in-system-programmable flash, 8-KB RAM with retention in all power modes and with hardware debug support.

Regarding the radio interface, it uses a 2.4-GHz IEEE 802.15.4 compliant RF transceiver with an excellent receiver sensitivity and robustness to interference. Only a single crystal is needed for configurable resolution asynchronous networks. The idea is that the device is suitable for systems targeting compliance with several serial protocols with worldwide radio-frequency regulations: ETSI EN 300 328 and EN 300 (19 × 4 mA, 2 × 20 mA) 440 (Europe), FCC CFR47 Part 15 (US) and ARIB STD-T-66 (Japan).

The CC2530 comes with five-channel DMA and an integrated High-Performance Op-Amp and ultralow-power comparator. Moreover, it provides an IEEE 802.15.4 MAC Timer, general-purpose timers (one 16-Bit, two 8-Bit), 32-kHz sleep timer with capture and IR generation circuitry. The mote provides Accurate Digital RSSI/LQI Support, battery monitor and temperature sensor, 12-Bit ADC with eight channels and configurable resolution, AES Security Coprocessor and 21 general purpose I/O pins.

A.4   Networks and Protocols

This section describes different network and protocol used throughout the book in order to give the reader a broad context for them.

A.4.1   ZigBee

The term ZigBee is intensively used in this book assuming the reader will already know about it. However, this section describes the ZigBee protocol stack for those readers who are not familiar with it. ZigBee is a set of protocols and layer specifications, which cover the first three levels of the OSI stack. Thus, at the bottom, the IEEE802.15.4 PHY physical layer accommodates high levels of integration by using direct sequence to permit simplicity in the analogue circuitry and enable cheaper implementations. On top of it, the IEEE802.15.4 MAC media access control layer permits use of several topologies without introducing complexity and is meant to work with large numbers of devices. Moreover, the network and application support layer permits growth of network without high power transmitters. This layer can handle huge numbers of nodes. This level in the ZigBee architecture includes the ZigBee Device Object (ZDO), user-defined application profile(s) and the Application Support (APS) sub-layer.

The APS sub-layer's responsibilities include maintenance of tables that enable matching between two devices and communication among them, and also discovery, the aspect that identifies other devices that operate in the operating space of any device.

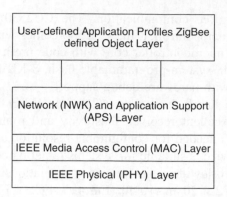

Figure A.24   Protocol Stack of the ZigBee specification.

The responsibility of determining the nature of the device (Coordinator/FFD or RFD) in the network, commencing and replying to binding requests and ensuring a secure relationship between devices rests with the ZDO (ZigBee Define Object). The user-defined application refers to the end device that conforms to the ZigBee Standard. Figure A.24 shows the protocol stack provided as part of the ZigBee specification.

There are three different ZigBee device types that operate on these layers in any self-organising application network. These devices have 64-bit IEEE addresses, with the option of enabling shorter addresses to reduce packet size, and work in either of two addressing modes – star and peer-to-peer. The first device is the ZigBee coordinator node, only one in each network to act as the router to other networks, and can be likened to the root of a (network) tree. It is designed to store information about the network. The second device type is the full function device FFD, acting as intermediary router transmitting data from other devices. It needs lesser memory than the ZigBee coordinator node, and entails lesser manufacturing costs. It can operate in all topologies and can act as a coordinator. Finally, the reduced function device RFD is just capable of talking in the network; it cannot relay data from other devices. It requires even less memory, (no flash, very little ROM and RAM), and thus is cheaper than an FFD. This device talks only to a network coordinator and can be implemented very simply in star topology. Figure A.25 shows an overview of the network topology available in ZigBee specification.

Regarding the physical layer, 2.4 GHz and 868/915 MHz are dual PHY modes. This represents three license-free bands: 2.4–2.4835 GHz, 868–870 MHz and 902–928 MHz. The number of channels allotted to each frequency band is fixed at 16 (numbered 11–26), one (numbered 0) and ten (numbered 1–10) respectively. The higher frequency band is applicable worldwide, and the lower band in the areas of North America, Europe, Australia and New Zealand.

Low power consumption with battery life ranging from months to years is a clear design objective of ZigBee. In the ZigBee standard, longer battery life is achievable

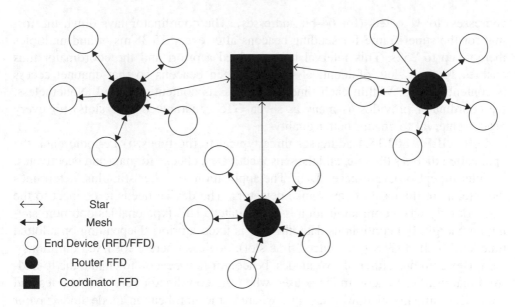

Figure A.25   Network topology associated with ZigBee specification.

by either of two means: continuous network connection and slow but sure battery drain, or intermittent connection and even slower battery drain. The maximum data rates allowed for each of these frequency bands are fixed as 250 kbps for the 2.4 GHz, 40 kbps for 915 MHz, and 20 kbps for 868 MHz frequencies.

The MAC layer accesses to the medium using carrier sense multiple access with collision avoidance (CSMA-CA) algorithm and using an address space of 64 bit length, 65,535 networks nodes. The typical wireless coverage range is about 50 m and the communication between nodes is based on a fully reliable 'hand-shaked' data transfer protocol.

The functions of the coordinator, which usually remain in the receptive mode, encompass network set-up, beacon transmission, node management, storage of node information and message routing between nodes. The network node, however, is meant to save energy (and so 'sleeps' for long periods) and its functions include searching for network availability, data transfer, checks for pending data and queries for data from the coordinator. For the sake of simplicity without jeopardising robustness, this particular IEEE standard defines a quartet frame structure and a super-frame structure used optionally only by the coordinator. The four frame structures are: beacon frame for transmission of beacons, data frame for all data transfers, acknowledgement frame for successful frame receipt confirmations and MAC command frame. These frame structures and the coordinator's super-frame structure play critical roles in security of data and integrity in transmission.

All protocol layers contribute headers and footers to the frame structure, such that the total overheads for each data packet range are from 15 octets (for short addresses) to 31 octets (for 64-bit addresses). The coordinator lays down the format for the super-frame for sending beacons after every 15.38 ms or/and multiples thereof, up to 252s. This interval is determined a priori and the coordinator thus enables 16 time slots of identical width between beacons so that channel access is contention-less. Within each time slot, access is contention-based. Nonetheless, the coordinator provides as many as seven GTS (guaranteed time slots) for every beacon interval to ensure better quality.

ZigBee/IEEE 802.15.4 addresses three typical traffic types. i) Periodic data, the application dictates the rate, and the sensor activates, checks for data and deactivates. II) Intermittent or Event-based Data. The application, or other stimulus, determines the rate, as in the case of say smoke detectors. The device needs to connect to the network only when communication is necessitated. This type enables optimum saving on energy. III) continuous data, the rate is fixed a priori. Depending on allotted time slots, called GTS (guaranteed time slot), devices operate for fixed durations.

ZigBee employs either of two modes, beacon or non-beacon to enable the to-and-from data traffic. Beacon mode is used when the coordinator runs on batteries and thus offers maximum power savings, whereas the non-beacon mode finds favour when the coordinator is mains-powered.

In the beacon mode, a device watches out for the coordinator's beacon that gets transmitted periodically, locks on and looks for messages addressed to it. If message transmission is complete, the coordinator dictates a schedule for the next beacon so that the device goes to sleep; in fact, the coordinator itself switches to sleep mode. While using the beacon mode, all the devices in a mesh network know when to communicate with each other. In this mode, necessarily, the timing circuits have to be quite accurate, or wake up sooner to be sure not to miss the beacon. This in turn means an increase in power consumption by the coordinator's receiver, entailing an optimal increase in costs.

The non-beacon mode will be included in a system where devices are asleep nearly always, as in smoke detectors and burglar alarms. The devices wake up and confirm their continued presence in the network at random intervals. On detection of activity, the sensors spring to attention, as it were, and transmit to the ever-waiting coordinator's receiver (since it is mains-powered). However, there is the remotest of chances that a sensor finds the channel busy, in which case the receiver unfortunately would miss a call.

A.4.2   RFID and Wireless Sensor Integration

The combination of RFID-equipped sensor and working for low proximity identification and the wireless sensor communication technologies is being used in some examples in this book. For example, the research work proposed by Tapia et al. (2010) in Section 4.5, describes a combined usage of RFID-equipped sensors with either ZigBee-based and Bluetooth-based wireless sensor devices. Concretely,

ZigBee devices employed are composed of a C8051F121 microcontroller and a CC2420 IEEE 802.15.4 radio frequency transceiver, whereas the Bluetooth biomedical sensors use a BlueCore4-Ext chip with a reduced instruction set computer (RISC) microcontroller with 48 kB of RAM and 1024 kB of external flash memory, and are compatible with the Bluetooth 2.0 standard. The integration between RFID-equipped devices and sensor-equipped devices has been analysed by (Zhang and Wang, 2006). In summary, they propose three different ways of integration between technologies.

One trend of the development of RFID is integrating it into the network. An RFID network is very mature now such as Real time locating System (RTLS), which implies integrating sensor nodes into RFID to get more environment information. A mix of tags and sensor nodes are deployed in any detected area. Smart stations gather information from tags and sensor nodes then transmit it to the base station. Here RFID and WSN information can be integrated in the base station.

As no network stack is embedded into the reader in a RFID system at present, the reader can only be operated passively and all its behaviour is controlled by the local control system. Moreover, the position of antennas of an RFID reader must be computed carefully to cover all the tags in range and not to conflict with other antennas or readers. All of these disadvantages limit the applications of RFID. If functions of a reader are cut short, an RFID reader might get much smaller, less expensive and easy to deploy. So, a new smart sensor containing less functional reader is provided by authors in order to be applied into applications without strict real-time requirement. A wide application in industry, such as maintenance and inventory security is a good example. Another example is the usage of smart nodes with temperature sensors deployed densely in the smart warehouse. Proper deployment algorithms should be used to adopt as few as nodes as possible to make all the tags in the range of readers. The maintenance of inventory can be identified by a smart node tracking system. An alarm could be triggered when an asset leaves a facility without authorisation. Furthermore, the temperature of some place in the warehouse can be controlled and can put out fire immediately before it gets out of control if it is much higher than usual.

A.4.3 Wireless Sensors for Industrial Environments

Throughout the book there have been an intensive number of references to ZigBee-based sensing platform developments. However, there are other solutions to be considered as alternatives to such technologies for the industry. They are especially expected to work efficiently in industrial scenarios in which hard environmental conditions may affect the efficiency of the wireless performance. This is the case of: *WirelessHART* (Song et al., 2008), *WIA-PA* (IEC) and *ISA100.11a* (International Society of Automation). The three networks are based on IEEE 802.15.4 and ZigBee technologies, albeit after various adaptations. These adaptations are directly related to the main features required from industry in order to provide an efficient utilisation of wireless networks in industrial environments:

Reliability: Industrial communication has to be highly reliable, not only when talking about the device itself, but also talking about communication's reliability. It may be increased by redundant communication paths, fault tolerance devices and communication protocols, and so on.

High Interference: A significant risk of electromagnetic interference and noise has to be taken into account for the industrial environment. There should be interference avoidance or mitigation measures in the wireless smart devices.

Security: Information about the process has to be protected from possible attacks by an insider or an outsider. These attacks could be a physical attack, sniffing, introduction of malicious packets, deliberate exposure and traffic analysis. Smart sensors have to be secure enough to avoid all of these.

Real-time: Real-time critical processes need to update information at fixed regular short intervals. The smart devices should be able to cope with asynchronous notifications of process safety in a guaranteed delivery time.

WirelessHART (Highway Addressable Remote Transducer Protocol) is an IEC 62591 standard defined as the extension of the wired HART protocol. It uses the IEEE 802.15.4 physical and MAC layers, but injects many unique features in these layers to cope with the stringent requirements for industrial purposes. It introduces channel hopping through frequency hopping spread spectrum (FHSS) and channel blacklisting into the MAC layer to minimise persistent interferences. It uses Time Division Multiple Access (TDMA) technology to arbitrate and coordinate communications between network devices. The TDMA establishes links specifying the time slot as 10 ms and frequency to be used for communications. These links are organised as superframes. This reduces the power consumption and provides collision free deterministic communications. It also provides Message Integrity Code (MIC) and authentication services providing confidentiality at both network and MAC layer.

ISA 100.11a is an industrial standard proposed by the International Society of Automation (ISA). It provides authentication methods hampering nodes to be joined in the network until the authentication has been successfully accomplished. The sensors or actuators provide data to routers but they cannot route data themselves. ISA 100.11a exposes many interesting features like interoperability, coexistence and determinism. The physical layer is built upon IEEE 802.15.4 operating in the 2.4 GHz band with 16 channels. The data link layer consists of the IEEE 802.15.4 MAC layers. It develops the MAC layers with an extended sub-layer to enhance features. It uses superframes such as WirelessHART and TDMA mechanism with CSMA-CA and the network can be configured to operate in different channel hopping schemes. The network layer is responsible for inter-network routing. In mesh

routing there is support for end-to-end network reliability and also special characteristics allow the network to adapt frequencies which can automatically suppress
coexistence issues. The transport layer is responsible for transparent transfer of
data between end systems. It includes the transport management entity, the transport
security entity and the transport data entity. ISA 100.11a also introduces a tunnelling
protocol. Tunnelling protocol allows the network to implement other existing protocols like HART, Modbus, Profibus and so on, which are already widely used in
the industry.

Finally, the WIA-PA (Wireless Networks for Industrial Automation-Process
Automation) is designed to meet the needs of process automation applications
as an IEC 62601standard. This protocol comprises seven logical roles: gateway,
redundant gateway, network manager (NM), security manager(SM), cluster head,
redundant cluster head and cluster member.

To guarantee real-time and reliable communication, WIAPA network uses the
beacon-enabled IEEE 802.15.4 superframe structure and TDMA to avoid packet
collisions. Frequency hopping is done using different techniques such as Adaptive Frequency Switch (AFS), Adaptive Frequency Hopping (AFH) and Timeslot
Hopping (TH). It is power efficient and has a very reliable self-organising network.

It works in a two-layer network topology that is hybrid star and mesh. The first
level of the network is mesh subnet where routing devices and gateway devices are
deployed. The second level of the network is star subnet where a routing device
and field devices are deployed. The routing devices play the role of cluster headers
and the field devices play the role of cluster members. Cluster members can only
forward data to the cluster head or accept data from the cluster head, however,
there is no direct communication among cluster members. The network manager
forms the mesh subnet, monitors the network and provides mechanisms for devices
joining and leaving the network.

A.5 Systems

CodeBlue[1] (Lorincz et al., 2004) is a new coding structure allowing for building
up wireless monitoring and tracking of patients and associates for emergency conditions. Its infrastructure requires a wide range of wireless devices with varying
capabilities to be integrated into medical, disaster response and emergency care
scenarios for new challenges to build the following requirements:

- tracking device locations;
- discovery and naming;
- robust routing;
- security;
- prioritisation of critical data.

---

[1] Available at http://fiji.eecs.harvard.edu/CodeBlue.

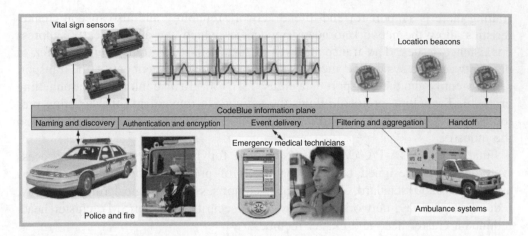

Figure A.26 The CodeBlue infrastructure for tracking of patients and associates for emergency condition. © 2004 IEEE. Reprinted, with permission, from Lorincz et al (2004) Sensor Networks for emergency responses: Challenges and opportunities.

Table A.2 Medical context of the patients monitored using telemedicine system

	Cardiac	Respiratory	With Mental Disease	Healthy over 40
Blood Pressure	YES	YES	YES	
Heart Rate	YES	YES	YES	YES
Temperature		YES	YES	
SpO$_2$	YES	YES		
ECG	YES			YES
Dyspnea	YES	YES		YES
Dizziness	YES		YES	YES
Pain	YES	YES	YES	

© 2010 IEEE. Reprinted, with permission, from Fengon et al (2010) A new telemedicine framework handling the emergency room overhead

Figure A.26 shows the structure of CodeBlue and the associated suite of protocols and services. Under the pervasive systems the community should work towards the seamless integration of computational devices for some time, but integrating low-power, low-capability wireless sensors into emergency and disaster response demands new approaches. For example, CodeBlue must run on sensor network devices with extremely limited resources then virtual machines and mobile agents generally become inappropriate for this domain.

To demonstrate wireless sensor nodes in disaster response, (Lorincz et al., 2004) have developed two mote-based vital sign monitors (Figure A.27), a pulse oximeter and a two-lead ECG monitor. The pulse oximeter captures a patient's heart rate and

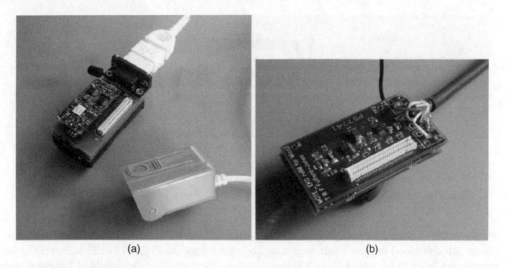

(a)   (b)

Figure A.27 Mote (a) Pulse oximeter and (b) Two-lead ECG. © 2004 IEEE. Reprinted, with permission, from Lorincz et al (2004) Sensor Networks for emergency responses: Challenges and opportunities.

SpO_2 by measuring the amount of light transmitted through a non-invasive sensor attached to the patient's finger. In practice emergency medical technicians (EMT) then use these standard vital signs to determine a patient's general circulatory and respiratory status, which are among the first vital signs.

If pulse oximeter is attached to the Mica2 mote, which transmits the heart rate and SpO_2 data periodically, about once a second and EKG continually monitors the heart's electrical activity in more severely injured patients. Then, a patient with internal bleeding requires cardiac monitoring. You can use ECG signals to detect arrhythmia or ischemia, and so on, then the monitoring PDA, Figure A.28, being used for multiple patient control shows real time activities of three patients running in such a crowded status.

With continuously growing demand on the limited resources, staff and space the need for a better controlled visiting systems and saving the resources for real emergencies and critical illnesses becomes apparent. To address this problem (Fengou et al., 2010) propose a telemedicine framework, a state controlled discipline infrastructure to build a consistent system based upon four structural elements:

- citizen's electronic medical record (EMR);
- citizen's profile;
- GP/MDPS profile;
- ERiOC profile.

The proposal under an expanded EMR and telemedicine requirement indicates changes to the systems. To clarify the idea and priorities they provide an example.

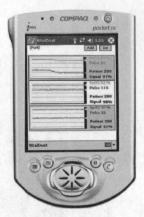

Figure A.28 PDA-based multiple-patient triage application. The screen shows real-time vital sign (heart rate and blood oxygen saturation) data from three patients. © 2004 IEEE. Reprinted, with permission, from Lorincz et al (2004) Sensor Networks for emergency responses: Challenges and opportunities.

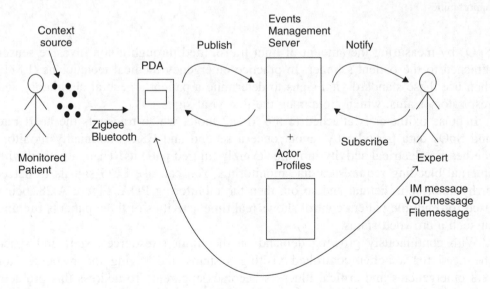

Figure A.29 General structure for telemedicine framework. © 2010 IEEE. Reprinted, with permission, from Fengon et al (2010) A new telemedicine framework handling the emergency room overhead.

The cases indicated in Table A.2 representing the parameters of the medical context of each of the four categories of conditional patients (c-patients) and the proposed framework are explained. For the deployment of the new system a new mechanism needs to be implemented, as shown in Figure A.29.

References

Ade M., Doulamis N., S. Wagle S., Ullah M.G. (2009) TeleHealth: Healthcare Technologies and TeleHealth Emergency (THE) System, in: IEEE (Ed.), 2nd International Conference on Wireless Communication, Vehicular Technology, Information Theory and Aerospace & Electronic Systems Technology, IEEE, Chennai. pp. 1–4.

Ahmed M.F., Vorobyov S.A. (2009) Node Selection For Sidelobe Control in Collaborative Beamforming for Wireless Sensor Networks, in: IEEE (Ed.), IEEE 10th Workshop on Signal Processing Advances in Wireless Communications, IEEE, Perugia. pp. 519–523.

Akkaya K., Younis M. (2005) A Survey on Routing Protocols for Wireless Sensor Networks. Ad Hoc Networks 3: 325–349.

Al-Karaki J.N., Kamal A.E. (2004a) A Survey on Sensor Networks. IEEE Wireless Communications 40: 1–37.

Al-Karaki J.N., Kamal A.E. (2004b) Routing Techniques in Wireless Sensor Networks: A Survey. IEEE Wireless Communications 11: 6–28.

Al-Karaki J.N., Ul-Mustafa R., Kamal A.E. (2004) Data Aggregation in Wireless Sensor Networks-Exact and Approximate Algorithms, in: IEEE (Ed.), IEEE Workshop on High Performance Switching and Routing, IEEE, Phoenix, Arizona.

Allen M.G. (2005) Micromachined Endovasculary-implantable Wireless Aneyrysm pressure sensors: from concept to clinic, in: IEEE (Ed.), 13th International Conference on Solid-State Sensors, Actuators and Microsystems, IEEE, Seoul, Korea. pp. 275–278.

Baghyalakshmi D., Ebenezer J., Satyamurty S.A.V. (2011) WSN based Temperature Monitoring for High Performance Computing Cluster, in: IEEE (Ed.), IEEE-International Conference on Recent Trends in Information Technology, IEEE, MIT, Anna University, Chennai. pp. 1105–1110.

Barr R., Bicket J.C., Dantas D.S., Du B., Kim T.W.D., Zhou B., Sirer E.G. (2002) On the Need for System-level Support for Ad Hoc and Sensor Networks. SIGOPS Operating System Review 36: 1–5.

Batalin M.A., Sukhatme G.S. (2004) Using a Sensor Network for Distributed Multi-robot Task Allocation, in: IEEE (Ed.), IEEE International Conference on Robotics and Automation, IEEE, New Orleans, LA, USA. pp. 158–164.

Benito-Lopez F., Coyle S., Byrne R., O'Toole C., Barry C., Diamond D. (2010) Simple Barcode System Based on Inonogels for Real Time pH-Sweat Monitoring, in: IEEE (Ed.), 2010 International Conference on Body Sensor Networks, IEEE, Singapor. pp. 291–296.

Benjamin S., Horan H.A., Marich M. (2009) Integrated Patient Health Information Systems to Improve Traffic Crash, in: IEEE (Ed.), 42nd Hawaii International Conference on System Sciences, IEEE, Big Island, HI pp. 1–10.

Bhatti S., Xu J. (2009) Survey of Target Tracking Protocols Using Wireless Sensor Network, in: IEEE (Ed.), 2009 Fifth International Conference on Wireless and Mobile Communications, IEEE, Cannes/La Bocca, French Riviera, France pp. 110–116.

Distributed Sensor Systems: Practice and Applications, First Edition.
Habib F. Rashvand and Jose M. Alcaraz Calero.
© 2012 John Wiley & Sons, Ltd. Published 2012 by John Wiley & Sons, Ltd.

Bhatti S., Carlson J., Dai H., Deng J., Rose J., Sheth A., Shucker B., Gruenwald C., Torgerson A., Han
  R. (2005) Mantis OS: An embedded multithreaded operating system for wireless micro sensor platforms.
  Mobile Networks and Applications 10: 563–579.

Biswas S., Tatchikou R., Dion F. (2006) Vehicle-to-Vehicle Wireless Communication Protocols for Enhancing
  Highway Traffic Safety. IEEE Communication Magazine 44: 74–82.

Borges L.M., Araújo P., Lebres A.S., Rente A., Salvado R., Velez F.J., Martinez-de-Oliveira J., Barroca N.,
  Ferro J.M. (2010) Wearable Sensors for Foetal Movement Monitoring in Low Risk Pregnancies, in: Springer
  (Ed.), Wearable and Autonomous Systems. pp. 115–136.

Braginsky D., Estrin D. (2002) Rumor Routing Algorithm in Sensor Networks, in: ACM (Ed.), ACM WSNA
  in conjunction with ACM MobiCom'02, ACM, Atlanta. pp. 22–31.

Brandl M., Grabner J., Kellner K., Seifert F., Nicolics J., Grabner S., Grabne G. (2009) A Low-Cost Wireless
  Sensor System and Its Application in Dental Retainers. IEEE Sensors Journal 9: 255–262.

Burrell J., Brooke T., Beckwith R. (2004) Vineyard Computing: Sensor Networks in Agricultural Production.
  IEEE Pervasive Computing 3: 38–45.

Byun J., Park S. (2011) Development of a Self-Adapting Intelligent Sensor for Building Energy Saving and
  Smart Services, in: IEEE (Ed.), IEEE International Conference on Consumer Electronics, IEEE, Las Vegas,
  NV. pp. 789–790.

Cao Q., Abdelzaher T., Stankovic J., He T. (2008) The LiteOS Operating System: Towards Unix Like Abstraction for Wireless Sensor Networks, in: IEEE (Ed.), 7th International Conference on Information Processing
  in Sensor Networks, IEEE, St. Louis, Missouri, USA.

Carullo A., Corbellini S., Parvis M., Vallan A. (2009) Wireless Sensor Network for Cold-Chain Monitoring.
  IEEE Transactions on Instrumentation and Measurement 58: 1405–1411.

Ceriotti M., Mottola L., Picco G.P., Murphy A.L., Guna S., Corrà M., Pozzi M., Zonta D., Zanon P. (2009)
  Monitoring Heritage Buildings with Wireless Sensor Networks: The Torre Aquila Deployment, in: ACM
  (Ed.), 8th ACM/IEEE International Conference on Information Processing in Sensor Networks ACM, San
  Francisco, CA, USA. pp. 1–12.

Compagna L., Khoury P.E., Massacci F., Saidane A. (2010) A Dynamic Security Framework for Ambient
  Intelligent Systems: A Smart-Home Based eHealth Application, in: M. L. G. e. al. (Ed.), Transaction on
  Computer Science (LNCS), Springer. pp. 1–24.

Corchado J.M., Bajo J., Tapia D.I., Abraham A. (2010) Using Heterogeneous Wireless Sensor Networks in a
  Telemonitoring System for Healthcare. IEEE Transactions on Information Technology in Biomedicine 14:
  234–240.

Costa P., Mottola L., Murphy A.L., Picco G.P. (2007) ProgrammingWireless Sensor Networks with the
  TeenyLIME Middleware, in: ACM (Ed.), 8th ACM/IFIP/USENIX International Middleware Conference
  ACM, Newport Beach (CA, USA). pp. 1–20.

County H. (2011) 2011 Public Safety Operating Budget: $254 million, Minnesota.

Chan A., Chang-Liew S. (2006) Merit of PHY-MAC Cross-Layer Carrier Sensing: A MAC-Address-based
  Physical Carrier Sensing Scheme for Solving Hidden-Node and Exposed-Node Problems in Large-Scale
  Wi-Fi Networks, in: IEEE (Ed.), IEEE Conference on Local Computer Networks, IEEE, Tampa, FL.
  pp. 871–878.

Chanboun A., Raissouni N. (2011) Wireless Sensor Network; Calibration; Modular In Situ System; Ad hoc;
  Satellite Remote Sensing; Land Surface Temperature. International Journal of Engineering Science and
  Technology 3: 5346–5357.

Chang W.-R., Lin H.-T., Cheng Z.-Z. (2008) CODA: A Continuous Object Detection and Tracking Algorithm
  for Wireless Ad Hoc Sensor Networks, in: IEEE (Ed.), 5th IEEE Consumer Communications and Networking
  Conference, IEEE, Las Vegas. pp. 168–174.

Chang W., Cao G., Porta T.L. (2004) Dynamic Proxy Tree-based Data Dissemination Schemes for Wireless
  Sensor Networks, in: IEEE (Ed.), IEEE International Conference on Mobile Ad-hoc and Sensor Systems,
  IEEE, Fort Lauderdale. pp. 21–30.

Chaudhary D.D., Nayse S.P., Waghmare L.M. (2011) Application of Wireless Sensor Network for Greenhouse Parameter Control in Precision Agriculture. International Journal of Wireless & Mobile Networks 3:
  140–149.

Chebrolu K., Raman B., Mishra N., Valiveti P.K., Kumar R. (2008) BriMon: A Sensor Network System for Railway Bridge Monitoring, in: ACM (Ed.), Sixth International Conference on Mobile Systems, Applications, and Services ACM, Breckenridge, CO, USA pp. 1–13.

Chen M., Gonzalez S., Vasilakos A., Cao H., Leung V.C.M. (2011) Body Area Networks: A Survey. Mobile Network Applications 16: 171–193.

Chen W.-T., Chen P.-Y., Lee W.-S., Huang C.-F. (2008) Design and Implementation of a Real Time Video Surveillance System with Wireless Sensor Networks, in: IEEE (Ed.), IEEE Vehicular Technology Conference, IEEE, Singapor. pp. 218–222.

Chen W., Chen L., Chen Z., Tu S. (2006) WITS: A Wireless Sensor Network for Intelligent Transportation System, in: IEEE (Ed.), First International Multi-Symposiums on Computer and Computational Sciences, IEEE, Hangzhou, Zhejiang, China. pp. 1–7.

Chen Y., Cheng L., Chen C., Ma J. (2009) Wireless Sensor Network for Data Sensing in Intelligent Transportation System, in: V. Spring (Ed.), IEEE 69th Vehicular Technology Conference, VTC Spring Barcelona. pp. 1–5.

Chintalapudi K., Fu T., Paek J., Kothari N., Rangwala S., Caffrey J., Govindan R., Johnson E., Masri S. (2006) Monitoring Civil Structures with a Wireless Sensor Network. IEEE Internet Computing 10: 26–34.

Chu M., Haussecker H., Zhao F. (2002) Scalable Information-driven Sensor Querying and Routing for Ad Hoc Heterogeneous Sensor Network. International Journal of High Performance Computing Applications 16: 293–313.

Chung W.Y. (2008) A Wireless Sensor Network Compatible Wearable U-healthcare Monitoring System Using Integrated ECG, Accelerometer and SpO$_2$, in: IEEE (Ed.), 30th Annual International Conference of the IEEE Engineering in Medicine and Biology Society, IEEE, Vancouver, BC pp. 1529–1532.

Daoxin D., Sailing H. (2011) Ultracompact Silicon Nanowire Circuits for Optical Communication and Optical Sensing, in: IEEE (Ed.), 2011 IEEE Winter Topicals, IEEE, Keystone, CO pp. 25–26.

Desourdis R.I., Smith D.R., Speights W.D., Dewey R.J., DiSalvo J.R. (2002) Emerging Public Safety Wireless Communication Systems Artech House, Inc., Massachusetts.

Dinh C., Tantinger D., Struck M. (2009) Automatic Emergency Detection Using Commercial Accelerometers and Knowledge-Based Methods. Computers in Cardiology 36: 485–488.

Dlouhy J., Cizek M., Rozman J. (2007) Patients Head Position Monitoring in: IEEE (Ed.), 17th International Conference Radioelektronika, IEEE, Brno pp. 1–3.

Duckham M., Nittel S., Worboys M. (2005) Monitoring Dynamic Spatial Fields Using Responsive Geosensor Networks, in: ACM (Ed.), 13th International Symposium of ACM on Advances in Geographical Information Systems, ACM, Bremen, Germany. pp. 1–10.

Dulman S., Havinga P. (2002) Operating System Fundamentals for the EYES Distributed Sensor Network, in: IEEE (Ed.), Proceedings of Progress, IEEE, Utrecht, the Netherlands.

Dunkels A., Gronvall B., Voigt T. (2004) Contiki - a lightweight and exible operating system for tiny networked sensors, in: IEEE (Ed.), 29th Annual IEEE International Conference on Local Computer Networks, IEEE, Washington, DC, USA. pp. 455–462.

Elmenreich W. (2002) Sensor Fusion in Time-Triggered Systems, Technique Institute for Informatik, Wien Techniscal University, Wien. pp. 173.

Elson J., Girod L., Estrin D. (2002) Fine-grained Network Time Synchronization Using Reference Broadcasts. ACM SIGOPS Operating Systems Review 38: 147–163.

Erol-Kantarci M., Mouftah H.T. (2010) Wireless Sensor Networks for Domestic Energy Management in Smart Grids, in: IEEE (Ed.), 25th Biennial Symposium on Communications, IEEE, Ontario, Canada. pp. 63–66.

Errachid A., Zine N., Samitier J., Bausells J. (2004) FET-based Chemical Sensor Systems Fabricated with Standard Technologies. Electroanalysis 16: 1843–1851.

Eswaran A., Rowe A., Rajkumar R. (2005) Nano-RK: An Energy-aware Resource-centric Ratios for Sensor Networks, in: IEEE (Ed.), The Real-Time Systems Symposium 2011, IEEE, Miami, Florida, USA. pp. 256–265.

Fengou M.-A.S., Panagiotakopoulos T.C., Fengos S.L., Lazarou N.G., Lymberopoulos D.K. (2010) A New Telemedicine Framework Handling the Emergency Room Overload, in: IEEE (Ed.), IEEE International Conference on Information Technology and Applications in Biomedicine, IEEE, Corfu. pp. 1–4.

Festag A., Hessler A., Baldessari R., Le L., Zhang W., Westhoff D. (2008) Vehicle-to-Vehicle and Road-side Sensor Communication for Enhnced Road Safety, in: IEEE (Ed.), 9th International Conference on Intelligent Tutoring Systems, IEEE, Montréal, Canada.

Fischer G.S., Akinbiyi T., Saha S., Zand J., Talamini M., Marohn M., Taylor R. (2006) Ischemia and Force Sensing Surgical Instruments for Augmenting Available Surgeon Information, in: IEEE (Ed.), The First IEEE/RAS-EMBS International Conference on Biomedical Robotics and Biomechatronics, IEEE, Pisa. pp. 1030–1035.

Flora D.F.J., Kavitha V., Muthuselvi M. (2011) Survey on Congestion Control Techniques in Wireless Sensor Networks, in: IEEE (Ed.), 2011 International Conference on Emerging Trends in Electrical and Computer Technology, IEEE, Tamil Nadu. pp. 1146–1149.

Fraile J.A., Bajo J., Corchado J.M., Abraham A. (2010) Applying Wearable Solutions in Dependent Environments. IEEE Transactions on Information Technology in Biomedicine 14: 1459–1467.

Franceschinis M., Gioanola L., Messere M., Tomasi R., Spirito M.A., Civera P. (2009) Wireless Sensor Networks for Intelligent Transportation Systems, in: IEEE (Ed.), 69th Vehicular Technology Conference, IEEE, Barcelona, Spain. pp. 1–5.

Ganeriwal S., Kumar R., Srivastava M.B. (2003) Timing-sync Protocol for Sensor Networks, in: ACM (Ed.), 1st International Conference on Embedded Networked Sensor Systems ACM, Los Angeles, California, USA.

Gautam G.C., Sharma T.P. (2011) A Comparative Study of Time Synchronization Protocols in Wireless Sensor Networks. International Journal of Applied Engineering Research 8.

Genova F., Bellifemine F., Gaspardone M., Beoni M., Cuda A., Fici G.P. (2009) Thermal and energy management system based on low cost Wireless Sensor Network Technology, to monitor, control and optimize energy consumption in Telecom Switch Plants and Data Centres, in: IEEE (Ed.), 4th International Conference on Telecommunication-Energy Special Conference IEEE, Vienna, Austria pp. 1–8.

Goh K.N., Chen Y.Y., Lin E.S. (2011) Developing a Smart Wardrobe System, in: IEEE (Ed.), IEEE Workshop on Personalized Networks, IEEE, Las Vegas, Nevada. pp. 303–307.

Goldoni E., Gamba P. (2010) W-TREMORS, a Wireless Monitoring System for Earthquake Engineering, in: IEEE (Ed.), IEEE Workshop on Environmental, Energy, and Structural Monitoring Systems, IEEE, Taranto, Italy. pp. 26–31.

Gong J.B. (2011) PDhms: Pulse Diagnosis via Wearable Healthcare Sensor Network, in: IEEE (Ed.), IEEE International Conference on Communications, IEEE, Kyoto pp. 1–5.

Grüner K., Reuter R., Smid H. (1991) A New Sensor System for Airborne Measurements of Mritime Pollution and of Hydrographic Parameters GeoJournal 24: 15. DOI: 0.1007/BF00213062

Han C.-C., Kumar R., Shea R., Kohler E., Srivastava M. (2005) A Dynamic Operating System for Sensor Nodes, in: ACM (Ed.), 3rd International Conference on Mobile Systems, Applications, and Services, ACM Press, New York, NY, USA. pp. 163–176.

Hartung C., Han R., Seielstad C., Holbrook S. (2006) FireWxNet: A MultiTiered Portable Wireless System for Monitoring Weather Conditions in Wildland Fire Environments, in: ACM (Ed.), The Fourth International Conference on Mobile Systems, Applications and Services, ACM, Uppsala, Sweden. pp. 28–41.

Hasswa A., Hassanein H. (2010) Using Heterogeneous and Social Contexts to Create a Smart Space Architecture, in: IEEE (Ed.), IEEE Symposium on Computers and Communications, IEEE, Riccione, Italy pp. 1838–1842.

Hefeeda M., Bagheri M. (2009) Forest Fire Modeling and Early Detection usingWireless Sensor Networks. Ad Hoc & Sensor Wireless Networks 7: 169–224.

Heinzelman W., Chandrakasan A., Balakrishnan H. (2000) Energy-efficient Communication Protocol for Wireless Sensor Networks, in: IEEE (Ed.), Hawaii International Conference System Sciences, IEEE, Hawaii.

Helal S., Mann W., El-Zabadani H., King J., Kaddoura Y., Jansen E. (2005) The Gator Tech Smart House: a programmable pervasive space. IEEE Computer 38: 50–60.

Helmis C.G., Tzoutzas J., Flocas H.A., Halios C.H., Stathopoulou O.I., Assimakopoulos V.D., Panis V., Apostolatou M., Sgouros G., Adam E. (2007) Indoor Air Quality in a Dentistry Clinic. Science of the Total Environment 377: 349–365.

Hill J., Szewczyk R., Woo A., Hollar S., Culler D.E., Pister K.S.J. (2000) System Architecture Directions for Networked Sensors. Architectural Support for Programming Languages and Operating Systems: 35:11: 93–104.

Hochmuth P. (2005) GM Cuts the Cords to Cut Costs. Mobility and Wireless Article of Techworld.

Hsiao C.-C., Sung Y.-J., Lau S.-Y., Chen C.-H., Hsiao F.-H., Chu H.-H., Huang P. (2011) Towards Long-Term Mobility Tracking in NTU Hospital's Elder Care Center, in: IEEE (Ed.), IEEE International Conference on Pervasive Computing and Communications Workshops, IEEE, Seattle, WA pp. 649–654.

Hu F., Xiao Y., Hao Q. (2009) Congestion-Aware, Loss-Resilient Bio-Monitoring Sensor Networking for Mobile Health Applications. IEEE Journal on Selected Areas in Communications 27: 450–465.

Hu W., Le T.D., Corke P., Jha S. (2010) Design and Deployment of Long-term Outdoor Sensornets: Experiences from a Sugar Farm. IEEE Pervasive Computing (first on-line).

Huang Y.M., Hsieh M.Y., Chao H.C., Hung S.H., Park J.H. (2009) Pervasive, Secure Access to a Hierarchical Sensor-Based Healthcare Monitoring Architecture in Wireless Heterogeneous Networks. IEEE Journal on Selected Areas in Communications 27: 400–412.

Huffman D.A. (1952) A Method for the Construction of Minimum-redundancy Codes. Proceedings of the Institute of Radio Engineers 40: 1098–1101.

Hwang J., Yoe H. (2010) Study of the Ubiquitous Hog Farm System Using Wireless Sensor Networks for Environmental Monitoring and Facilities Control. Sensors 10: 10752–10777. DOI: 10.3390/s101210752.

IEC. (2009) IEC/PAS 62601: Industrial Communication Networks–Fieldbus specifications–WIA-PA communication network and communication profile, in: IEC (Ed.), IEC.

Intanagonwiwat C., Govindan R., Estrin D., Heidemann J., Silva F. (2003) Directed Diffusion for Wireless Sensor Networking. IEEE/ACM Transactions on Networking 11: 2–16.

International Society of Automation. (2008) ISA 100-Wireless Systems for automation, in: ISA (Ed.), ISA.

Jarlskog C.Z., Jiang H., Paganetti H. (2006) Proton Monte Carlo in the Clinic, in: IEEE (Ed.), IEEE Nuclear Science Symposium Conference Record, IEEE, San Diego, CA pp. 3891–3893.

Jo K., Bak C., Lee J.-W., Cho W.-D. (2010) Service-Oriented Actuator for Ubiquitous Smart Space, in: IEEE (Ed.), IEEE International Conference on Sensor Networks, Ubiquitous, and Trustworthy Computing, IEEE, Newport Beach, California pp. 333–339.

Jung W., Li C., Kim D.-S., Ahn C.H. (2009) A Sensing Tube with an Integrated Piezoelectric Flow Sensor for Liver Transplantation, in: IEEE (Ed.), 31st Annual International Conference of the IEEE Engineering in Medicine & Biology Society, IEEE, Minneapolis, Minnesota, USA. pp. 4469–4472.

Junnila S., Kailanto H., Merilahti J., Vainio A.-M., Vehkaoja A., Zakrzewski M., Hyttinen J. (2010) Wireless, Multipurpose In-Home Health Monitoring Platform: Two Case Trials. IEEE Transactions on Information Technology in Biomedicine 14: 447–455.

Karp B., Kung H.T. (2000) GPSR: Greedy Perimeter Stateless Routing for Wireless Networks, in: ACM (Ed.), The Annual International Conference on Mobile Computing and Networking, ACM, Boston. pp. 243–254.

Kawashima T., Ma J., Apduhan B.O., Huang R., Rong C. (2008) A System Prototype with Multiple Robots for Finding u-Objects in a Smart Space, in: IEEE (Ed.), International Conference on Embedded Software and Systems, IEEE, Sichuan pp. 229–236.

Kawoos U., Tofighi M.-R., Warty R., Kralick F.A., Rosen A. (2008) In-Vitro and In-Vivo Trans-Scalp Evaluation of an Intracranial Pressure Implant at 2.4GHz. IEEE Transactions on Microwave: The Theory and Technique 56: 2256–2265.

Kemp J., Gaura E.I., Brusey J. (2008) Using Body Sensor Networks for Increased Safety in Bomb Disposal Missions, in: IEEE (Ed.), IEEE International Conference on Sensor Networks, Ubiquitous, and Trustworthy Computing, IEEE, Taiwan. pp. 81–89.

Kiess W., Rybicki J., Mauve M. (2007) On the Nature of Inter-Vehicle Communication, in: IEEE (Ed.), 4th Workshop on Mobile Ad-Hoc Networks, IEEE, Bern. pp. 493–502.

Kijewski-Correa T., Henderson A., Montestruque L., Rager J. (2009) Real-Time Sensor Fusion to Enhance Plume Detection in Urban Zones, in: IEEE (Ed.), Prooceding at 11th Americas Conference on Wind Engineering, International Association for Wind Engineering, San Juan, Puerto Rico.

Kim B.-K., Hong S.-H., Jeong Y.-S., Eom D.-S. (2008) The Study of Applying Sensor Networks to a Smart Home, in: IEEE (Ed.), Fourth International Conference on Networked Computing and Advanced Information Management, IEEE Computer Society, Gyeongju, Korea. pp. 676–681.

Kim S., Pakzad S., Culler D., Demmel J., Fenves G., Glaser S., Turon M. (2007) Health Monitoring of Civil Infrastructures Using Wireless Sensor Networks, in: ACM (Ed.), International Conference on Information Processing in Sensor Networks, ACM, Cambridge, MA. p. 10.

Kim Y., Kim J.H., Basoglu C., Winter T.C. (1997) Programmable Ultrasound Imaging Using Multimedia Technologies: A Next-Generation Ultrasound Machine. IEEE Transactions on Information Technology in Biomedicine 1: 19–29.

Kim Y.W., Lee S.J., Kim G.H., Jeon G.J. (2009) Wireless Electronic Nose Network for real-time Gas Monitoring System, in: IEEE (Ed.), IEEE International Workshop on Robotic and Sensors Environments, IEEE, Lecco. pp. 169–172.

Kimoto A., Shida K. (2008) A NewMultifunctional Sensor Using Piezoelectric Ceramic Transducers for Simultaneous Measurements of Propagation Time and Electrical Conductance. IEEE Transactions on Instrumentation and Measurement 57: 2542–2547.

King J. (2005) BP Pioneers Large-scale Use of Wireless Sensor Networks. ComputerWorld. http://www.computerworld.com/s/article/99962/BP_Pioneers_Large_Scale_Use_of_Wireless_Sensor_Networks, last viewed 2 Dec. 2011.

Ko J., Gao T., Rothman R., Terzis A. (2010) Wireless Sensing Systems in Clinical Environments. IEEE Engineering in Medicine and Biology Magazine March/April: 103–110.

Koo I., Jung K., Koo J., Nam J.-D., Lee Y., Choi H.R. (2006) Wearable Tactile Display Based on Soft Actuator, in: IEEE (Ed.), IEEE International Conference on Robotics and Automation, 2006, IEEE, Orlando, FL pp. 2220–2225.

Koshy J., Pandey R. (2005) VMSTAR: Synthesizing Scalable Runtime Environments for Sensor Networks, in: ACM (Ed.), 3rd international conference on Embedded Networked Sensor Systems ACM, San Diegom, CA, USA. pp. 243–254.

Kosucu B., Irgan K., Kucuk G., Baydere S. (2009) FireSenseTB: A Wireless Sensor Networks TestBed for Forest Fire Detection, in: ACM (Ed.), International Conference on Wireless Communications and Mobile Computing: Connecting the World Wirelessly ACM, Caen, France. pp. 1173–1177.

Kulik J., Heinzelman W., Balakrishnan H. (2002) Negotiation-based Protocols for Disseminating Information in Wireless Sensor Networks. Wireless Networks 8: 169–185.

Lee B.-G., Do K.-H., Chung W.-Y. (2009) WSN Based 3D Mobile Indoor Multiple User Tracking, in: IEEE (Ed.), IEEE Sensors, IEEE, Christchurch, New Zealand. pp. 1598–1603.

Lee H.J., Chen K. (2007) PingPong-128, A New Stream Cipher for Ubiquitous Application, in: IEEE (Ed.), International Conference on Convergence Information Technology, IEEE, Gyeongju. pp. 1893–1899.

Levis P., Culler D. (2002) Mate: A Tiny Virtual Machine for Sensor Networks, in: IEEE (Ed.), International Conference on Architectural Support for Programming Languages and Operating Systems, IEEE, San Jose, CA, USA,.

Li-min Y., Anqi L., Zheng S., Hui L. (2008) Design of Monitoring System for Coal Mine Safety Based on Wireless Sensor Network, in: IEEE (Ed.), IEEE/ASME International Conference on Mechtronic and Embedded Systems and Applications, IEEE, Beijin. pp. 409–414.

Li L., Halpern J.Y. (2001) Minimum Energy Mobile Wireless Networks Revisited, in: IEEE (Ed.), IEEE International Conference on Communications, IEEE, Helsinki, Finland. pp. 278–283.

Li M., Liu Y. (2007) Underground Structure Monitoring with Wireless. Information Processing in Sensor Networks., in: ACM (Ed.), International Symposium on Information Processing in Sensor Networks, ACM, Cambridge, MA, USA. pp. 69–78.

Li M., Liu Y. (2009) Underground Coal Mine Monitoring with Wireless Sensor Networks. ACM Transactions on Sensor Networks 5:10:1–10:29.

Li X., Nayak A., Stojmenovic I. (2010) Sink Mobility in Wireless Sensor Networks, in: Wiley (Ed.), Wireless Sensor and Actuator Networks: Algorithms and Protocols for Scalable Coordination and Data Communication, Wiley.

Lim J.S., Kim J., Friedman J., Lee U., Vieira L., Rosso D., Gerla M., Srivastava M.B. (2011a) SewerSnort: A Drifting Sensor for in situ Wastewater Collection System Gas Monitoring. Ad Hoc Networks (on-line first). DOI: 10.1016/j.adhoc.2011.01.016.

Lim Y.G., Hong K.H., Kim K.K., Shin J.H., Lee S.M., Chung G.S., Baek H.J., Jeong D.-U., Park K.S. (2011b) Monitoring Physiological Signals Using Nonintrusive Sensors Installed in Daily Life Equipment. Biomed Eng Lett 1: 11–20. DOI: 10.1007/s13534-011-0012-0.

Lindsey S., Raghavendra C.S. (2002) PEGASIS: Power Efficient Gathering in Sensor Information Systems, in: IEEE (Ed.), IEEE Aerospace Conference, IEEE, Big Sky, Montana. pp. 1125–1130.

Lindsey S., Raghavendra C.S., Sivalingam K.M. (2002) Data Gathering Algorithms in Sensor Networks using Energy Metric. IEEE Transactions on Parallel and Distributed Systems 13: 924–935.

Lopez G., Custodio V., Moreno J.I. (2010) LOBIN: E-Textile and Wireless-Sensor-Network-Based Platform for Healthcare Monitoring in Future Hospital Environments. IEEE Transactions on Information Technology in Biomedicine 14: 1446–1458.

López M., Gómez J.M., Sabater J., Herms A. (2010) IEEE 802.15.4 Based Wireless Monitoring of pH and Temperature in a Fish Farm, in: IEEE (Ed.), 15th IEEE Mediterranean Electrotechnical Conference IEEE, Valletta pp. 575–780.

Loriga G., Taccini N., Rossi D.D., Paradiso R. (2005) Textile Sensing Interfaces for Cardiopulmonary Signs Monitoring, in: IEEE (Ed.), 2005 IEEE Engineering in Medicine and Biology IEEE, Shagai, China. pp. 1–4.

Lorincz K., Malan D.J., Fulford-Jones T.R.F., Nawoj A., Clavel A., Shnayder V., Mainland G., Welsh M. (2004) Sensor Networks for Emergency Response: Challenges and Opportunities. IEEE Pervasive Computing 4: 16–23.

Low K.S., Win W.N.N., Er M.J. (2005) Wireless Sensor Networks for Industrial Environments, in: IEEE (Ed.), International Conference on Computational Intelligence for Modelling, Control and Automation, IEEE, Vienna. pp. 1–6.

Luinge H., Veltink P., Baten C. (1999) Estimating Orientation with Gyroscopes and Accelerometers. Technol Health Care 7.

Madan R., Cui S., Lall S., Goldsmith A. (2005) Cross-Layer Design for Lifetime Maximization in Interference-Limited Wireless Sensor Networks, in: IEEE (Ed.), IEEE International Conference on Computer Communications, IEEE, Miami. pp. 1964–1973.

Maróti M., Kusy B., Simon G., Lédeczi Á. (2004) The Flooding Time Synchronization Protocol, in: ACM (Ed.), 2nd International Conference on Embedded Networked Sensor Systems, ACM, Baltimore, MD, USA pp. 39–49.

Martinez K., Hart J.K. (2010) Glacier Monitoring: Deploying Custom Hardware in Harsh Environments, in: S. Science+BusinessMedia (Ed.), Wireless Sensor Networks: Deployments and Design Frameworks, Springer. pp. 245–258.

Marzuki A., Aziz Z.A.A., Manaf A.A. (2011) A Review of CMOS Analog Circuits for Image Sensing Application in: IEEE (Ed.), IEEE International Conference on Imaging Systems and Techniques, IEEE, Penang pp. 180–184.

MayoClinic (2011) Diseases and Treatments Alphabetically. http://www.mayoclinic.com/

Maza I., Kondak K., Bernard M., Ollero A. (2010) Multi-UAV Cooperation and Control for Load Transportation and Deployment. Journal of Intelligent Robot Systems 57: 417–449.

McKeever E., Pavuluri S.K., Lopez-Villarroya R., Goussetis G., Kavanagh D.M., Mohammed M.I., Desmulliez M.P.Y. (2010) Label-free Chemical/biochemical Sensing Device Based on an Integrated Microfluidic Channel Within a Waveguide Resonator in: IEEE (Ed.), 3rd Electronic System-Integration Technology Conference, IEEE, Berlin pp. 1–3.

Melodia T., Vuran M.C., Pompili D. (2006) State of the Art in Cross-Layer Design for Wireless Sensor Networks, in: L. F. M. Cesana (Ed.), Wireless Systems/Network Architectures, Springer-Verlag, Berlin. pp. 78–92.

Mishra A., Gondal F.M., Afrashteh A.A., Wilson R.R. (2006) Embedded Wireless Sensors for Aircraft/Automobile Tire Structural Health Monitoring, in: IEEE (Ed.), 2nd IEEE Workshop on Wireless Mesh Networks, IEEE, Reston, VA. pp. 163–165.

Misra P., Kanhere S., Ostry D., Jha S. (2010) Safety Assurance and Rescue Communication Systems in High-Stress Environments: A Mining Case Study. IEEE Communication Magazine 48: 66–73.

Mohammed A., Al-Kuwari A.H., Ortega-Sanchez C., Sharif A., Potdar V. (2011) User Friendly Smart Home Infrastructure: BeeHouse, in: IEEE (Ed.), IEEE International Conference on Digital Ecosystems and Technologies, IEEE, Daejeon, Korea. pp. 1–6.

Mulder S.D., Hoesel L.v., Having P. (2003) Peeros-system Software for Wireless Sensor Networks. System Software for Wireless Sensor Networks Preprint.

Murad M., Yahya K.M., Hassan G.M. (2009) Web Based Poultry Farm Monitoring System Using Wireless Sensor Networks, in: ACM (Ed.), 7th International Conference on Frontiers of Information Technology, ACM, Abbottabad, Pakistan pp. 1–5.

Nachtigall J., Redlich J.-P. (2011) Wireless Alarming and Routing Protocol for Earthquake Early Warning Systems, in: IEEE (Ed.), 4th IFIP International Conference on New Technologies, Mobility and Security, IEEE, Paris, France. pp. 1–6.

Nasser N., Chen Y. (2010) Anytime and Anywhere Monitoring For the Elderly, in: IEEE (Ed.), IEEE 6th International Conference on Wireless and Mobile Computing, Networking and Communications, IEEE, Canada. pp. 429–433.

Nathy S., Gibbons P.B., Seshany S., Anderson Z.R. (2004) Synopsis Diffusion for Robust Aggregation in Sensor Networks, in: ACM (Ed.), 2nd ACM International Conference on Embedded Networked Sensor Systems, ACM, Sydney, Australia. pp. 1–13.

Niziałek A., Zabierowski W., Napieralski A. (2008) Application of JEE 5 Technologies for a System to Support Dental Clinic Management, in: IEEE (Ed.), International Conference on Modern Problems of Radio Engineering, Telecommunications and Computer Science, IEEE, Lviv-Slavsko, Ukraine. pp. 565–568.

O'Rourke J. (1998) Chapter 3. Convex Hulls in 2D, in: C. U. Press (Ed.), Computational Geometry in C, Cambrige University Press, New York.

Ohta J., Tagawa A., Minami H., Noda T., Sasagawa K., Tokuda T., Hatanaka Y., Ishikawa Y., Tamura H., Shiosaka S. (2009) A Multimodal Sensing Device for Fluorescence Imaging and Electrical Potential Measurement of Neural Activities in a Mouse Deep Brain, in: IEEE (Ed.), 31st Annual International Conference of the IEEE Engineering in Medicine & Biology Society, IEEE, Minneapolis, Minnesota, USA. pp. 5887–5890.

Open University (1996) Block 3-Global Patterns in Technological Innovation, in: T. C. Team (Ed.), Innovation Design Environment and Strategy, Open University, UK.

Palomo J.M., Wolf G.R., Hans M.G. (2004) Use of Digital Photography in the Case Orthodontic Clinic. American Journal of Orthodontics and Dentofacial Orthopedics 126: 381–385.

Pan M.-S., Yeh L.-W., Chen Y.-A., Lin Y.-H., Tseng Y.-C. (2008) WSN-Based Intelligent Light Control System Considering User Activities and Profiles. IEEE Sensors 8: 1710–1721.

Pandian P.S., Mohanavelu K., Safeer K.P., Kotresh T.M., Shakunthala D.T., Gopal P., Padak V.C. (2008) Smart Vest: Wearable Multi-parameter Remote Physiological Monitoring System. Medical Engineering & Physics 30: 466–477.

Pang Z., Chen Q., Zheng L. (2009) A Pervasive and Preventive Healthcare Solution for Medication Non-compliance and Daily Monitoring, in: IEEE (Ed.), 2nd International Symposium on Applied Sciences in Biomedical and Communication Technologies, IEEE, Bratislava. pp. 1–6.

Pantelopoulos A., Bourbakis N.G. (2010) A Survey on Wearable Sensor-Based Systems for Health Monitoring and Prognosis. IEEE Transactions on Systems, Man, and Cybernetics, Part C: Applications and Reviews 40: 1–12.

Park C., Chou P.H., Bai Y., Matthews R., Hibbs A. (2006) An Ultra-Wearable, Wireless, Low Power ECG Monitoring System, in: IEEE (Ed.), IEEE Biomedical Circuits and Systems Conference, IEEE, London. pp. 241–244.

Pei Z., Deng Z., Yang B., Cheng X. (2008) Application-Oriented Wireless Sensor Network Communication Protocols and Hardware Platforms: a Survey, in: IEEE (Ed.), IEEE International Conference on Industrial Technology, IEEE, Chengdu pp. 1–6.

Pei Z., Deng Z., Xu S., Xu X. (2009) Anchor-Free Localization Method for Mobile Targets in Coal Mine Wireless Sensor Networks. Sensors 9: 2836–2850.

Piran M.J., Murthy G.R. (2010) A Novel Routing Algorithm for Vehicular Sensor Networks. Wireless Sensor Network 2: 919–923. DOI: 10.4236/wsn.2010.212110.

Qureshi M.S., Basu A., Bicen B., Degertekin L., Hasler P. (2010) Integrated Low Voltage and Low Power CMOS Circuits for Optical Sensing of Diffraction Based Micromachined Microphone in: IEEE (Ed.), IEEE International Symposium on Circuits and Systems, IEEE, Paris. pp. 2031–2034.

Rashvand H.F., Salcedo V.T., Sanchez E.M., Iliescu D. (2008) Ubiquitous Wireless Telemedicine. IET Communications 2: 237–254.

Rashvand H.F., Salah K., Calero J.M.A., Harn L. (2010) Distributed Security for Multi-Agent Systems–Review and Applications. IET Information Security 4: 188–201.

Rawi M.I.M., Al-Anbuky A. (2009) Passive House Sensor Networks: Human Centric Thermal Comfort Concept, in: IEEE (Ed.), International Conference on Intelligent Sensors, Sensor Networks and Information Processing, IEEE, Melburne, Australia. pp. 255–260.

Rienzo M.D., Rizzo F., Parati G., Ferratini M., Brambilla G., Castiglioni P. (2005) A Textile-Based Wearable System for Vital Sign Monitoring: Applicability in Cardiac Patients. Computers in Cardiology 32: 699–712.

Rienzo M.D., Rizzo F., Meriggi P., Bordoni B., Brambilla G., Ferratini M., Castiglioni P. (2006) Applications of a Textile-based Wearable System in Clinics, Exercise and Under Gravitational Stress, in: IEEE (Ed.), Medical Devices and Biosensors, 2006, IEEE, Cambridge. pp. 8–10.

Robb R.A. (2001) The Biomedical Imaging Resource At Mayo Clinic. IEEE Transactions on Medical Imaging 20: 854–867.

Rodoplu V., Ming T.H. (1999) Minimum energy mobile wireless networks. IEEE Journal of Selected Areas in Communications 17: 333–1344.

Sain M., Kumar P., Lee Y.-D., Lee H.J. (2010) Secure Middleware in Ubiquitous Healthcare, in: IEEE (Ed.), 5th International Conference on Computer Sciences and Convergence Information Technology, IEEE, Seoul. pp. 1072–1077.

Sathik M.M., Mohamed M.S., Balasubramanian A. (2010) Fire Detection Using Support Vector Machine in Wireless Sensor Network and Rescue Using Pervasive Devices. International Journal of Advanced Networking and Applications 2: 636–639.

Schiewe J. (2005) Status and Future Perspectives of the Application Potential of Digital Airborne Sensor Systems. International Journal of Applied Earth Observation and Geoinformation 6: 13.

Seok M., Hanson S., Wieckowski M., Chen G.K., Lin Y.-., Blaauw D., Sylvester D. (2010) Circuit Design Advances to Enable Ubiquitous Sensing Environments in: IEEE (Ed.), IEEE International Symposium on Circuits and Systems, IEEE, Paris. pp. 285–288.

Sha K., Shi W., Watkins O. (2006) Using Wireless Sensor Networks for Fire Rescue Application: Requirements and Challenges, in: IEEE (Ed.), 6th IEEE International Conference on Electro/Information Technology, IEEE, East Lansing, MI, USA. pp. 1–6.

Shah R.C., Rabaey J. (2002) Energy Aware Routing for Low Energy Ad Hoc Sensor Networks, in: IEEE (Ed.), IEEE Wireless Communications and Networking Conference, IEEE, Orlando, Florida.

Shah R.C., Roy S., Jain S., Brunette W. (2003) Data MULEs: Modeling a Three-tier Architecture for Sparse Sensor Networks, in: IEEE (Ed.), IEEE International Workshop on Sensor Network Protocols and Applications, IEEE, Anchorage, AK. pp. 30–41.

Shnayder V., Chen B., Lorincz K., FulfordJones T.R.F., Welsh M. (2005) Sensor Networks for Medical Care, in: H. University (Ed.), Technical Report Harvard University.

Shnayder V., Chen B., Lorincz K., FulfordJones T.R.F., Welsh M. (2010) Ubiquitous Access to Cloud Emergency Medical Services, in: IEEE (Ed.), 10th IEEE International Conference on Information Technology and Applications in Biomedicine, IEEE, Corfu. pp. 1–4.

Sikka P., Corke P., Overs L. (2004) Wireless Sensor Devices for Animal Tracking and Control, in: IEEE (Ed.), 29th Annual IEEE International Conference on Local Computer Networks, IEEE. pp. 446–454.

Singh S.K., Singh M.P., Singh D.K. (2010a) Routing Protocols in Wireless Sensor Networks–A Survey. International Journal of Computer Science & Engineering Survey 1: 63–83.

Singh S.K., Singh M.P., Singh D.K. (2010b) A Survey of Energy-Efficient Hierarchical Cluster-Based Routing in Wireless Sensor Networks. International Journal of Advanced Networking and Applications 2: 570–580.

Sleman A., Moeller R. (2011) SOA Distributed Operating System for Managing Embedded Devices in Home and Building Automation, in: IEEE (Ed.), IEEE International Conference on Consumer Electronics, IEEE, Las Vegas, NV pp. 629–630.

Smith G. (1980) The Contract Net Protocol: High-level Communication and Control in a Distributed Problem Solver. IEEE Transaction on Computing 29.

Sohrabi K., Gao J., Ailawadhi V., Pottie G.J. (2000) Protocols for Self-organization of a Wireless Sensor Network. IEEE Personal Communications 7: 16–27.

Soliman H., Sudan K., Mishra A. (2010) A Smart Forest-Fire Early Detection Sensory System: Another Approach of Utilizing Wireless Sensor and Neural Networks, in: IEEE (Ed.), IEEE 2010 International Conference on SENSORS, IEEE, Waikoloa, USA. pp. 1900–1904.

Somov A., Spirjakin D., Ivanov M., Khromushin I., Passerone R. (2010) Combustible Gases and Early Fire Detection: an Autonomous System for Wireless Sensor Networks, in: ACM (Ed.), 1st International Conference on Energy-Efficient Computing and Networking ACM, Passau, Germany pp. 85–93.

Song J., Welch D., Christen J.B. (2011) A Fully-adjustable Dynamic Range Capacitance Sensing Circuit in a 0.15 μm 3D SOI Process in: IEEE (Ed.), IEEE International Symposium on Circuits and Systems, IEEE, Rio de Janeiro, Brazil pp. 1708–1711.

Song J., Han S., Mok A.K., Chen D., Lucas M., Nixon M. (2008) WirelessHART: Applying Wireless Technology in Real-Time Industrial Process Control, in: 2008 (Ed.), IEEE Real-Time and Embedded Technology and Applications Symposium, IEEE, St Louise, MO, USA. pp. 377–386.

Song W.-Z., Shirazi B., Huang R., Xu M., Peterson N., LaHusen R., Pallister J., Dzurisin D., Moran S., Lisowski M., Kedar S., Chien S., Webb F., Kiely A., Doubleday J., Davies A., Pieri D. (2010) Optimized Autonomous Space In-Situ Sensor Web for Volcano Monitoring. IEEE Journal of Selected Topics in Applied Earth Observations and Remote Sensing 3: 541–546.

Stark D., Davis J. (2004) Friendly Object Tracking and Foreign Object Detection and Localization with an SDAC Wireless Sensor Network, in: IEEE (Ed.), 10th IEEE International Workshop on Future Trends of Distributed Computing Systems, IEEE, Suzhou, China. pp. 30–36.

Su W., Akyildiz I.F. (2005) Time-diffusion Synchronization Protocol for Wireless Sensor Networks. IEEE/ACM Transactions on Networking 13: 384–397.

Tan W., Wang Q., Huang H., Guo Y., Zhang G. (2007) Mine Fire Detection System Based on Wireless Sensor Network, in: IEEE (Ed.), International Conference on Information Acquisition, IEEE, Jeju City, Korea. pp. 148–151.

Tao Z., Yajuan Q., Deyun G., Hongke Z. (2010) Environmental Monitoring and Air-conditioning Automatic Control with Intelligent Building Wireless Sensor Network, in: IEEE (Ed.), International Conference on Control, Automation, Robotics and Vision, IEEE, Singapore. pp. 2431–2436.

Tapia D.I., Rodríguez S., Bajo J., Corchado J.M., García O. (2010) Wireless Sensor Networks for Data Acquisition and Information Fusion: A Case Study, in: IEEE (Ed.), 13th Conference on Information Fusion (FUSION), IEEE, Edinburgh. pp. 1–8.

Tian H., Stankovic J.A., Chenyang L., Abdelzaher T. (2003) SPEED: A Stateless Protocol for Real-time Communication in Sensor Networks in: IEEE (Ed.), International Conference on Distributed Computing Systems, IEEE, Reno, Nevada. pp. 46–55.

Tsujita W., Ishida H., Moriizumi T. (2005) Dynamic Gas Sensor Network for Air Pollution Monitoring and Its Auto-Calibration. IEEE Sensors 1: 56–59. DOI: 10.1109/ICSENS.2004.1426098

Tubaishat M., Shang Y., Shi H. (2007) Adaptive Traffic Light Control with Wireless Sensor Networks, in: IEEE (Ed.), 4th Ieee Consumer Communications And Networking Conference, IEEE, Las Vegas, NV, USA. pp. 187–191.

Vasilescu I., Kotay K., Rus D., Dunbabin M., Corke P. (2005) Data Collection, Storage, and Retrieval with an Underwater Sensor Network, in: ACM (Ed.), 3th ACM Conference on Embedded Networked Sensor Systems ACM, San Diego, California, USA. pp. 1–12.

Vincen D., Stampouli D., Powell G. (2009) Foundations for System Implementation for a Centralised Intelligence Fusion Framework for Emergency Services, in: IEEE (Ed.), 12th International Conference on Information Fusion, IEEE, Seattle, Washington.

Wang C., Li B., Sohraby K., Daneshmand M., Hu Y. (2007) Upstream Congestion Control in Wireless Sensor Networks Through Cross-Layer Optimization. IEEE Journal of Selected Areas in Communications 25: 786–795.

Wang Z., Yang R., Wang L. (2010) Multi-agent Control System with Intelligent Optimization for Smart and Energy-efficient Buildings, in: IEEE (Ed.), 36th Annual Conference on IEEE Industrial Electronics Society IEEE, Glendale, AZ pp. 1144–1149.

Weingärtner E., Kargl F. (2007) A Prototype Study on Hybrid Sensor-Vehicular Networks, Kommunikation und Verteilte Systeme, Technischer Bericht der RWTH Aachen, Aachen, Germany. pp. 79–82.

Wenning B.-L., Pesch D., Timm-Giel A., Görg C. (2008) Environmental Monitoring Aware Routing in Wireless Sensor Networks, in: Springer (Ed.), IFIP joint conference on Mobile and Wireless Communications Networks (MWCN 2008) and Personal Wireless Communications, Springer, Toulouse, France. pp. 5–16.

Werner-Allen G., Lorincz K., Welsh M., Marcillo O., Johnson J., Ruiz M., Lees J. (2006) Deploying a Wireless Sensor Network on an Active Volcano. IEEE Internet Computing 10: 18–25.

Woo A., Tong T., Culler D. (2003) Taming the Underlying Challenges of Reliable Multihop Routing in Sensor Networks, in: ACM (Ed.), 1st International Conference on Embedded Networked Sensor Systems ACM, Los Angeles (CA), USA. pp. 14–27.

Woolard D.L., Brown E.R., Pepper M., Kemp M. (2005) Terahertz Frequency Sensing and Imaging: A Time of Reckoning Future Applications? Proceedings of the IEEE 93: 1722–1743.

Wu M.-Y., Huang W.-Y. (2011) Health Care Platform with Safety Monitoring for Long-Term Care Institutions, in: IEEE (Ed.), 7th International Conference on Networked Computing and Advanced Information Management, IEEE, Gyeongju. pp. 313–317.

Xu Y., Heidemann J., Estrin D. (2001) Geography-informed Energy Conservation for Ad-hoc Routing, in: IEEE (Ed.), Seventh Annual ACM/IEEE International Conference on Mobile Computing and Networking, IEEE, Atlanta, Georgia. pp. 70–84.

Yang H., Kuang B., Mouazen A.M. (2011) Wireless Sensor Network for Orchard Management, in: IEEE (Ed.), Third International Conference on Measuring Technology and Mechatronics Automation, IEEE, Shanghai,China. pp. 1162–1165.

Yannakopoulos J., Bilas A. (2005) Cormos: A Communication-oriented Runtime System for Sensor Networks, in: IEEE (Ed.), The Second European Workshop on Wireless Sensor Networks, IEEE, Istanbul, Turkey.

Yanying G., Lo A., Niemegeers I. (2009) A Survey of Indoor Positioning Systems for Wireless Personal Networks. IEEE-CST 11: 13–32.

Yao F., Yang S.-H. (2010) Mitigating Interference Caused by IEEE 802.11 b in the IEEE 802.15.4 WSN within the Environment of Smart House, in: IEEE (Ed.), International Conference on Systems Man and Cybernetics, IEEE, Istanbul pp. 2800–2807.

Yedavalli K., Krishnamachari B. (2008) Sequence-based Localization in Wireless Sensor Networks. IEEE Transaction on Mobile Communications 7: 81–94.

Younis M., Akkaya K. (2008) Strategies and Techniques for Node Placement in Wireless Sensor Networks: A Survey. Journal Ad Hoc Networks 4: 621–655.

Younis O., Fahmy S. (2004) Heed: A Hybrid, Energy-efficient, Distributed Clustering Approach for Ad-hoc Networks. IEEE Transactions on Mobile Computing 3: 366–369.

Yu L., Wang N., Meng X. (2005) Real-time Forest Fire Detection with Wireless Sensor Networks, in: IEEE (Ed.), International Conference on Wireless Communications, Networking and Mobile Computing, IEEE, Shanghai, China. pp. 1214–1217.

Yu W., Yuan J. (2005) Joint Source Coding, Routing, and Resource Allocation for Wireless Sensor Networks, in: IEEE (Ed.), EEE International Conference on Communications, IEEE, Seoul, 2: 737–741.

Yu Y., Estrin D., Govindan R. (2001) Geographical and Energy-Aware Routing: A Recursive Data Dissemination Protocol for Wireless Sensor Networks, in: UCLA (Ed.), Computer Science Department Technical Report, UCLA, California, Los Angeles.

Zhang B., Sukhatne G.S., Requicha A.A.G. (2004) Adaptive Sampling for Marine Microorganism Monitoring, in: IEEE (Ed.), International Conference on Intelligent Robots and Systems, IEEE, Sendai, Japan. pp. 1115–1122.

Zhang J., Anwen Q. (2010) The Application of Internet of Things(IOT) in Emergency Management System in China, in: IEEE (Ed.), International Conference on Technologies for Homeland Security, IEEE, Waltham, MA. pp. 139–142.

Zhang L., Wang Z. (2006) Integration of RFID into Wireless Sensor Networks: Architectures, Opportunities and Challenging Problems, in: IEEE (Ed.), Fifth International Conference on Grid and Cooperative Computing Workshops, IEEE, Changsha, Hunan, China. pp. 1–7.

Index

Distributed Sensor Systems: Practice and Applications, First Edition.
Habib F. Rashvand and Jose M. Alcaraz Calero.
© 2012 John Wiley & Sons, Ltd. Published 2012 by John Wiley & Sons, Ltd.